Nanotechnology for Light Pollution Reduction

Light is essential for living organisms; however, excessive light causes adverse health conditions. This book covers the most recent progress on nanotechnology for reducing light pollution, discussing many approaches and technologies for controlling light pollution.

The book explores the fundamentals of light and the causes of light pollution, delving into light pollution's social, economic, and ecological impacts, its effects on living beings and the environment, as well as possible solutions and methods of control. The text discusses smart lighting technology, covering the various smart nanomaterials, nanosensors, and nanodevices involved. It also explores smart lighting involving natural light from the sun, artificial skydomes, shadow-free/secondary light sources, and the basics of many emerging devices such as light-emitting diodes and photosensors.

Nanotechnology is key to providing a new route for the next generation of lighting devices and systems with reduced light pollution. This essential reference illuminates emerging technologies and their applications, providing new directions to scientists, researchers, and students to better understand the principles, technologies, and applications of nanotechnology in light pollution.

Nanotechnology for Light Pollution Reduction

Edited by

Tuan Anh Nguyen
Head of Microanalysis Department at the Institute for Tropical Technology,
Vietnam Academy of Science and Technology

Ram K. Gupta
Associate Professor at Pittsburg State University, Pittsburg, Kansas, USA

CRC Press
Taylor & Francis Group
Boca Raton London New York

CRC Press is an imprint of the
Taylor & Francis Group, an **informa** business

First edition published 2023
by CRC Press
6000 Broken Sound Parkway NW, Suite 300, Boca Raton, FL 33487-2742

and by CRC Press
4 Park Square, Milton Park, Abingdon, Oxon, OX14 4RN

CRC Press is an imprint of Taylor & Francis Group, LLC

© 2023 selection and editorial matter, Tuan Anh Nguyen and Ram K. Gupta; individual chapters, the contributors

Library of Congress Cataloging-in-Publication Data
Names: Anh Nguyen, Tuan, editor. | Gupta, Ram K., editor.
Title: Nanotechnology for light pollution reduction / edited by Tuan Anh
Nguyen, Head of Microanalysis Department at the Institute for Tropical
Technology, Vietnam Academy of Science and Technology, Ram K. Gupta,
Associate Professor at Pittsburg State University.
Description: First edition. | Boca Raton, FL : CRC Press, [2023] |
Includes bibliographical references and index. | Identifiers: LCCN 2022009566 (print) |
LCCN 2022009567 (ebook) | ISBN 9781032021201 (hardback) |
ISBN 9781032027722 (paperback) | ISBN 9781003185109 (ebook)
Subjects: LCSH: Light pollution. | Nanotechnology.
Classification: LCC TD195.L52 N36 2023 (print) | LCC TD195.L52 (ebook) |
DDC 363.738—dc23/eng/20220603
LC record available at https://lccn.loc.gov/2022009566
LC ebook record available at https://lccn.loc.gov/2022009567

ISBN: 9781032021201 (hbk)
ISBN: 9781032027722 (pbk)
ISBN: 9781003185109 (ebk)

DOI: 10.1201/9781003185109

Typeset in Minion
by codeMantra

Access the Support Material: www.routledge.com/9781032021201

Contents

Editors

Dr. Tuan Anh Nguyen completed his BSc in Physics from Hanoi University in 1992, and his Ph.D. in Chemistry from Paris Diderot University (France) in 2003. He was a Visiting Scientist at Seoul National University (South Korea, 2004) and the University of Wollongong (Australia, 2005). He then worked as a Postdoctoral Research Associate & Research Scientist at Montana State University (USA), 2006–2009. In 2012, he was appointed as Head of the Microanalysis Department at the Institute for Tropical Technology (Vietnam Academy of Science and Technology). He has managed 4 Ph.D. theses as thesis director and 3 are in progress. He is Editor-In-Chief of *Kenkyu Journal of Nanotechnology & Nanoscience* and Founding Co-Editor-In-Chief of *Current Nanotoxicity & Prevention*. He is the author of 4 Vietnamese books and Editor of 32 Elsevier books in the Micro & Nano Technologies Series.

Dr. Ram K. Gupta is an Associate Professor at Pittsburg State University. Dr. Gupta's research focuses on conducting polymers and composites, green energy production and storage using biowastes and nanomaterials, optoelectronics and photovoltaics devices, organic-inorganic hetero-junctions for sensors, bio-based polymers, flame-retardant polymers, bio-compatible nanofibers for tissue regeneration, scaffold and antibacterial applications, corrosion inhibiting coatings, and bio-degradable metallic implants. Dr. Gupta has published over 250 peer-reviewed articles, made over 300 national, international, and regional presentations, chaired many sessions at national/international meetings, edited many books, and written several book chapters. He has received several million dollars for research and educational activities from many funding agencies. He is serving as Editor-in-Chief, Associate Editor, and editorial board member of numerous journals.

Contributors

Magdalene Asare
Department of Chemistry
Pittsburg State University
Pittsburg, KS, USA
Kansas Polymer Research Center
Pittsburg State University, Pittsburg
KS, USA

Humira Assad
Department of Chemistry, Faculty of
    Technology and Science
Lovely Professional University
Phagwara, India

Irfan Ayoub
Department of Physics
National Institute of Technology Srinagar
Jammu and Kashmir, India

Anjali Banger
Department of Chemistry
Banasthali Vidyapith
Rajasthan, India

Reyhaneh Barzegar
Babol Noshirvani University of Technology
Babol, Iran

Ajit Behera
Department of Metallurgical & Materials
    Engineering
National Institute of Technology
Rourkela, India

Ramakrishna Biswal
Department of Humanities and Social
    Sciences
National Institute of Technology
Rourkela, India

Ederson Dias Pereira Duarte
Department of Chemistry
UEM - State University of Maringa
Maringá, Brazil

Ali Farmani
Lorestan University
Khorramabad, Iran

Ishrat Fatma
Department of Chemistry, Faculty of
    Technology and Science
Lovely Professional University
Phagwara, India

Luiz Fernando Gorup
School of Chemistry and Food Science
Federal University of Rio Grande
Rio Grande, Brazil
Materials Engineering
Federal University of Pelotas
Pelotas, Brazil
Institute of Chemistry
Federal University of Alfenas
Alfenas, Brazil

Hadis Foladi
Lorestan University
Khorramabad, Iran

Richika Ganjoo
Department of Chemistry, Faculty of
    Technology and Science
Lovely Professional University
Phagwara, India

M. Barzegar Gerdroodbary
Babol Noshirvani University of Technology
Babol, Iran

Ram K. Gupta
Department of Chemistry
Pittsburg State University
Pittsburg, KS, USA
Kansas Polymer Research Center
Pittsburg State University
Pittsburg, KS, USA

Elizângela Hafemann Fragal
Department of Chemistry
UEM - State University of Maringa
Maringá, Brazil

Vanessa Hafemann Fragal
Department of Chemistry
UEM - State University of Maringa
Maringá, Brazil

Essia Hannachi
Institute for Research and Medical
    Consultations (IRMC), Imam
    Abdulrahman Bin Faisal University,
    Dammam, Saudi Arabia

Eduardo José de Arruda
Faculty of Exact Sciences and Technology
    (FACET)
Federal University of Grande Dourados
Dourados, Brazil

Mahfuz Kabir
Bangladesh Institute of International and
    Strategic Studies (BIISS)
Dhaka, Bangladesh

Zobaidul Kabir
School of Environmental and Life Sciences
University of Newcastle
Callaghan, Australia

Halima Khatun
Bangladesh Livestock Research Institute
    (BLRI)
Savar, Dhaka, Bangladesh

Ashish Kumar
Department of Science and Technology,
    Government of Bihar, Nalanda College
    of Engineering, India

Vijay Kumar
Department of Physics
National Institute of Technology Srinagar
Jammu and Kashmir, India
Department of Physics
University of the Free State
Bloemfontein, South Africa

Santos Kumar Das
Department of Electrical and Electronics
    Engineering
National Institute of Technology
Rourkela, India

Mantu Kumar Mahalik
Department of Humanities and Social
    Sciences
Indian Institute of Technology
Kharagpur, Medinapur, India

Sidney A. Lourenço
Post-graduation Program in Material
  Science and Engineering
Federal University of Technology (UTFPR)
Londrina, Brazil

Bharti Mehlawat
Department of Chemistry
Banasthali Vidyapith
Rajasthan, India

Kwadwo Mensah-Darkwa
Kwame Nkrumah University of Science
  and Technology
Kumasi, Ghana

Ali Mir
Lorestan University
Khorramabad, Iran

Edvani C. Muniz
Department of Chemistry
UEM - State University of Maringa
Maringá, Brazil
Post-graduation Program in Material
  Science and Engineering
Federal University of Technology (UTFPR)
Londrina, Brazil
Department of Chemistry
Federal University of Piauí
Piauí, Brazil

Tuan Anh Nguyen
Vietnam Academy of Science and
  Technology
Hanoi, Vietnam

Elisangela Pacheco da Silva
Department of Chemistry
UEM - State University of Maringa
Maringá, Brazil

Rakesh Sehgal
Department of Mechanical Engineering
National Institute of Technology Srinagar
Jammu and Kashmir, India

Rishabh Sehgal
Department of Electrical and Computer
  Engineering
University of Texas at Austin
Austin, TX, USA

Hendrik C. Swart
Department of Physics
University of the Free State
Bloemfontein, South Africa

Thiago Sequinel
Faculty of Exact Sciences and Technology
  (FACET)
Federal University of Grande Dourados
Dourados, Brazil

Rafael Silva
Department of Chemistry
UEM - State University of Maringa
Maringá, Brazil

Yassine Slimani
Institute for Research and Medical
  Consultations (IRMC), Imam
  Abdulrahman Bin Faisal University,
  Dammam, Saudi Arabia

Anamika Srivastava
Department of Chemistry
Banasthali Vidyapith
Rajasthan, India

Manish Srivastava
Department of Chemistry
Banasthali Vidyapith
Rajasthan, India

Felipe M. de Souza
Kansas Polymer Research Center
Pittsburg State University
Pittsburg, KS, USA

Abhinay Thakur
Department of Chemistry, Faculty of
    Technology and Science
Lovely Professional University
Phagwara, India

Anjali Yadav
Department of Chemistry
Banasthali Vidyapith
Rajasthan, India

CHAPTER 1

Light Pollution and Prevention: An Introduction

Abhinay Thakur and Richika Ganjoo
Lovely Professional University

Ashish Kumar
Nalanda College of Engineering

CONTENTS

DOI: 10.1201/9781003185109-1

1.1 INTRODUCTION

Lighting has now become a critical component in developing cities appropriate for livable environments in all climates. Lighting is important from an economic, ecological, and social perspective. Lighting needs for widely used outdoor working areas are regulated by standards in European Union countries. However, because the lighting of environmental and architectural elements is not standardized, technical papers like the Commission Internationale de l'Eclairage (CIE) study have offered recommendations [1–3]. Unnecessary or insufficient lighting has detrimental effects on the ecosystem and human situations. Due to population expansion and increased outdoor lighting per person, this condition has resulted in light pollution, which is characterized as an undesired skylight. Light pollution has been a problem for astronomers and scientists in this era. Many scientific and technological issues have arisen as a result of the smart city model, notably light pollution. Light pollution, like sound, waste, and chemical pollutants, has a wide range of detrimental consequences on persons, ecosystems, astronomy, and energy usage. All of this is the product of human activity. Hormonal imbalance, drowsiness, malignancy, risky driving, and other harmful impacts are only a few instances.

Furthermore, light pollution harms certain plants and animals. Current research shows that urbanized surroundings have proceeded to become much brighter and much more lighted at nighttime in modern times, and also that man-made light pollution through misguided, intrusive, and undesired outdoor lighting has recently increased by more than 2% every year. H.S. Lim et al. [4] conducted a field assessment in South Korea to determine lighting environmental control regions, which they used to discuss the realities of light pollution. They discovered that outdoor illumination at night is a regular occurrence in Korea. Furthermore, as Korea's economy develops, the usage of artificial lighting is likely to rise, putting it susceptible to the negative impacts of artificial lighting. Collected from field measurements taken in Seoul, South Korea, they explore the topic of light pollution in this paper. The measurements were collected in terms of perception and analyze light pollution. Following our research, we noticed that some types of light pollution are mostly present in established metropolitan and heavily populated areas. Experiments to measure light pollution are now being conducted in Korea. It'll be interesting to see how the whole of South Korea fares in concerns of light pollution levels in comparison to Seoul. Figure 1.1

FIGURE 1.1   Eoul City, South Korea's Environmental Management Regions. Adapted with permission [4], MDPI, Distributed under Creative Commons Attribution License 4.0 (CC BY).

depicts the Seoul City Environmental Management Regions, which were formed based on a field survey.

However, lighting has now become a requirement in human life for addressing aesthetic goals and psychological comforts. Lighting is the process of transmitting light from a specified source of light to an item or environment to provide visibility. Light is considered the foundation of the lighting concept; nonetheless, the goal of lighting is to visualize the illuminated area using a light source [5]. In many ways, urban lighting influences the comfort of illumination as well as the health of living creatures (humans, animals, and plants). In outdoor lighting, the incomplete understanding and misleading of installations and bulbs can result in light intrusion, dazzle, vertical light, and overbearing light. Today, both quite poor lighting systems and bad practices are prevalent in Turkey and around the world, and these bad practices are growing increasingly common. Light pollution occurs when light is used in the incorrect location, in the incorrect amount, in the incorrect direction, and at the incorrect time. Although not as harmful as air and water pollution, excess and inappropriate lighting results in poor illumination; as a result, a large percentage of the energy spent to produce light is lost. The exploitation of light in a manner that wastes energy inhibits astronomical investigations and produces consequences that harm natural life is known as light pollution. It is possible to change the design of the luminaires.

Light pollution is primarily caused by artificial lighting used outside. Lighting should be conducted at the correct location, in the correct direction, with the appropriate amount of light, and at the appropriate time. The issue isn't with the lighting, but with how it's being used. The most important component in preventing light pollution and energy usage is proper lighting. When eye quality and relaxation are neglected, illumination can create suffering rather than help. Light pollution can occur when improper lighting sources are used in outdoor areas, or when the strength, location, and color of the light are incorrectly adjusted [6–8]. Light pollution is caused by architecture that does not permit uniform illumination, resulting in visual pollution related to visual health, convenience, and aesthetics. M.S. Faid et al. [9] addressed the hazards of light pollution to the environment's resilience. They concluded that light pollution is a risk to planetary exploration, global health, ecosystems, and economics, rendering it one of the most serious threats to the concept of sustainability. To save energy and maximize the use of renewable energy sources, W.A. Jabbar et al. [10] designed the SGStreet-LS smart and sustainable sidewalk illumination technology. The suggested system combines powerful thoughts and ideas to effortlessly and inexpensively govern the working of street lightings based on sunlight accessibility and motion sensors using an Arduino-based controller using radio frequency (RF) wireless transmission technology. It also replaces traditional high-powered lighting with low-power LEDs powered by solar panels. To turn on the lights, two conditions must be met: a Passive infrared (PIR) motion detector can detect the location of objects in the roadway and a Light-dependent resistor (LDR) sensor can sense reduced light intensity concentrations (darkness circumstances). If this does not happen, the street lights will be shut off. The use of electricity for street lights might be reduced as a result of smart and green street light system (SGStreet-LS) deployment, while CO_2 levels could be reduced by utilizing renewable energy sources. The lights switch on before pedestrians and cars arrive, then turn off or dim when no one is present. In this chapter, we will be discussing the basic introduction of light pollution and its prevention by employing the significance of the smart city, smart lighting (such as LEDs), and the adverse impact of light pollution on our ecosystem. Figure 1.2 shows the various examples of light pollution at night time.

1.1.1 Various Sources Responsible for Light Pollution

Light pollution is one of the unintended consequences of modern civilization. It is derived from a variety of sources, including exterior and interior illumination of buildings, advertising, commercial properties, offices, industries, streetlights, and lit sports arenas as shown in Figure 1.3. The truth is that most nighttime lighting systems are inefficient, overly brilliant, improperly directed, improperly shielded, and, in many cases, completely unnecessary. This light, and the energy required to generate it, is being squandered by not concentrating it on the things and places that people want lit. Luminous pollution occurs when outdoor lights are utilized in excess and where they are not required. Residential, industrial, and commercial outdoor lighting that is poorly planned also contributes substantially to light pollution. Unshielded light fixtures direct more than half of their illumination upward or laterally. Often, just 40% of the light projected illuminates the ground.

FIGURE 1.2   Examples of light pollution. Adapted with permission [11], MDPI, Distributed under Creative Commons Attribution License 4.0 (CC BY).

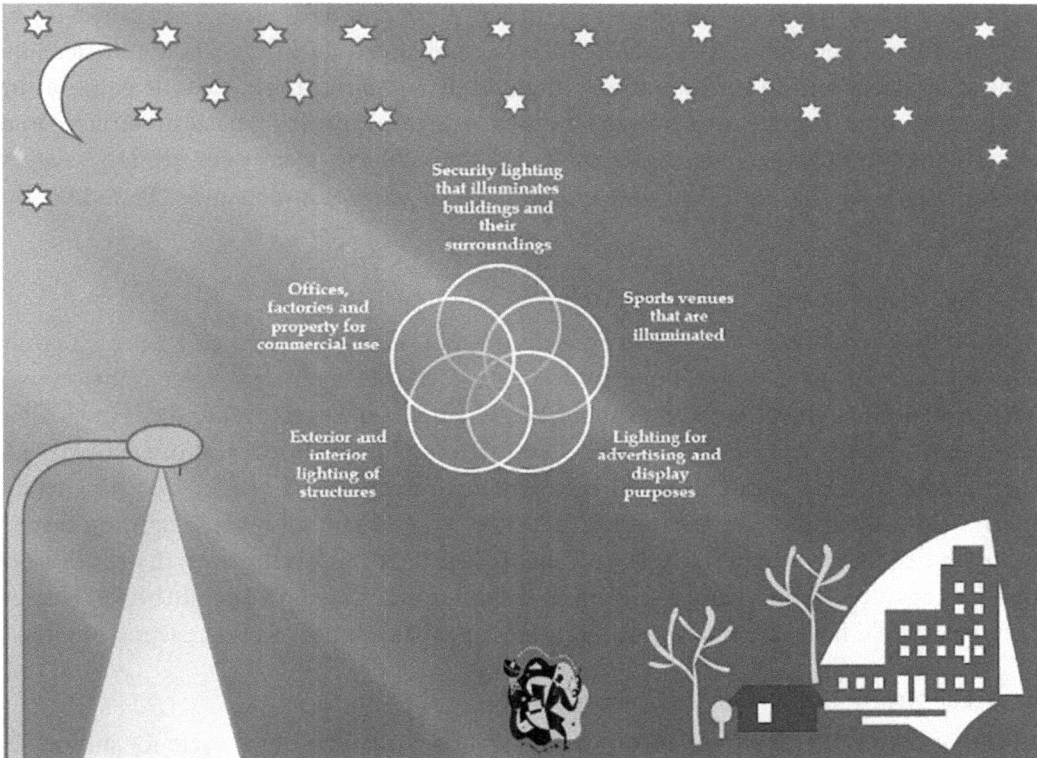

FIGURE 1.3   Various sources of light pollution.

Nearly 30% of outdoor lighting is projected to be wasted as a result of this bad design. Each year, wasteful lighting contributes 1.7 million tons of carbon dioxide and $2.2 billion in lost energy in the United States alone [12,13]. Light pollution may be caused by a variety of factors, including the following shown in Figure 1.3.

Environmental stewardship necessitates energy saving and efficiency. Installing high-quality outdoor lighting may result in a 60%–70% reduction in energy use, billions of dollars in savings, and a reduction in carbon emissions. Outside lights should be completely covered as well as light should be directed downwards, not above. Fully safeguarded lights can offer the same degree of brilliance as unshielded lamps on the ground while using less energy and money. Indoor lighting that is not required – especially in darkened office buildings – should be switched off.

1.2 THE ADVERSE IMPACT OF LIGHT POLLUTION

Although light is a significant source of pollution in the environment, it may have a negative effect on the human body, animals, and plants, as well as degrade the night sky and raise energy costs. Artificial lighting increases sky glow, limiting astronomical observations and altering the night sky. Skyglow occurs as a result of inadequately protected illumination, poor adjustment, and the addition of superfluous light fixtures. Different kinds of light may have psychological and physiological effects on humans. Certain types of illumination may enhance mood, alleviate tiredness, and relieve eye strain [14]. For 3 billion years, life on Earth was governed by a rhythm of light and dark produced exclusively by the Sun, Moon, and Stars. Artificial lights have supplanted darkness, and our cities now shine at night, upsetting the natural day-night cycle and upsetting the subtle equilibrium of our ecosystem. The negative consequences of losing this inspiring natural resource may seem intangible. However, a rising amount of data establishes a clear connection between the brightness of the night sky and quantifiable negative consequences, including the following:

1.2.1 Increasing Energy Consumption

Lighting that produces an excessive amount of light or shines when and where it is not required is inadequate. Wasted energy has severe environmental and economic consequences. Outside lighting consumes about 120 TW-hours of energy annually in the United States alone, mostly to brighten roadways and parking lots. This amount of energy is sufficient to satisfy New York City's entire power requirements for 2 years. According to the International Development Association (IDA), at least 30% of all outside lighting system in the United States is squandered, with unsecured lights accounting for the majority of this. This costs $3.3 billion and results in 21 million tons of Carbon dioxide being released every year. We'd have to plant 875 million trees per year to offset all that carbon dioxide.

1.2.2 Ecosystem and Wildlife Disruption

All life on Earth has been reliant on the Earth's constant day-night cycle for billions of years. All plants and animals have it encoded in their deoxyribonucleic acid (DNA). By illuminating the night, we have drastically disturbed this cycle. The daily cycle of light and darkness on Earth regulates life-sustaining processes including sleep, nourishment, reproduction, predator defense for plants and animals. Artificial light at night seems to have a detrimental and even lethal impact on a variety of species, including birds, mammals, amphibians, plants, and insects. Artificial light has a negative impact on the world's ecosystems [15,16]. Nocturnal animals sleep throughout the day and are active at night. Light pollution profoundly alters their nocturnal environment by turning night into day. Glare from artificial lights may also have an effect on wetland environments that are home to amphibians such as frogs and toads, whose breeding ritual includes nocturnal croaking. Artificial lights disrupt this nocturnal activity, interfering with reproduction, and reducing populations.

1.2.3 The Demise of Baby Turtles

Artificial light disrupts this nighttime activity, interfering with reproduction and contributing to population decline. Artificial lights can lead to the death of baby sea turtles. Sea turtles spend most of their lives in the water, but they lay their eggs on the beach at night. Hatchlings determine their location in the water by sensing the bright horizon above the ocean. Artificial lights are used to lure them away from the water. Thousands of hatchlings die in this way each year in Florida alone. Female turtles avoid nesting on lit beaches, which results in nests being concentrated in less illuminated and shady areas. This may result in the selection of an inefficient nesting environment or a cluster of nests, affecting the quantity and sex ratio of hatchlings produced, as well as increased hatchling mortality.

1.2.4 Impact on Aquatic Life

The reaction of fish to artificial light (attraction or avoidance) varies by species but has an effect on their natural behavior in both directions. There have been numerous researches on using artificial light in fish farming and deep-sea fish. The vast majority of studies suggest that fish resist white light sources. Nonetheless, certain species are drawn to light, and this is how sport fishermen and commercial fisheries capture them. Mukene is captured via a technique of light attraction. Anglers often employ light attraction to capture fish in the dark. Fishing using floating lights is employed to catch Mukene in Lake Victoria, as per the Food and Agriculture Organization (FAO). Mukene is caught with scoop-nets and nets dragged from canoes (lampara nets) and beaches (beach seines) [17,18]. Because this technique is employed in shallow waters along coasts, it may jeopardize nursery sites for juvenile Mukene, Nile perch, and Tilapia. In Atlantic salmon production cages, underwater light increases swimming depth and reduces the fish population. These synthetic photoperiods are used to delay sexual maturity and promote development. According to research conducted in these farms, salmon retain schooling behavior by positioning themselves in respect to the artificial light gradient. The amount of light utilized in Halibut farms affects their swimming behavior. Halibut swim less and develop more when exposed to artificial light. The fish may be especially susceptible to UV harm. Proof of damage (dermatitis, etc.) has already been found in Halibut. It's especially true for fish that have been raised indoors and are moved outside in the spring when the light is at its most intense. Shade netting can help farmers protect their cattle. White light disrupts deep-sea fish's normal behavior, as shown in the research of illumination systems used in deep-sea fish monitoring. The average number of fish spotted on camera under red light was substantially higher than under white light, as per investigations. The reasons for this are that deep-sea fish's eyes have adapted to the gloomy environment and that strong lights may cause eye injury.

1.2.5 Devastating Effects of Artificial Light on Bird Species

Migratory and night-hunting birds use moonlight and stars to navigate. Artificial light may lead them to deviate from their intended path and into dangerous urban nightscapes. Millions of birds are killed each year when they collide with ineffectively lit skyscrapers and buildings. Migratory birds rely on accurate seasonal cycles to guide their migration. Artificial light may induce them to migrate too early or too late, resulting in missed opportunities for breeding, eating, and other behaviors.

1.2.6 Devastating Effects of Artificial Lights on Plants

Excessive light may be harmful to plants and can alter their life cycle. Numerous plants rely on the duration of night to determine their rate of development and the timing of flowering and fruiting. Light pollution may alter the metabolism of plants to the point where they cannot develop flowers or colors if they are not kept in a dark location for an extended length of time. Using lights at night may prevent the photochromic hormone from being released, thereby eradicating the plants. Low-pressure sodium lights have the potential to disturb a plant's photoperiodic control of growth and development.

1.2.7 Devastating Effects of Artificial Lights on Insects

Light attracts many insects, yet artificial light may be deadly. Insect population deficits affect any animal that relies on insects for food or pollination. Certain predators take advantage of this attraction, with unintended consequences for food webs.

1.2.8 Impact on Human Health

Humans have evolved to follow the natural day-night cycle of light and darkness. Because of the widespread use of artificial lighting, the majority of us no longer get genuinely black evenings. Artificial light at night time has been proven to be harmful to the health of a human being, depression, and sleep issues, increasing the risk of obesity, breast cancer, diabetes, and prostate cancer, among other health concerns. Humans, like the majority of species on Earth, have a circadian rhythm – our biological clock – a sleep-wake cycle regulated by the day-night cycle. At night, artificial light may interrupt this cycle. Our bodies generate melatonin in response to our circadian cycle. Melatonin is beneficial to our overall health [18,19]. It includes antioxidants, improves the immune system, promotes sleep, supports the proper operation of the thyroid, pancreas, ovaries, testes, and adrenal glands, and lowers cholesterol.

Artificial light exposure throughout the night inhibits melatonin synthesis. Fluorescent lighting in a normal office environment is enough to increase blood pressure by eight points. Some research suggests that everyday exposure to moderately bright light decreases sexual performance. Numerous published research also indicates a connection between exposure to light at night and an increased risk of breast cancer, perhaps as a result of melatonin synthesis being suppressed during the nocturnal period. Cohen et al. [3] suggested in 1978 that decreased melatonin production may raise the risk of breast cancer.

1.3 PREVENTION TECHNIQUES

1.3.1 Smart Cities

A smart city is a community that uses information and communication technology (ICT) to improve operations efficiency, community knowledge exchange, federal program reliability, and resident well-being. While definitions differ, the ultimate aim of a city of the future is to leverage smart technologies and digital insights to improve inhabitants' well-being also while optimizing city activities and stimulating economic development.

A City's Smartness is Influenced by a Multitude of Things. These Characteristics Involve:

- A technologically based architecture;
- environmental activities;
- a well-functioning mass transit system;
- a robust sense of town development;
- the ability for humans to live, function, and use the city's assets.

Internet of things (IoT) devices, user interfaces (UI), professional support, and data transmission infrastructure are all used in smart cities. IoT is a framework of interlinked items which can share and exchange information, like cars, smartphones, and household devices. IoT uses detectors and appliances to acquire sufficient information, that is then saved on databases [20,21]. The use of data analytics (DA) to connect these devices allows for the integration of traditional and cyber city characteristics, resulting in increased private and public sector productivity, financial advantages, and improved resident lifestyles. Smart cities focus heavily on energy-saving and productivity. When there are no automobiles or pedestrians on the roads, smart streetlights dim utilizing smart sensors. Smart grid technology could be used to identify features, repair, and forecasting, as well as provide on-demand electricity and track energy outages.

K.H. Bachanek et al. [22] discussed and studied the use of intelligent lighting as part of the smart energy and smart city concepts. The investigation was carried out in two stages. In the first step, statistics data on power usage in each country during the last 30 years were analyzed. This research allows for the drawing of conclusions about variations in power usage as well as an indicator of energy consumption on metropolitan lighting. The attention then shifted to evaluating technologies and beliefs connected to street lighting, with an emphasis on LEDs. The next phase in the study process was to determine what the smart city notion meant in terms of smart energy. The presentation of case studies concludes the second track of research. According to the author, introducing appropriate solutions, such as dynamic lighting management, backed up by IT tools from the lighting management system, ensures the quality and effectiveness of smart lighting applications while obtaining the agreement of all urban stakeholders. One of the conditions for a concept's success and compelling execution is the use of energy efficiency indicators and the optimization of LED lighting in cities [9,23].

G. Gagliardi et al. [20] demonstrated an urban smart lighting network that can adjust the lighting brightness of street lamps independently using data from vehicles (buses, cars, motorcycles, and bikes) and/or walkers in a given region. The system can adjust the illumination brightness based on the demands of the user, lowering energy consumption. Regional controllers, video cameras, motion sensors, and electronic equipment for video processing are used to achieve this. The sensor inputs are stored in this way so that a 1–10 V control input voltage can be applied to the luminaire or controller to fade the lights optimally. Monitoring can be done on a decentralized basis on each street lamp or a group of them. Experiments show that the proposed architecture can save up to 65% of energy when compared to a normal street lamp system.

1.3.2 Smart Lighting

As part of the smart city paradigm, smart lighting is an essential part of energy administration. Smart lighting refers to a diverse area of lighting management that allows for the use of a wide range of approaches and technology in the development of concepts. It is no more merely lighting connected with a lamppost, but rather a device that permits the installation and usage of information and communications technology to achieve the highest efficiency while minimizing negative impacts on the environment [23,24]. The primary function of a smart lighting design is to develop the appropriate amount of light at any given time: where and when it is required. It should adjust the amount and luminance to improve the viewing experience depending on the type of task being performed. It must ensure the end-users health, security, and protection. It does not squander our Earth's resources quietly, but rather proactively restricts the consequences of light pollution on the biotope, as well as any other environmental implications. Through Visual Light Communication protocols, the platform might theoretically provide additional services (geo-localization, data connectivity) to end-users. The following are the features of smart lighting:

- Adapt the amount of light, the dispersion of light in space, and the quality of light interactively to improve viewing experience while adhering to any normative criteria.
- Prevent any visual disturbances that could jeopardize the security and well-being of the end-user at any time.
- Always lower the installation's energy consumption without jeopardizing the aforesaid parameters.
- Actively restrict the impacts of light pollution on the environment and biotope, while adhering to new sky-protection guidelines and policies.

K. Brock et al. [24] published a model based on an in-depth examination of four smart city scenarios (Amsterdam, Eindhoven, Stratumseind, and Veghel) conducted by Philips

Lighting over 5 years, demonstrating four distinct business framework that enables existing firms to enter the smart city sector. They compare and contrast the four types of business models on individual and mutual value generation and value capture aspects, and explain how, based on the scope and environment, each business model might be beneficial to an incumbent. They propose precise implementation advice for incumbent firms and demonstrate these business models with experiences from Philips Lighting's move from public lighting to smart cities.

1.3.3 Use of LED Light

Only warm-colored bulbs must be utilized with LEDs and compact fluorescents (CFLs) to help save the environment for future generations. When you switch to LED lights, you may lower the luminance without sacrificing visibility. Only warm light sources should be utilized for outdoor lighting, according to the IDA. Low-pressure sodium (LPS low-color-temperature LEDs), high-pressure sodium (HPS) are all examples. To reduce blue emission, use "warm" or filtered LEDs (CCT 3,000 K; S/P ratio 1.2). N. Schulte-Römer et al. [25] emphasized the decrease of light pollution by discussing sustainable lighting with the LED light paradox in their paper.

Human-centric lighting (HCL) is the next great thing in lighting. Biologically effective lighting refers to illumination that is meant to promote the biological organism and hence improve cognitive efficiency. Lighting systems that are psychologically efficient are intended to generate emotionally stimulating settings and appealing atmospheres. Visual signals in space design assist us in interacting with our surroundings. It's useful to use natural and artificial lighting to illuminate products in a place while limiting negative effects like glare and low contrast.

The benefits of the smart lighting system can be listed as follows:

- To reduce energy costs,
- Save time and energy (helping the maintenance team in the field solve problems quickly),
- To reduce maintenance costs,
- To provide efficient business,
- Remote monitoring of electrical parameters,
- To eliminate the negative psychological effect of incorrect lighting,
- To reduce light pollution caused by incorrect lighting,

Following are some of the advantages of a smart lighting system:

To save time and energy (by assisting the field maintenance staff in quickly resolving problems), to decrease the cost of maintenance, to provide effective business, provide the monitoring system of electrical parameters, to eliminate the negative psychological impact of incorrect lighting, to reduce light pollution caused by incorrect lighting. Lamp on-off, lowering, and other operations can be controlled remotely, and fixtures can be managed remotely. Defect identification can be performed remotely, and the fault location can be determined promptly and intervened immediately.

1.4 INNOVATION BASED ON SMART LIGHTING TO REDUCE LIGHT POLLUTION

Blue light exposure at night is very dangerous. Regrettably, the majority of LEDs used in, computer screens, outdoor lighting televisions, and other electronic displays, produce an excessive amount of blue light [26–28]. As LEDs improve inefficiency, a luminaire's performance is measured not only in lumens but also in the quality of light it produces. Innovative technologies such as tunable liquid lenses have enabled the development of luminaires with changeable beam patterns, focus, and color temperature. GaN substrate LEDs now provide full-spectrum light, which means that colors are not created by mixing monochromatic red, green, and blue (RGB) LEDs. This enables even more human-centric natural illumination, which benefits both productivity and sleep quality throughout the day. Control electronics are also refining. Power factor correction is now standard on low power control gear, resulting in increased full-load efficiency and decreased standby usage. These advancements are necessary to decrease the lighting load in big buildings by combining natural and artificial lighting for maximum efficiency while being environmentally friendly.

1.4.1 Light-Emitting Diode

Each day, new technologies are developed to enhance lighting performance. Organic LEDs are being developed at a rapid pace as a possible substitute for conventional LEDs. These are equivalent to high-intensity discharge lights in terms of efficiency. Motion sensor/passive infrared motion sensing-based lighting, as well as task/ambient/distribution system lighting, is changing the world of lighting. Avnet India has established a cost-effective PIR motion sensor for ceiling and wall mounting that detects motion correctly and automatically switches lights on/off based on motion detection and visibility [29]. To ease the burden and provide more people with the opportunity to make their houses more sustainable, companies have developed plug sockets that can convert standard lights to smart lighting. This may assist in offsetting the expense of implementing smart technology. Using sockets that convert conventional lights to smart lighting results in the same amount of long-term savings. A. Abdullah et al. [23] devised a mechanism for controlling street lights to reduce energy consumption. LDR, Infrared Sensor (IR), Battery, and LED were used to create the prototype. To save energy, the brightness of the bulbs was adjusted in this project. The lamps fade depending on the pace of detected item motion, such as walkers, bicycles, and autos. The greater the degree of a moving thing, the higher the frequency. The development of street lights for this concept is not the same as traditional street lights, which are regulated by a timed switch or a light sensor that constantly switches the light on at sunset and off at sunrise. According to the research, motion detection systems can save up to 40% of energy per month.

1.4.2 Smart Street Lighting

Streetlights have long been a source of concern because of how much energy they use and how they contribute to light pollution in cities by shining luminously when no one is present as shown in Figure 1.4. These are only a few of the issues that some smart lighting solution companies are attempting to solve with their products. They seek to improve city safety,

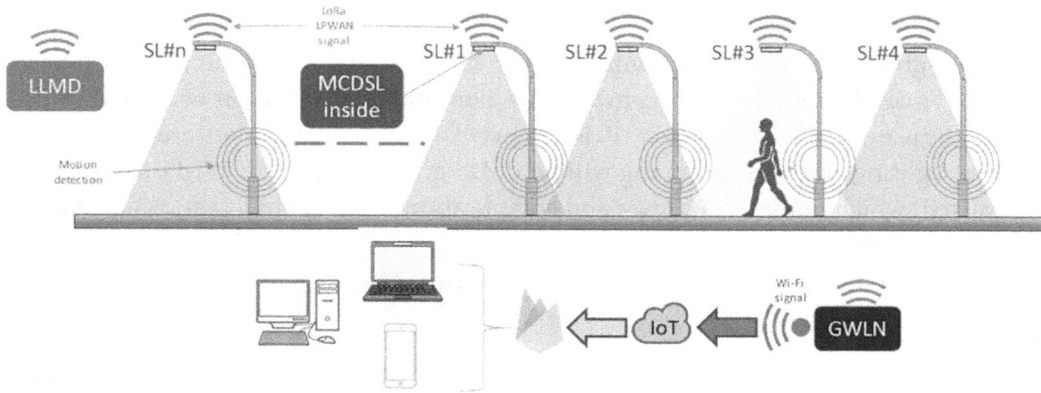

FIGURE 1.4 Lighting level measurement device (LLMD) system being used in smart lighting detection technology. Adapted with permission [30], MDPI, Distributed under Creative Commons Attribution License 4.0 (CC BY).

make lights more effective, make poles multi-functional, and reduce costs of maintenance and energy, amid other things, in a variety of methods. Maksat Technologies' smart street lighting system includes streetlight control and analysis, as well as image surveillance and tracking and data transfer. To accomplish multi-node low data transfer, ZigBee technology is utilized to gather data. Temperature, humidity, carbon dioxide, carbon monoxide, air quality, air pressure, and other environmental sensors are also included.

A single light control device and data collection unit are connected to the lamp post for each streetlight. The light and data control units are managed via the remote unit gateway. Up to 200 light control and 50 data control units may be managed by a single remote unit gateway. A smart street lighting system that uses two types of sensors: wireless sensor networks (WSN) and pyroelectric infrared sensors (PIR) were developed [26]. The design of an intelligent street light system employing a Zigbee device is included in this paper. WSNs and a monitoring center make up this system type. The monitoring center's primary function is to interact with sensor networks via wired or wireless methods. Information regarding pedestrian and vehicular mobility can be collected by analyzing the characteristics of each street lamp. Illumination sensors, PIR sensors, and other sensors perform tasks such as detecting and controlling.

ZigBee Technology's Benefits:

- Lower Power Score
- The size is little
- Affordability
- Long-lasting battery
- Can handle a high number of nodes
- No licensing fees for an open standard procedure
- A variety of sources are available
- Easy to maintain

1.4.3 Wireless Communication Technology Lighting

Explosive growth in wireless communication technologies, for instance, the Internet Protocol Smart Object (IPSO) organization, a non-profit organization dedicated to promoting the use of Internet Protocol (IP) for Smart Object communication, are vital enablers of smart lighting. The next big thing will be LED lights with built-in Wi-Fi. They are not centralized, which minimizes clutter and complications. Rather than that, Wi-Fi-enabled lights link to your home's network through a smartphone, a smart TV, or many supplementary smart devices, and the lights may be switched off even while we are not at home.

1.4.4 Luminaries

This sector is advancing at a breakneck pace in terms of technology. Currently, a diverse array of creative as well as lucrative smart lights and equipment are accessible, some of which are eligible for government subsidies. Manufacturers of luminaires provide new solutions that address customer demands, for example, human-centric lighting and linked or IoT lighting. Innovations of the modern time, such as the IoT, have a plethora of uses in the energy industry, namely in the supply, distribution, and transmission of energy, as well as demand. The IoT may be used to increase energy efficacy, increase the proportion of renewable energy, and mitigate the ecological effect of energy consumption. The IoT is a relatively new technology that makes use of the Internet to link physical objects or "things." Physical gadgets include household appliances and industrial machinery. These gadgets, when equipped with the proper sensors and communication networks, may provide important data and allow the provision of a variety of services to individuals. For example, intelligently managing the energy use of buildings provides cost savings in the long run. The IoT has a broad variety of applications, including manufacturing, logistics, and the construction sector. Additionally, the IoT is extensively used in monitoring the environment, as well as healthcare systems and services, well-organized building energy management, and drone-based applications. Figure 1.5 shows the utilization of smart home networks to reduce light pollution.

IoT plays a crucial part in the construction of smart houses, which may aid in the efficient use of light and the avoidance of light pollution. Frugal Labs IoT Platform (FLIP) is indeed an open-source software IoT system constructed by Frugal Labs in Bangalore, India for anyone interested in studying and working with IoT. The FLIP device is linked to detectors, lighting, climate control, cameras, window and gate systems, and other equipment in the planned smart home system. The gateway connects the flip gadget to the Internet. The proposed smart home network's gateway is critical because it adds a layer of security to the network, keeping it more secure. The presented smart home system can monitor the environment and control home appliances, air quality and security, doors, locks, and window frames from a distant area, generate alert notifications based on predetermined situations, regulate internal temperature by detecting the intensity of light and heat extent in the cabin, and therefore auto-adjust lighting system, and generate alerts and notifications based on preset conditions, and generate alerts and notifications based on preset conditions. Toma et al. [31] explained the benefits of using an IoT system to reduce numerous types of pollution, particularly light pollution. They looked into the essential components

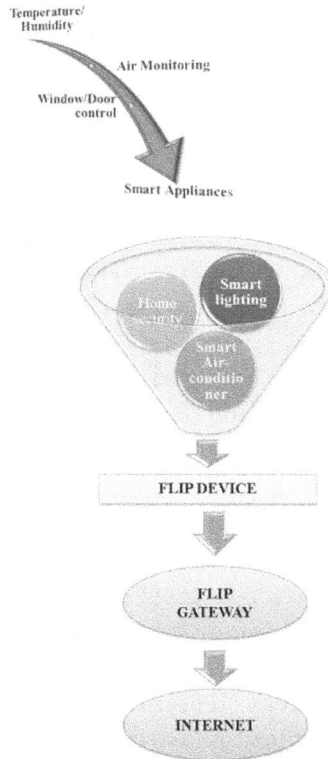

FIGURE 1.5   Smart Home Network to reduce light pollution.

of a real-time pollution surveillance network, such as IoT connection methods, sensors, data collection and transfer over communication channels, data protection and reliability, and data acquisition and transfer over communication channels.

1.4.5 Artificial Intelligence-Based Lighting

Lighting design often includes novel techniques, and control and artificial intelligence have established themselves as evolutionary forces, demonstrating their capacity to radically change existing processes. The potential for artificial intelligence (AI) in the lighting industry is intriguingly broad, impacting the many stages of the lighting life cycle such as design, implementation, certification, and customization. In the future, a mix of LED lights and AI may be used to improve the illumination in the buildings we live in while also reducing energy usage. AI can give light the same quality as sunshine – changes are so subtle that they are nearly imperceptible. This offers many advantages since our eyes don't want to stay in the same place all of the time. People can see in a range of light spectrums and appreciate the variation in lighting, but the manner light has been conceptualized till now – as either on or off – overlooks the fact that they can see in a variety of light spectrums and enjoy diversity in illumination. LED lights differ from previous generations of bulbs in that they allow precise color and brightness management and may be incorporated into materials such as walls and ceilings. This opens up many possibilities for creating artificial light that mimics natural light. By monitoring and analyzing the indoor environment, a lighting system powered by AI can optimize and modify light settings to enhance the user experience and well-being. Beyond end-users and tenants, the use of such a system extends to other stakeholders like property owners as well as facility managers. A network of data-driven lighting components generates data continuously, which is collected and stored centrally. AI algorithms may be programmed to operate on the source component, for instance, a sensor, for decentralized, real-time decision making, or on a server for centralized decision making. Additionally, the information collected may be utilized for different Building Management Systems.

Nevertheless, the technique does have certain limitations. While cameras enhanced with AI can identify exact occupancy and movement in a room or area, the visual stream would have to conform to stringent privacy regulations. Another major issue is the poor human comprehension of AI, which hinders its pace of adoption too. In summary, there are many possibilities for bringing AI to the lighting sector, with the potential to significantly enhance user experience, well-being, production, and eventually cost-effectiveness.

Researchers in Europe have developed an AI system capable of tricking office workers into believing that all of the building's lights are on even when they're off. "SceneUnderLight is a European Union-funded project that brings together Osram, the University of Verona, and the Italian Institute of Technology (IIT) in Genova." Its mission was to devise a technique for providing optimum comfort and security while achieving significant energy savings." It has successfully integrated sophisticated computer vision research and cutting-edge technology to create a lighting control paradigm in which each employee sees the whole workplace as "all lighted," while lights that are not seen are turned off. The program continuously assesses the propagation of light in workplaces and calculates how much of it is seen by individuals. Then, intelligent lighting management automatically changes the illumination depending on the presence of individuals and their location inside the workplace. The idea is especially well-suited for big open-plan workplaces since it allows for dimming of luminaires located farther away without impairing employee comfort or feeling of security.

Another successful research endeavor, OpenLicht, stands for "smart design" and was launched in September 2016 to provide tailored lighting solutions via the use of open source and novel resources. The German Ministry of Education and Research has supported ten research projects as part of the "Open Photonics" program. The Open Photonics initiatives are exploring a variety of objectives, including open innovation approaches for improving photonic component and system usage, as well as open-source methods for encouraging their widespread use. Additionally, they embrace methods for more directly engaging the public in scientific endeavors. It automatically adjusts the lighting in the room to the user's position and activities, like watching television or reading, remembers the user's preferences, and it can also respond to unexpected circumstances to a certain extent. AI seems to be the unifying evolutionary force in all industries, with the potential to fundamentally alter the way things are done. While the uproar around technology is very thrilling and hopeful, it remains an elusive dimension for the average population.

1.5 ROLE OF NANOTECHNOLOGY IN THE PREVENTION OF LIGHT POLLUTION

Several researchers have been exploring nanotechnology for solutions to problems in medical, computer programming, the environment, or even athletics for the past two decades. Nanoparticles are pushing the boundaries of technology's possibilities, resulting in the development of new and improved pollution management strategies. Nanoparticles, which are defined as particles with a length of 1–100 nm (1 nm equals 1 billionth of a meter), have immense potential for the future of research. Their small size allows them to target highly specific places in the body, such as sick cells, without damaging healthy cells. Furthermore, at the nanoscale scale, elemental qualities can drastically change: many become better conductors for heat or reflecting light, others change color, some become sturdier, and others change or gain magnetic properties. The durability of certain plastics in the nanoscale range is comparable to that of steel. Nano-silicon dioxide crystals are already used by tennis racquet manufacturers to increase equipment performance. Because microscale defects between molecules are removed at the nanoscale, super-strength and other remarkable qualities arise. Nanoparticles without these faults enable materials to achieve their maximum chemical bond strength. Nanoparticles' unique features, as well as their vast surface area, make them ideal for developing efficient energy management and pollution control approaches. For instance, if super-strong polymers could have been used to substitute metal in automobiles, tractors, airplanes, and other industrial equipment, huge energy savings, and pollution reductions would result. Nanoscale materials are also being used to enhance batteries, allowing them to provide more power faster. Batteries have been recharged using nanomaterials that collect enough light to convert them into electrical energy. Other ecologically friendly innovations comprise non-thermal white LEDs that use less energy and SolarStucco, a self-cleaning covering that uses photocatalysts to dissolve organic contaminants.

Using a new nanoscale structure, the researchers, led by electrical engineering professor Stephen Chou, enhanced the luminosity and efficacy of LEDs made of organic materials (flexible carbon-based sheets) by 57%. This method should improve LEDs manufactured of inorganic (silicon-based) materials, which are now the most common. The concept improves the visual clarity of LED displays by 400% when compared to standard methods. The researchers detail how they achieved this by devising a technology that alters light on a wavelength smaller than a single wavelength in an article published online on August 19 in the journal Advanced Functional Materials. Remote phosphorous semiconductors are being employed by firms like Nanosys to make LEDs that convert blue light into a warmer shade of white, closer to the classic white of fluorescent bulbs. The phosphorous in this experiment was made from "nano-materials." As human eyes are sensitive to the color green, the LEDs have a higher degree of green to offer us a false sensation of brightness without really increasing the display's brightness. This aids in the creation of outstanding picture quality while consuming very little energy, making it the predominant method used in devices with display panels. This is encouraging news for people interested in the applications of nanotechnology in numerous industries. Nanosys claims that through this effort, it will be able to build LEDs in almost any color, which would be a major step forward into the existing LED displays and other electrical gadgets that emit powerful light.

1.6 MITIGATING LIGHT POLLUTION BY ALTERING OUTDOOR LIGHTING

1.6.1 Employ LPS or LED Lights

For the amount of light they provide, both LED and LPS bulbs are extremely energy-efficient. LPS bulbs emit a very warm light that has little impact on the environment. When picking LEDs, make sure they're low-power, well insulated, and have a warm (yellow or amber) correlated color temperature (CCT) of 3,000 K or less (2,700 K is preferable). The aim is to keep blue and green light outputs to a minimum.

1.6.2 The Light Should Be Directed Where It is Required

Outside, use reflectors to assist illuminate any walkways. This prevents you from having to buy an entire row of lights to illuminate the same road. To minimize glare, choose the more directed, lower light options whenever practical.

1.6.3 Lights with Bulb Caps and Shields Should Be Purchased

Bulb caps and shields keep the light from shining directly into the sky. Lights ranging from "non-cutoff" to "full cutoff" are available. Full cutoff barriers minimize the most "bleeding" of light between the shields.

1.6.4 Direct the Light From Your Outdoor Lights Downward

This reduces the amount of stray light that enters the environment. Even at night, light reflected off the ground offers ample light for nighttime activities.

1.6.5 Install Lights that Detect Movement

Only when they detect movement in their field of view do these lights turn on. These can be used on walkways, near garages, in gloomy areas, and even in halls. Because the lights are only turned on when they are needed, they soon pay for themselves.

1.6.6 Smart Lighting Should Be Installed

Smart lighting systems are an excellent way to keep greater control over your lights. Smart lights can be controlled and configured using your smart device, allowing you to achieve pinpoint efficiency in your house. They can also help you save money on your electric costs.

Adjust the brightness of your lights to the job at hand, and save the configuration for subsequent use. At the touch of a button, you can get the bare minimum of lights. Set your smart lights to turn on exclusively during specific hours of the day or to turn off at specific intervals. This assures that when you leave for work, all of your lights are turned off, even if you forget to do so. Remote control of smart lighting is also possible. You can turn off lights from your smart device if you've left the house and they've been left on.

1.7 INITIATIVES TO PREVENT LIGHT POLLUTION

- The Campaign to Protect Rural England (CPRE) is among the UK's older and most well-known environmentalist groups. It was founded in 1926 and presently has a membership of approximately 40,000 people. One of their goals is to maintain and enhance black skies by avoiding the growth of light pollution that is unwanted. The CPRE provides a boundary between rural and urban zones since darkness at night-time is among the most distinctive characteristics of rural communities. CPRE provides local governments with competent planning assistance and is known for advocating for European and International policies.

- The New England Light Pollution Advisory Group (NELPAG) was established in 1993 in the United States. It provides expert advice and funds to help resolve conflicts caused by obstructive light sources, as well as assisting in the integration of municipal and state lighting legislation and rules and regulations as well as explaining to the public about the benefits of using effective, glare-free external night light sources to reduce light pollution. This organization is credited with popularizing and promoting conferences in the New England territory for illumination specialists, astrophysicists, lawmakers, reporters, and the public in general to discuss the progress of exterior nighttime lighting and to evaluate what is suitable for illumination and when it should be used.

- The IDA is a grassroots group that is one of the world's older and most well-known. Its goal is to promote awareness about the significance of darkness at nighttime and deliver awareness and advice on adequate outdoor lighting in order to preserve and protect the nighttime environment and dark sky history. It was formed in 1988 in the United States. Local chapters of the IDA sprang up rapidly, followed by international chapters. There are presently 50 formal IDA chapters worldwide, with over 20 international branches spanning five continents. This worldwide network of IDA activists for teaching decision-makers, municipal officials, and the general public about the need of safeguarding the night sky in their communities. They also help with the development of lighting regulation concepts and also the implementation of lighting laws to eliminate light pollution. A technical group within the IDA organization recommends the executive committee on the expanding concerns of outdoor lighting practices which affect human health as well as the survival of flora and fauna and works to draught rules to address these issues.

- The International Dark-Sky Places (IDSP) initiative was established by the IDA in 2001 to recognize and protect areas with extraordinary nighttime features for future generations, as well as encourage awareness of the issues involved. It encourages cities, parks, and conserved places all across the world to preserve and protect dark locations by promoting smart lighting policy and public awareness. Candidates must go through a rigorous application process that includes demonstrating strong community support for dark sky protection. There are currently 130 IDSPs certified working globally.

- The European Union/European Commission supported STARS4ALL, a project to encourage investigations in Europe and build a collaborative attention infrastructure for fostering dark skies, as part of the Horizon 2020 Program between January 2016 and January 2018. Among the breakthroughs was the Dark Sky Meter app that monitors sky luminance and exchanges data with researchers all over the world, as well as a TESS-W photometer capable of measuring and providing a continual assessment of the night sky illumination for light pollution investigations.

1.8 CHALLENGES AND FUTURE OUTLOOKS

Although LEDs and compact fluorescent light bulbs (CFLs) may assist decrease energy consumption and preserving the environment, they should be used in conjunction with warm-colored lights. Dimmers, motion sensors, and timers all contribute to lowering the average lighting level and thus saving more energy. Global solutions for smart cities based on Low Power Wide Area (LPWA) networks have the potential to substantially decrease energy consumption, carbon emissions, and operational complexity. Wireless connection innovations lead to cost savings and improved maintenance procedures, benefiting both the environment and governments' budgets. Streetlights are now equipped with LED lights that are centrally controlled and monitored. The brightness of the lights may be adjusted to suit individual requirements. Additionally, voice commands may be used to operate these smart lights. The ecological consequences are only partly known at the present, and the area of light pollution urgently needs further study. Collaboration between physical scientists, engineers, medical specialists, biologists, and ecologists should be explicitly interdisciplinary. Environmental issues, particularly the assessment of light properties and the comprehension of tropical and aquatic habitats, need further study.

1.9 CONCLUSION

Power consumption is increasing rapidly in today's world due to urbanization in both established and emerging cities. Increased demand must be addressed, which may pose issues with existing systems and infrastructure. Furthermore, some negative effects of human activities inflict substantial harm to the environment (therefore energy resources), humans, and nature, as well as disrupting economic, societal, and environmental balances. For the environment, selecting the appropriate lighting product is critical. The most common reasons for streetlight lighting nowadays are investment rates and energy conservation. However, making the best decision is significantly more difficult. Though there's an ongoing argument about the life term of LED tech and maintenance fees under various real-world settings, the technology's energy effectiveness and the relatively minimal maintenance demands of long-lasting LED modules appear to substantiate the sustainability claim. However, LED devices are rapidly being offered and used for outdoor illumination as well as indoor lighting. Short-wavelength illuminants, such as metal halide or LEDs, can drastically enhance skyglow, inhibit melatonin, and interrupt sleep cycles. Furthermore, this blue-rich light causes glare, which reduces vision for the aged. To maximize the energy consumption of cold-white LEDs, filters that modify the spectral electric grid and produce warm-white light are required.

An additional investigation could point to strategies to improve the spectrum and energy efficiency of LEDs. In fluctuating situations, it is critical to make efficient use of available resources and to avoid undesirable repercussions in order to avoid jeopardizing comfortability and adversely impacting the environment. It is not appropriate to create a system just based on necessities. The requirements should be accomplished at the lowest possible cost and with the lowest amount of negative impact while providing the most benefit. As a result, after taking into account all of the characteristics of the region to be created; it must be developed with the intent in mind. Smart systems spring to mind at this moment. Outdoor lighting is one of the most basic requirements of humans in general and cities in particular. As a result, all circumstances must be considered while addressing this need. In this regard, smart technologies that cut energy consumption and eliminate light pollution by guaranteeing proper outdoor lighting are required to avoid unwanted repercussions while developing outdoor lighting.

A smart outdoor lighting system contains a luminaire and network infrastructure that consumes very little energy and operates properly to avoid light pollution. Given that outdoor lighting accounts for 14.75% of the grid's energy consumption, the reductions in this area will have a considerable impact on energy production. Furthermore, when light pollution-induced dark sky shortage is alleviated through proper and intelligent outdoor lighting, individuals, the ecosystem, and scientific studies will benefit. This is a simple problem to tackle. It's just too straightforward for someone to comprehend, and it doesn't require the development of new technologies. Because of so many falsehoods that have been perpetuated over the years, it does require a shift of mindset. Now is the time to make a change!

Aim the light down to where it is needed, use only what is necessary, and when done, turn it off.

REFERENCES

[1] D. Henderson, Valuing the stars: on the economics of light pollution, *Environ. Philos.* 7 (2010) 17–26. https://doi.org/10.5840/envirophil2010712.

[2] J.S. Cha, J.W. Lee, W.S. Lee, J.W. Jung, K.M. Lee, J.S. Han, J.H. Gu, Policy and status of light pollution management in Korea, *Light. Res. Technol.* 46 (2014) 78–88. https://doi.org/10.1177/1477153513508971.

[3] R. Rajkhowa, Light pollution and impact of light pollution, *Int. J. Sci. Res.* 3 (2012) 2319–7064.

[4] H.S. Lim, J. Ngarambe, J.T. Kim, G. Kim, The reality of light pollution: a field survey for the determination of lighting environmental management zones in South Korea, *Sustainability.* 10 (2018) 374. https://doi.org/10.3390/su10020374.

[5] N.A. Kerenyi, E. Pandula, G. Feuer, Why the incidence of cancer is increasing: the role of "light pollution," *Med. Hypotheses.* 33 (1990) 75–78. https://doi.org/10.1016/0306-9877(90)90182-E.

[6] I. Zatsepin, M. Svitek, Night earth observation for Smart Cities, *2015 Smart Cities Symp. Prague, SCSP.* 20 (2015) 34–37. https://doi.org/10.1109/SCSP.2015.7181556.

[7] K.J. Gaston, T.W. Davies, J. Bennie, J. Hopkins, Reducing the ecological consequences of night-time light pollution: options and developments, *J. Appl. Ecol.* 49 (2012) 1256–1266. https://doi.org/10.1111/j.1365-2664.2012.02212.x.

[8] R.K. Ibrahim, M. Hayyan, M.A. AlSaadi, A. Hayyan, S. Ibrahim, Environmental application of nanotechnology: air, soil, and water, *Environ. Sci. Pollut. Res.* 23 (2016) 13754–13788. https://doi.org/10.1007/s11356-016-6457-z.

[9] M.S. Faid, N.N.M. Shariff, Z.S. Hamidi, The risk of light pollution on sustainability, *ASM Sci. J.* 12 (2019) 134–142.

[10] W.A. Jabbar, M.A. Bin Yuzaidi, K.Q. Yan, U.S.B.M. Bustaman, Y. Hashim, H.T. Alariqi, Smart and green street lighting system based on arduino and RF wireless module, *2019 8th Int. Conf. Model. Simul. Appl. Optim. ICMSAO 2019.* (2019) 0–5. https://doi.org/10.1109/ICMSAO.2019.8880451.

[11] K.M. Zielińska-Dabkowska, K. Xavia, K. Bobkowska, Assessment of citizens' actions against light pollution with guidelines for future initiatives, *Sustainability.* 12 (2020) 4997. https://doi.org/10.3390/su12124997.

[12] K.H. Kim, J.W. Choi, E. Lee, Y.M. Cho, H.R. Ahn, A study on the risk perception of light pollution and the process of social amplification of risk in Korea, *Environ. Sci. Pollut. Res.* 22 (2015) 7612–7621. https://doi.org/10.1007/s11356-015-4107-5.

[13] J. Lyytimäki, P. Tapio, T. Assmuth, Unawareness in environmental protection: the case of light pollution from traffic, *Land Use Policy.* 29 (2012) 598–604. https://doi.org/10.1016/j.landusepol.2011.10.002.

[14] Z. Song, X. Li, Hazards, causes and legal governance measures of China's urban light pollution, *Nat. Environ. Pollut. Technol.* 16 (2017) 975–980.

[15] K.T. Yeun, S.-H. Bae, A study on development direction of smart pole for smart city construction, *J. Chosun Nat. Sci.* 12 (2019) 1–8.

[16] Z. Karagöz Küçük, N. Ekren, Light pollution and smart outdoor lighting, *Balk. J. Electr. Comput. Eng.* 9 (2021) 191–200. https://doi.org/10.17694/bajece.874343.

[17] J. Škvareninová, M. Tuhárska, J. Škvarenina, D. Babálová, L. Slobodníková, B. Slobodník, H. Středová, J. Minďaš, Effects of light pollution on tree phenology in the urban environment, *Morav. Geogr. Rep.* 25 (2017) 282–290. https://doi.org/10.1515/mgr-2017-0024.

[18] I.A. Vinogradova, V.A. Ilyukha, E.A. Khizhkin, L.B. Uzenbaeva, T.N. Ilyina, A. V. Bukalev, A.I. Goranskii, Y.P. Matveeva, V.D. Yunash, T.A. Lotosh, Light pollution, desynchronosis, and aging: state of the problem and solutions, *Adv. Gerontol.* 4 (2014) 260–263. https://doi.org/10.1134/S2079057014040213.

[19] K.T. Yeun, P.N. Hong, Design of path prediction smart street lighting system on the Internet of Things, *J. Chosun Nat. Sci.* 12 (2019) 14–19.

[20] G. Gagliardi, A. Casavola, M. Lupia, G. Cario, F. Tedesco, F. Lo Scudo, F.C. Gaccio, A. Augimeri, A smart city adaptive lighting system, *2018 3rd Int. Conf. Fog Mob. Edge Comput. FMEC 2018.* 23 (2018) 258–263. https://doi.org/10.1109/FMEC.2018.8364076.

[21] T.W. Davies, T. Smyth, Why artificial light at night should be a focus for global change research in the 21st century, *Glob. Chang. Biol.* 24 (2018) 872–882. https://doi.org/10.1111/gcb.13927.

[22] K.H. Bachanek, B. Tundys, T. Wiśniewski, E. Puzio, A. Maroušková, Intelligent street lighting in a smart city concepts—a direction to energy saving in cities: an overview and case study, *Energies.* 14 (2021) 1–19. https://doi.org/10.3390/en14113018.

[23] A. Abdullah, S.H. Yusoff, S.A. Zaini, N.S. Midi, S.Y. Mohamad, Energy efficient smart street light for smart city using sensors and controller, *Bull. Electr. Eng. Inf.* 8 (2019) 558–568. https://doi.org/10.11591/eei.v8i2.1527.

[24] K. Brock, E. den Ouden, K. van der Klauw, K. Podoynitsyna, F. Langerak, Light the way for smart cities: lessons from philips lighting, *Technol. Forecast. Soc. Change.* 142 (2019) 194–209. https://doi.org/10.1016/j.techfore.2018.07.021.

[25] N. Schulte-Römer, J. Meier, M. Söding, E. Dannemann, The LED paradox: how light pollution challenges experts to reconsider sustainable lighting, *Sustain.* 11 (2019) 6160. https://doi.org/10.3390/su11216160.

[26] P. Yadav, E. Studies, smart street lighting system, A platform.pdf, *Int. J. Sci. Res.* 4 (2021) 2013–2016.

[27] J. Rodrigo-Comino, S. Seeling, C. Egner-Duppich, M. Palm, M. Seeger, J.B. Ries, Challenges and opportunities facing light pollution: smart light-hub interreg, *Proceedings.* 30 (2020) 63. https://doi.org/10.3390/proceedings2019030063.

[28] S. Garcia-Segura, X. Qu, P.J.J. Alvarez, B.P. Chaplin, W. Chen, J.C. Crittenden, Y. Feng, G. Gao, Z. He, C.H. Hou, X. Hu, G. Jiang, J.H. Kim, J. Li, Q. Li, J. Ma, J. Ma, A.B. Nienhauser, J. Niu, B. Pan, X. Quan, F. Ronzani, D. Villagran, T.D. Waite, W.S. Walker, C. Wang, M.S. Wong, P. Westerhoff, Opportunities for nanotechnology to enhance electrochemical treatment of pollutants in potable water and industrial wastewater-a perspective, *Environ. Sci. Nano.* 7 (2020) 2178–2194. https://doi.org/10.1039/d0en00194e.

[29] M. Dahan, A.A. Mbacké, O. Iova, H. Rivano, M. Dahan, A.A. Mbacké, O. Iova, H. Rivano, D. Smart, A. Aziz, Challenges of designing smart lighting, *EWSN 2020-Int. Conf. Embed. Wirel. Syst. Networks. ACM.* 1 (2020) 1–6.

[30] F. Sánchez Sutil, A. Cano-Ortega, Smart public lighting control and measurement system using lora network, *Electronics.* 9 (2020) 124. https://doi.org/10.3390/electronics9010124.

[31] C. Toma, A. Alexandru, M. Popa, A. Zamfiroiu, IoT solution for smart cities' pollution monitoring and the security challenges, *Sensors (Switzerland).* 19 (2019) 3401. https://doi.org/10.3390/s19153401.

CHAPTER 2

Freeform Optics/Nonimaging: An Introduction

Ajit Behera

National Institute of Technology

CONTENTS

2.1 INTRODUCTION

Generally, freeform optics can be categorized into two types (i.e. non-imaging and imaging freeform optics). Freeforms are optical shapes or optical surfaces that are designed with unsymmetrical configurations. Manufacturing a freeform is the same as that of a highly complex asphere, as shown in Figure 2.1. The surface form and local slope change in the asphere are all parameters that tend to the complexity of the shape. According to the 17,450-1:2011 ISO standard, the design of freeform optics can have at least one freeform surface with no rotational or translational symmetry about the axis which is normal to the mean plane [1]. The non-imaging type optical freeform surfaces include beam shaping, concentrators, and illumination. The optical freeform surfaces can control the light direction which helps in improvement in uniformity and efficiency of energy supply [2]. Development of non-imaging began in the mid-1960s in three different locations by V. K. Baranov (USSR), M. Ploke (Germany), and R. Winston (USA), leading the individual origin of the first non-imaging concentrators. Later these are also used for solar energy

DOI: 10.1201/9781003185109-2

Type of Asphere

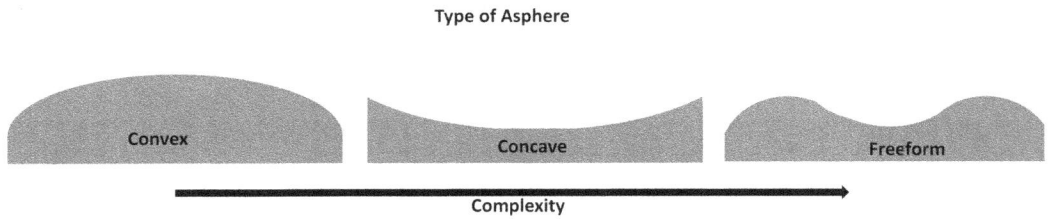

FIGURE 2.1 The free form asphere, showing the complexity surface as a comparison to the convex and concave surface.

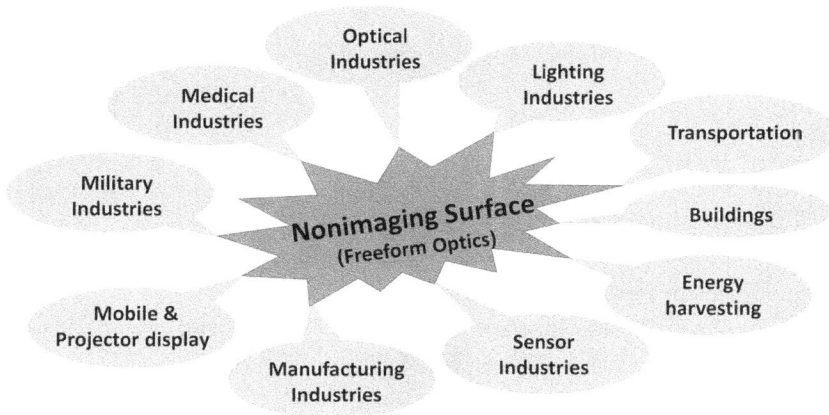

FIGURE 2.2 Use of freeform optics in various industries.

concentration [3,4]. Recently many industries and universities are investigating non-imaging optics. Currently, the CeDInt-UPM's Advanced Optics group is considered the largest research group related to this research area [5]. Non-imaging optics can concentrate the sunlight to 84,000 times the ambient intensity of sunlight. This can exceed the flux at the sun's surface and also can approach the theoretical limit of heating of an object to the temperature of the surface of the sun based on the second law of thermodynamics [6].

Non-imaging optics plays a great role to improve light quality, which is responsible to reduce light pollution. A good optical design determines the energy efficiency of illumination devices and the minimization of light pollution [7]. Figure 2.2 is showing the various type of industries that explores freeform optic devices. In addition, this technology, therefore, achieves much higher design freedom that can be applied for precise control of energy transmission.

2.2 DESIGN AND WORKING PRINCIPLE OF FREE FORM

In the case of high-end imaging of optics design, optical surfaces are changed from traditional spheres and aspheres to freeform design. Freeform applications are broadly accepted in a wide range of spectrums such as for infrared, visible, and ultraviolet rays. Therefore, easier shape tunability with high precision is needed. This increases the challenges of dynamic range and accuracy of their measurement technologies [8]. The easiest route to

Source

Receiver

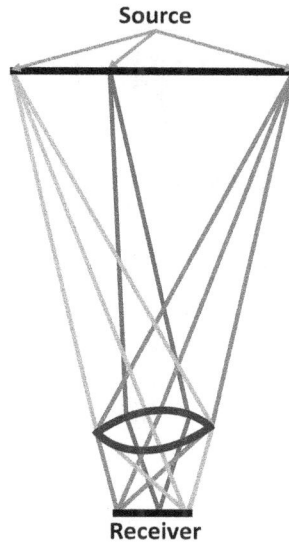

FIGURE 2.3   A simple design by edge ray principle.

make the freeform optics is the "string technique" based on the edge ray principle shown in Figure 2.3. The edge ray principle in the figure shows a lens receives light from a source and directs it toward the receiver. In the early 1990s, more sophisticated methods were developed which can handle the extended light sources in a better way compared to the edge-ray method [9]. These were primarily designed to give a solution to the optic structural complexity associated with solid-state automobile headlamps and complex lighting systems. The edge ray principle tells, if rays coming from the edge of the light source are redirected towards the edge of the receiver, this ensures that all rays coming from the inner point of the light source reach the receiver [10]. There are no conditions to form an image, the only purpose is to transfer the light from the light source to the target surface. So, based on the transfer of radiation from the source to the target, non-imaging optics becomes simpler with higher efficiency when compared to imaging optics.

Nonimaging optics equipment is constructed based on techniques such as a flow-line method or Winston-Welford design method, Simultaneous Multiple Surface (SMS) or Minano-Benitez design method, Minano design method, multi-parameter optimization, tailoring method, and Geometric construction method [11–14]. Flowline technology is generally the most widely used technology, although SMS is extremely versatile and has proven to produce a variety of optical systems. The 3D SMS design technique was made in 2003 by an optical scientific group at the Light Prescriptions Innovators Laboratory [15]. The abbreviation SMS is suggested as it tunes the simultaneous manufacture of multiple optical surfaces. The design technique in 2-D was initially developed by Miñano and Benitez, and later the method was developed for 3-D geometry [17]. The developed 3D method permits the optic design with freeform surfaces, which may not have any kind of symmetry. Later a third design method has been proposed, which is not applied to real-world application till today and remained as a theoretical optics. In multichannel or stepped flow-line optics,

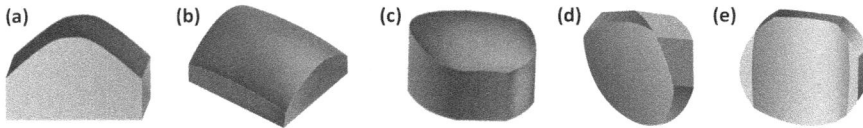

FIGURE 2.4 Common 3D freeform optic surfaces: (a) Acylinder, (b) Toroid, (c) Atoroid/biconic, (d) Off-axis parabola, and (e) Anamorph.

the light splits into several channels and then after recombination gives single output. The major utilization of this technique is found to design the ultra-compact optics [16]. Various types of 3D freeform surfaces are given in Figure 2.4.

Fabrication of a nonimaging surface is identical to that of a highly complex asphere. Surface form and local change in slope are the two major factors that influence the complexity of the shape and the fabrication process. For example, Optimax fabricates freeform surfaces having limits and tolerances to the fabrications for specific optical aspheres, prisms, cylinders, and spheres having a diameter around 0.010 mm, center thickness ±0.050, Irregularity-profilometry ±1.0 μm, and surface roughness 10 Å RMS [18]. During the designing of a freeform surface from scratch, for better convenience, we should assume that the source is a point. There are two major surfaces making techniques known as the flux mapping method and the support ellipsoid method [19,20]. The flux mapping method is defined by finding a corresponding map between the light source and the desired target of interest. The distributions of source and target flux are divided into bins that contain equal amounts of energy in them [21]. By comparing these source and destination grids, each destination location is related to the corresponding source origination direction. Next, the designer needs to create a surface that reflects (mirrors) or refracts (lenses) the direction of each ray to the corresponding target location. The simplicity of this method has some limitations [22]. Allocations can only be easily set for simple light distributions. Also, there is no guarantee that the constructed surface will be smooth and continuous. This may seem counter-intuitive, but some mappings cannot be achieved using continuous surfaces. A more sophisticated approach is needed to enforce surface continuity [23,24]. The method of supporting ellipsoids provides a general framework for solving the problem of discrete targets. The light distribution of the target is like a pixelated image. An ellipsoid is associated with each pixel in the image. Its first focus coincides with the point source and its second focus coincides with the pixel position [25]. Given the optical properties of the conic section, the light emitted by the light source that hits the ellipsoid is reflected at the second focal point. Therefore, the light is distributed over all pixels of the target. All that is left to do is to scale the ellipsoid so that the required amount of flux reaches each pixel. The concept is simple, but when a large number of target pixels are used, the scaling process is computationally intensive [26,27].

Several industries have been designed machines to make the freeform surface. For example, the industry, TNO, has recently acquired a grinding machine, a polishing machine, and a measurement machine for freeform surfaces. The new grinding machine (a Cranfield Precision TTG 350, Figure 2.5a) makes free forms in addition to convex, concave, and flat shapes. Instead of the common three-axis system, the machine has a unique

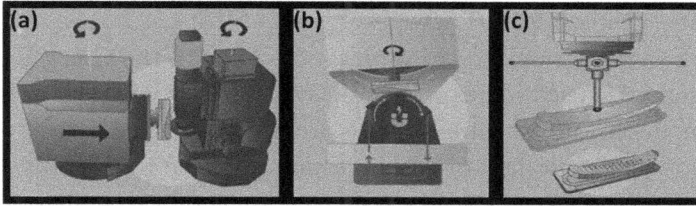

FIGURE 2.5 (a) Free form grinding machine, (b) Free form polishing machine, and (c) Free form surface precision measurement system.

two vertical rotary axes system. This two axes rotation capability enables the machine to grind with very large accuracy. The polishing machine (a Q-flex 300, Figure 2.5b) can be used to polish ceramic and glass lenses and mirrors. The surface roughness is measured by light. Then computer algorithms are used to correct the desired surface, to a very high degree of accuracy. Cerium oxide or diamond (fluid) is used for polishing known as the magnetorheological finishing technique. Optical components are measured in each stage that is before, during, and after production and finishing, checking the retained specifications. The ZEISS PRISMO ultra can measure free forms to a high degree of accuracy. The Coordinate Measuring Machine in Figure 2.5c is a fully automated measuring machine that has a probe to measure the fabricated components. Once again the measurement is taken after grinding.

2.3 NONIMAGING OPTICS FOR THE SOLUTION OF LIGHT POLLUTION

Nonimaging optics works on two design principles. That is, the concentration of solar energy (maximizing the energy quantity to the receiver, usually a solar cell or thermal receiver) and illumination (controlling the distribution of light in some areas and completely isolated from other areas) [28,29]. This illumination nonimaging plays a major role in reducing light pollution [30]. Common applications for nonimaging optical systems include many areas of lighting engineering. The current utilization of nonimaging optical designs includes automotive headlights, LCD, illuminated instrument panel displays, backlights, LED lights, optical fiber lighting devices, projection display systems, and luminaires [31–38]. Especially in lighting applications, it is to be focused on the intensity distribution. Nonimaging optics provides precise control of light, providing high quality and uniform lighting for LED applications. Designing an optical system for using LEDs as a high-intensity light source for projectors is more difficult than using a compact arc lamp. The high-brightness LED projection screen has improved the efficiency of the light engine by using nonimaging components that match the shape and emission pattern of the LED [39]. Freeform optics has been used to increase the angular color uniformity of a white LED and obtain a large angular range of uniform illumination. Currently, LEDs are also used in street lighting equipment to reduce light pollution. Freeform lenses are optimized to produce a controlled brightness distribution on street and enhance the brightness uniformity of the surface [40]. Freeform surface reduces the light pollution by shielding the light source to minimize glare and light trespass, and by reducing skyglow by optimizing the light for a confined area.

2.4 FREEFORM OPTICS IN OPTICAL INDUSTRIES

Freeform optics has revolutionized the optics industry with the Green Revolution. Freeform optical designs are used in beam shaping systems with a variety of lighting characteristics. Most modern beam shaping systems are tuned for fixed optical properties. That is, the distribution of output light of a beam shaping system is generally not changeable. However, some classes of beam shaping systems allow modifying the optics to meet the requirements of different applications. The beam shaping system consists of a freeform lens and a non-classical zoom system designed for ray aiming and energy savings rather than aberration control [41]. The freeform lens contains two elaborately designed freeform optical planes, whereby both the intensity distribution and the wavefront of the incident light beam are manipulated in the required manner. After propagating through a non-classical zoom system, the light beam produces an illumination pattern with a variable pattern size and an invariant irradiance distribution at various zoom positions on a fixed observation surface.

Due to the nature of the zoom system, the zoom system contains at least three lens elements. A divergent beam emitted from a point source is converted into a parallel beam parallel to the optical axis with a predefined irradiance distribution after refraction by the freeform input and output surfaces of the lens. The parallel beam is then propagated through the zoom system to produce a given illumination pattern on the target plane. Here the target plane's position is fixed. That is, the working distance between the target plane and the zoom system does not change. In traditional zoom systems, parallel incident beams emitted from a point on an object at infinity are known to converge to a point in the image plane [42].

For the Theatrical Spotlighting system, coupling of free form surface and the imaging optics plays a great role in the production of uniform light distribution along with a sharp cut angle. It also minimizes the loss of light through the optical pathway. In the case of stage performers, for a better view in theatres selective and consistent illumination is required. Historically lament bulbs are generally used for theatre lighting. Because of the production of a large amount of heat, actors and patrons feel discomfort. After the invention of LED lighting, the old lament bulbs of the theatres are getting replaced with LED lights as they provide the same quality of lighting with the reduction in energy requirement. However, the transition to LEDs has been in the spotlight because of the high standards of theatre experts. A typical spotlight uses a halogen-tungsten bulb coupled with an ellipsoid reflector to provide 10,000 lumens of illumination with 750 W power input. There is the coupling of a nonimaging optics reflector coupled with a 70 W LED texture and an imaging lens column to produce a new type of spotlight. This new spotlight can show up to 10,000 lumens as required for spotlighting in the theatres, with very less power consumption and heat generation during the production of uniform illumination. For the creation of an ideal spotlighting system incorporating nonimaging with imaging optics, the knowledge of two important concepts to be known are Kohler illumination and the concept of etendue. Both concepts are necessary to ensure the maximum throughput of light through an optical system. The design method for lens columns must use these two concepts which can be coupled with a nonimaging CPC (compound parabolic concentrator) for the production of uniform illumination [43].

Currently, there is a growing demand for oblique illumination in general lighting by freeform reflector array design. The design of illumination can be divided into axial-illumination design and oblique-illumination design. In the case of axial-illumination, the optics of the system and the illumination targets are placed on the same axis as that of the optical system. The chief ray of the central field of view and the target plane are perpendicular to each other. The problem solving of axial-illumination design incorporates methods like the differential equation method, the tailoring method, and the parameter optimization method. On the other hand, light is projected at the target sideways with a slanting angle for the oblique-illumination technique. A typical example of this method is the illumination system of a digital micro-mirror device projection. In comparison with axial-illumination, oblique-illumination has better optical performance due to the more flexible layout of the illumination system. This provides a wider application of oblique illumination in general lighting. To achieve the oblique-illumination traditionally, generally quadric surfaces like ellipsoidal, spherical, and parabolic surfaces are used to redirect the emitted light from a light source (e.g., halogen lamp) to the target surface. The mentioned quadric surfaces play an important role in the illumination design. But the low degree of freedom in the design of the quadric surfaces creates many challenges for the designers. To reduce light pollution along with the energy consumption in street lights and advertising board lights, specific illumination patterns like rectangular patterns are preferred because they can perfectly match the illumination plane. With the increasing focus on green and healthy lighting, high degrees of freedom to the freeform surface in design that can achieve a compact design with excellent optical performance, it is used for illumination designs [44]. Freeform surfaces help in the development of unique luminaires that can produce uniform rectangular illumination patterns that can match accurately to the city roads and the advertising screens, as shown in Figure 2.6. Figure 2.6a shows the quadric surfaces used to direct the emitted rays from the light source. Due to the low degree of freedom of the design of the quadric surfaces, the distribution of light is difficult to control and it causes energy loss along with the pollution of light. Instead of quadric surfaces, the luminaire with freeform surfaces shown in Figure 2.6b provides illumination, which matches perfectly to the lighting environment and helps to achieve energy-efficient good lighting.

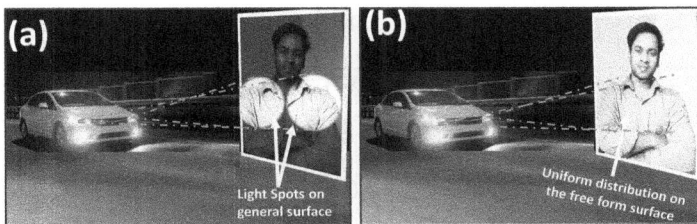

FIGURE 2.6   The effect of oblique illumination in (a) Quadric surfaces, and (b) Freeform surfaces.

2.5 FREEFORM OPTICS IN STREET LIGHT AND AUTOMOBILE TRANSPORTATION

When the optical radiation interacts with the surrounding environment for a large range of length scales, its utility as a remote sensing modality would be limited if the observed interactions at one length scale are difficult to transform to another through magnification or reduction. Fortunately, the light propagation follows a geometric ray model in many practical instances which enables the construction of optical systems made from lenses and/or mirrors. Such systems collect and focus light from distant objects by mapping points in object space to points on a photosensitive detector which forms an image. This mapping allows a great degree of magnification or reduction which permits high-resolution non-contact sensing of distant objects large and small. Gradually automobile industries are adopting the principle of nonimaging technique. Harnessing Multiscale Nonimaging optics for automotive is possible by integrating the Flash LiDAR with the heterogenous Semiconductor. There is a limitation on optical strain metrology with displacement resolution smaller than 100 nanometers in the semiconductor industries, which is capable of the measurement of strain fields between high-density interconnect lines. Despite being a cheaper technology, small attention is given to the perimeter sensing characteristics of scene-illuminating flash LiDAR in the field of automotive applications. The nonimaging optics provides intriguing instrument design and explanations of observed sensor performance for a large variation in length scales. By a novel approach in metrology for the next-generation semiconductor packages, an effective non-contact technique is established for mapping nanoscale mechanical strain fields and out-of-plane surface warping by using laser diffraction. In addition to that, the distance of object detection on the order of tens of meters of a low-cost automotive flash LiDAR is understandable through the principles of optical energy transfer from the surface of a remote object to an extended multi-segment detector. The design of an automotive perception system to identify various roadway objects in low-light situations can be the consequence of the mentioned information [45].

At night, street lighting becomes the key player in energy wastage and light pollution. Still, there are major issues with efficient lighting of curved and twisted roadways. The design and development of LED Street Lighting for freeform roads can overcome such issues, as shown in Figure 2.7. These LED lamps can deliver a roadway-shaped light pattern with a maximized performance of the illumination. The light is directed to the desired places efficiently and homogeneously. This reduces glare and improves the comfort of the eyes and the visual discrimination ability of the pedestrians and the car drivers. The proposed luminaire is very practical in which only the cover plate is to be replaced by a special

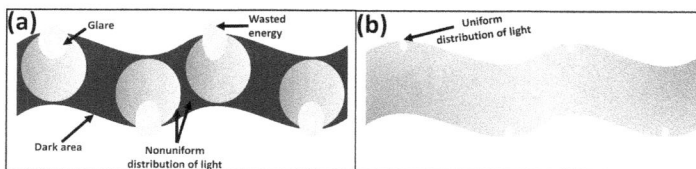

FIGURE 2.7 (a) Light distribution in a traditional street surface, and (b) Use of nonimaging technology in the advanced street surface.

microlens array sheet for the production of different shaped light patterns. This mentioned mechanism is so simple and effective that after being collimated the LED light can be distributed efficiently on a freeform roadway by the special microlens sheet [46].

2.6 FREEFORM OPTICS IN SPACE APPLICATION

Due to the extreme vastness of the space, putting optics in it is a very big challenge. Freeform optics give a new direction for space research providing a clear targeted vision. Optical components can act as a new form of Wi-Fi for us. Laser light is used to communicate between the ground station to satellite and one satellite to another satellite. Freeform optic manufactured transmitters and receivers are used here and are compact and very precise. Laser communication expanding its use with the help of freeform surfaces and is expected to illuminate data traffic in communication. In various optical applications, a single freeform lens can be replaced the multiple lenses as in the traditional optic practice [47].

Freeform mirror-based optical systems are used in nano-satellites used for urban monitoring, environmental protection disaster response, precision farming, defense, and security using high-resolution satellite imaging. State-of-the-art micro-satellites can apply in sub-meter ground resolution. The optic equipment is based on classical two, three, or four mirror optical systems and the surfaces are rotationally symmetric or sections of rotationally symmetric surfaces. Now, all these requirements have been replaced by the freeform surface facilitating the limited volume and mass budget [48].

The optics and optomechanical combination required to be higher mechanical strength to be sustained at forces experienced during rocket launch and are lightweight and compact as possible to reduce costs. Often the entire optomechanical assembly needs to be folded for launch and then reliably employed in an aligned configuration in space. Radiation effects, operation in a vacuum, and the potential for large thermal shifts are also required to be counted. Technological advances in materials, optics manufacturing/analysis, as well as active compensation are enabling lesser weight, higher performance instruments in a more compact structure. Future sensor advances offer even more direction. One of the great achievements for more compact, space-based optics is the use of freeform optic surfaces. Freeform optic surfaces will be used with the next-generation Infrared Atmospheric Sounding Interferometer (IASI-NG), as well as other structures. Freeform mirrors help to eliminate the aberration caused by constrained mirror locations to be within the packaging structure due to the folded telescope design architecture. Freeforms are applied to correct aberration over a wide field of view NASA Need: large cryo-stable freeform mirrors [49]. Freeform optics enable future NASA missions that are large astrophysics missions such as LUVOIR and OST and small, wide-field Earth Science and Planetary missions such as MiniSpec, TIMERS, or CubeSats. NASA has a strong objective to detect and estimate changes to Earth's ecosystems as described in Strategic and Science plans. Earth's vegetated ecosystems sustain life on Earth. Vegetation activities with respect to its surrounding vary with the time flow. Vegetation productivity is also related to 3D structure, as it is a major parameter to detect the light environment within the canopy. Productivity can be measured by light use efficiency models with remote sensing inputs that account for vegetation stress from soil moisture, disease, and insects. Vegetation responds to rates and magnitudes changes among vegetation types and seasonal stage, and with the nature and magnitude of environmental stresses. The magnitude, rate, and duration of these responses estimate ecosystem activities such as carbon balance and evapotranspiration. Existing approaches, such as Normalized Difference Vegetation Index (NDVI) from Advanced Very High-Resolution Radiometer (AVHRR) and Moderate Resolution Imaging Spectroradiometer (MODIS), and proposed missions like Hyspiri are useful for addressing the changes at weekly/monthly time scales. Multi-temporal calculation of structure and functioning of vegetation will give the entire and more precise measurements of productivity [50]. Due to the low Earth's orbit and high spatial resolution, it requires a very fast detector that can make an exposure in just a few tens of minutes of 1 μs while the satellite passes through the pixels on the ground. This requires a time delay and integral (TDI) detector, where a pixel size of 5–10 μm is considered a common value for such detectors, and a pixel size of 10 μm is a lower detector resolution. The application of freeform surfaces provides a very compact structure with a small spatial envelope and low mass, reduces geometric aberrations, improves balance and control, and optical performance (image quality, depth of field, field of view) can be modified [51]. To detect and characterize spacecraft in orbit, ground-based optical telescopes were adopted with both imaging and non-imaging sensors. For objects that are unresolved or poorly resolved due to their small size and/or long range from the observing station, nonimaging techniques are being explored for their potential to extract information for satellite identification and determination of status and function [52,53].

2.7 FREEFORM OPTICS FOR OTHER INDUSTRIES

Freeform optics have become widely used in several industries, including solar concentrating, optical fiber, remote sensing, manufacturing, depth estimation, etc. Non-imaging optics have the potential to revolutionize solar concentrator design. Designs for flexible, high-power-density, remote irradiation systems can be achieved using non-imaging methods. Applications include industrial infrared heaters, e.g. in semiconductor processing, alternatives to laser light for certain medical procedures, and general remote high-brightness lighting. The high power densities inherent in the small active radiant areas of conventional metal halide, halogen, xenon, microwave-sulfur, and related lamps can be restored with low power loss non-imaging concentrators. These high flux levels can be transported at high transmissivity with light channels such as fiber optics or light pipes and transformed into fixtures capable of delivering prescribed angular and spatial flux distributions to desired targets [54,55]. Maximum efficiency of fiber optic irradiation can be achieved with non-imaging designs. Rays emerging from a fiber over a restricted angular range (small numerical aperture) are required to illuminate a small near-field detector with maximum radiation efficiency. These designs range from pure reflectors (all mirrors) to pure dielectrics (refractive and based on total internal reflection) to lens-mirror combinations. Compared to traditional imaging solutions, non-imaging units offer significant practical advantages in terms of compactness and ease of alignment, as well as significantly higher radiation efficiency [56].

Depth estimation can directly calculate the volume and distance of the surrounding structure. Because the angle of view of a camera lens remains constant, no matter how far away an imaged object is. It follows that the apparent size of a structure in an image is simply a function of the angular occupancy of the field of view and not the distance from the camera lens. In addition, the objects that occupy a larger portion of the camera's angle of view appear larger, while the structure that occupies a smaller portion of the angle of view appear smaller, regardless of distance or actual size. However, it should be noted that the farther an object of fixed size is from a camera lens, the smaller it will appear. The size relationship between the object and the physical dimensions of an imaged scene is reduced with distance. Despite the basic ambiguities related to object size and distance in imaging systems, the cost, performance, angular resolution, and general utility of digital cameras make the idea of per-pixel depth estimation difficult to avoid [57].

Free-form optics are also used in medical diagnostics for engineers. It can help perform simple physical exams, construct a differential diagnosis and perform non-imaging tools used in medicine such as labs, ECG, cultures, and pathological samples. Non-imaging nuclear medicine methods are also available. Historically, nuclear medicine was largely an imaging specialty, employing modalities as diverse and increasingly sophisticated as rectilinear scanning, gamma camera (planar) imaging, photon emission tomography, and positron emission tomography. However, non-imaging radiation detection remains an essential part of nuclear medicine.

2.8 ADVANTAGES OF FREEFORM OPTIC SURFACE

In comparison to conventional optical designed components, freeform optics has several advantages, as given below [58–61]:

- Several lenses are used in traditional design are now replaced by only one (or a few) special freeform lenses.
- Freeform surfaces are the most energy-efficient, as around 100% of the incident light that transferred to the target plane without loss of the light.
- Having more compact optics.
- It is possible to combine several light sources as well as the light distribution to several target places.
- Mappings can be achieved by the freeform discontinuous surface
- Better handling of extended sources.
- Color mixing capabilities.
- Can enhance the performance of an optical structure to the maximum possible extent.
- Increased range of fabricated surfaces, allowing optical designers with more flexibility and scope for the development of new equipment.
- Shield the light source to minimize glare and light trespass.
- Reduce skyglow by optimizing the light for a confined area.
- Eliminate landscape lighting.
- Eliminate the intensive harm interference to the biological system.

2.9 LIMITATIONS OF FREEFORM OPTIC SURFACE

Different types of limitations observed in the case of freeform surface are [62–65]:

- Very narrow choice of suitable materials for making the optic components as well as less variety of light sources.
- Lack of smooth and continuous surface.
- Due to the unusual shape of freeform surfaces, it is not easy to fabricate, test, and measure such lenses and mirrors.

2.10 SUMMARY

Traditional used lenses and mirrors have a smooth shape, either convex or concave, and they have limitations due to loss of light beyond its periphery. They cannot produce certain light-beam paths, whereas lenses and mirrors with a more complex aspherical or freeform can solve this limitation. Freeform optics expanded successfully from the introduction of freeform surfaces into both imaging and nonimaging advanced optical applications. This chapter discussed the importance of the nonimaging/free form surfaces and their designs to be used for the application of light pollution control. Various type of advantages and limitations associated with the freeform surface has been discussed. As per the nonimaging technology demand on an industrial scale, light pollution can be controlled on a significant scale, globally.

REFERENCES

[1] J. Reimers, A. Bauer, K. Thompson, Freeform spectrometer enabling increased compactness, *Light Sci. Appl.* 6 (2017) e17026. https://doi.org/10.1038/lsa.2017.26.

[2] F. Fournier, J. Rolland, Optimization of freeform light pipes for light-emitting-diode projectors, *Appl. Opt.* 47 (2008) 957–966. https://www.osapublishing.org/ao/abstract.cfm?URI=ao-47-7-957.

[3] J. C. Miñano, R. Mohedano, P. Benítez, *Nonimaging Optics* (2015). https://doi.org/10.1002/9783527600441.oe1014.

[4] A. A Radu, John R. Mattox, Steven Ahlen, Design studies for nonimaging light concentrators to be used in very high-energy gamma-ray astronomy, *Nucl. Instrum. Methods Phys. Res. Sect. A* 446(3) (2000) 497–505. https://doi.org/10.1016/S0168-9002(99)01417-5.

[5] M. Tian, Y. Su, H. Zheng, G. Pei, G. Li, S. Riffat, A review on the recent research progress in the compound parabolic concentrator (CPC) for solar energy applications, *Renewable Sustainable Energy Rev.* 82(Part 1) (2018) 1272–1296. https://doi.org/10.1016/j.rser.2017.09.050.

[6] J. O'Gallagher, R. Winston, Atomic Optics: Nonimaging Optics on the Nanoscale. Chicago: N. p. (2005). https://doi.org/10.2172/838024.

[7] F. Falchi, P. Cinzano, C. D. Elvidge, D. M. Keith, A. Haim, Limiting the impact of light pollution on human health, environment and stellar visibility, *J. Environ. Manage.* 92(10) (2011) 2714–2722. https://doi.org/10.1016/j.jenvman.2011.06.029.

[8] A. Cygan, P. Wcisło, S. Wójtewicz, G. Kowzan, M. Zaborowski, D. Charczun, K. Bielska, R. S. Trawiński, R. Ciuryło, P. Masłowski, D. Lisak, High-accuracy and wide dynamic range frequency-based dispersion spectroscopy in an optical cavity, *Opt. Express* 27 (2019) 21810–21821. https://doi.org/10.1364/OE.27.021810.

[9] E. Chen, F. Yu, Design of LED-based reflector-array module for specific illuminance distribution, *Opt. Commun.* 289 (2013) 19–27. https://doi.org/10.1016/j.optcom.2012.09.082.

[10] P. Benitez, J. C. Minano, J. Blen, R. Mohedano, J. Chaves, O. Dross, M. Hernandez, J. L. Alvarez, W. Falicoff, SMS design method in 3D geometry: examples and applications, Proc. SPIE 5185, Nonimaging Optics: Maximum Efficiency Light Transfer VII (2004). https://doi.org/10.1117/12.506857.

[11] Y. Hu, H. Shen, Y. Yao, A novel sun-tracking and target-aiming method to improve the concentration efficiency of solar central receiver systems, *Renewable Energy* 120 (2018) 98–113. https://doi.org/10.1016/j.renene.2017.12.035.

[12] P. Benítez, Elliptic ray bundles in three-dimensional geometry for nonimaging optics: a new approach, *J. Opt. Soc. Am. A* 16 (1999) 2245–2252. https://doi.org/10.1364/JOSAA.16.002245.

[13] P. Gimenez-Benitez, J. C. Miñano, J. Blen, R. M. Arroyo, J. Chaves, O. Dross, M. Hernandez, W. Falicoff, Simultaneous Multiple Surface optical design method in three dimensions, *Opt. Eng.* 43(7) (2004). https://doi.org/10.1117/1.1752918.

[14] R. M. Arroyo, J. C. Minano, P. Benitez, J. L. Alvarez, M. Hernandez, J. Gonzalez, K. Hirohashi, S. Toguchi, Ultracompact nonimaging devices for optical wireless communications, *Opt. Eng.* 39(10) (2000). https://doi.org/10.1117/1.1308173.

[15] Julio Chaves, Manuel Collares-Pereira, Variable-geometry nonimaging optics devices, Proc. SPIE 4446, Nonimaging Optics: Maximum Efficiency Light Transfer VI (2001). https://doi.org/10.1117/12.448809.

[16] S. Thiele, T. Gissibl, H. Giessen, A. M. Herkommer, Ultra-compact on-chip LED collimation optics by 3D femtosecond direct laser writing, *Opt. Lett.* 41 (2016) 3029–3032. https://doi.org/10.1364/OL.41.003029.

[17] J. C. Minano, P. Benitez, Fermat's principle and conservation of 2D etendue, Proc. SPIE 5529, Nonimaging Optics and Efficient Illumination Systems (2004). https://doi.org/10.1117/12.560754.

[18] https://www.optimaxsi.com/capabilities/optimax-freeforms/, 30.12.2021.

[19] F. R. Fournier, W. J. Cassarly, J. P. Rolland, Designing freeform reflectors for extended sources, Proc. SPIE 7423, Nonimaging Optics: Efficient Design for Illumination and Solar Concentration VI, 742302 (2009). https://doi.org/10.1117/12.826021.

[20] J. Petrasch, P. Coray, A. Meier, M. Brack, P. Häberling, D. Wuillemin, A. Steinfeld, 2006, A novel 50kW 11,000 suns high-flux solar simulator based on an array of xenon arc lamps., ASME. *J. Sol. Energy Eng.* 129(4) (November 2007) 405–411. https://doi.org/10.1115/1.2769701.

[21] F. R. Fournier, W. J. Cassarly, J. P. Rolland, Fast freeform reflector generation using source-target maps, *Opt. Express* 18 (2010) 5295–5304. https://doi.org/10.1364/OE.18.005295.

[22] J. Meyron, Q. Mérigot, B. Thibert, Light in power: a general and parameter-free algorithm for caustic design. *ACM Trans. Graph.* 37(6), Article 224 (December 2018) 13. https://doi.org/10.1145/3272127.3275056.

[23] G. Captur, P. Gatehouse, K. E. Keenan, A medical device-grade T1 and ECV phantom for global T1 mapping quality assurance-the T1 Mapping and ECV Standardization in cardiovascular magnetic resonance (T1MES) program. *J. Cardiovasc. Magn. Reson.* 18 (2016) 58. https://doi.org/10.1186/s12968-016-0280-z.

[24] A. A. Mingazov, D. A. Bykov, E. A. Bezus, L. L. Doskolovich, On the use of the supporting quadric method in the problem of designing double freeform surfaces for collimated beam shaping, *Opt. Express* 28 (2020) 22642–22657. https://doi.org/10.1364/OE.398990.

[25] R. Perez-Enciso, A. Gallo, D. Riveros-Rosas, E. Fuentealba-Vidal, C. Perez-Rábago, A simple method to achieve a uniform flux distribution in a multi-faceted point focus concentrator, *Renewable Energy* 93 (2016) 115–124. https://doi.org/10.1016/j.renene.2016.02.069.

[26] J. Meyron, Initialization procedures for discrete and semi-discrete optimal transport, *Comput. Aided Des.* 115 (2019) 13–22. https://doi.org/10.1016/j.cad.2019.05.037.

[27] J. Burger, A. Gowen, Data handling in hyperspectral image analysis, *Chemom. Intell. Lab. Syst.* 108(1) (2011) 13–22. https://doi.org/10.1016/j.chemolab.2011.04.001.

[28] M. Abdelhamid, B. K. Widyolar, L. Jiang, R. Winston, E. Yablonovitch, G. Scranton, D. Cygan, H. Abbasi, A. Kozlov, Novel double-stage high-concentrated solar hybrid photovoltaic/thermal (PV/T) collector with nonimaging optics and GaAs solar cells reflector, *Appl. Energy* 182 (2016) 68–79. https://doi.org/10.1016/j.apenergy.2016.07.127.

[29] N. Bushra, T. Hartmann, A review of state-of-the-art reflective two-stage solar concentrators: technology categorization and research trends, *Renewable Sustainable Energy Rev.* 114 (2019) 109307. https://doi.org/10.1016/j.rser.2019.109307.

[30] E. Freniere, M. A. Gauvin, R. N. Youngworth, The different (yet similar) realms of illumination and stray light modeling, Proc. SPIE 9577, Optical Modeling and Performance Predictions VII, 957708 (2015). https://doi.org/10.1117/12.2190683.

[31] A. Cvetkovic, O. Dross, J. Chaves, P. Benítez, J. C. Miñano, R. Mohedano, Etendue-preserving mixing and projection optics for high-luminance LEDs, applied to automotive headlamps, *Opt. Express* 14 (2006) 13014–13020. https://doi.org/10.1364/OE.14.013014.

[32] D. G. Pelka, K. Patel, An overview of LED applications for general illumination, Proc. SPIE 5186, Design of Efficient Illumination Systems (2003). https://doi.org/10.1117/12.509164.

[33] J. F. V. Derlofske, Computer modeling of LED light pipe systems for uniform display illumination, Proc. SPIE 4445, Solid State Lighting and Displays (2001). https://doi.org/10.1117/12.450035.

[34] D. Feuermann, J. M Gordon, M. Huleihil, Solar fiber-optic mini-dish concentrators: first experimental results and field experience, *Sol. Energy* 72(6) (2002) 459–472. https://doi.org/10.1016/S0038-092X(02)00025-7.

[35] F. Chen, S. Liu, K. Wang, Z. Liu, X. Luo, Free-form lenses for high illumination quality light-emitting diode MR16 lamps, *Opt. Eng.* 48(12), 123002, 2009. https://doi.org/10.1117/1.3274677.

[36] W. J. Cassarly, Taming light using nonimaging optics, Proc. SPIE 5185, Nonimaging Optics: Maximum Efficiency Light Transfer VII (2004). https://doi.org/10.1117/12.508709.

[37] W. Zhang, Q. Liu, H. Gao, F. Yu, Free-form reflector optimization for general lighting, *Opt. Eng.* 49(6) (2010) 063003. https://doi.org/10.1117/1.3430576

[38] W. Monch, Micro-optics in lighting applications, *J. Adv. Opt. Tech.* 4(1) (2015). https://doi.org/10.1515/aot-2014-0061.

[39] F. Z. Fang, X. D. Zhang, A. Weckenmann, G.X. Zhang, C. Evans, Manufacturing and measurement of freeform optics, *CIRP Ann.* 62(2) (2013) 823–846. https://doi.org/10.1016/j.cirp.2013.05.003.

[40] D. Rui, Z. Lin, K. Qi, W. Chen, Optical design in illumination system of digital light processing projector using laser and gradient-index lens, *Opt. Eng.* 51(1) (2012) 013004. https://doi.org/10.1117/1.OE.51.1.013004.

[41] L. Yang, F. Shen, Z. Ding, X. Tao, Z. Zheng, F. Wu, Y. Li, R. Wu, Freeform optical design of beam shaping systems with variable illumination properties. *Opt. Express.* 2021, 29(20):31993–32005. doi: 10.1364/OE.436340. PMID: 34615279.

[42] R. Wu, L. Yang, Z. Ding, L. Zhao, D.Wang, K. Li, F. Wu, Y. Li, Z. Zheng, X. Liu, Precise light control in highly tilted geometry by freeform illumination optics, *Opt. Lett.* 44 (2019) 2887–2890.

[43] H. Ries, J. Muschaweck, Tailored freeform optical surfaces, *J. Opt. Soc. Am. A* 19 (2002) 590–595. https://doi.org/10.1364/JOSAA.19.000590.

[44] E. Chen, R. Wu, Design of freeform reflector array for oblique illumination in general lighting, *Opt. Eng.* 54(6) (2015) 065103. https://doi.org/10.1117/1.OE.54.6.065103.

[45] T. Houghton, Harnessing multiscale nonimaging optics for automotive flash LiDAR and heterogenous semiconductor integration, Dissertation or Thesis, Arizona, State University. ProQuest Dissertations Publishing (2020) 28030956.

[46] C. Sun, Design of LED street lighting adapted for free-form roads, *IEEE Photonics J.* 9(1) (February 2017) 1–13, Art no. 8200213. https://doi.org/10.1109/JPHOT.2017.2657742.

[47] https://www.tno.nl/en/tno-insights/articles/optics-2-0-the-future-of-freeform-optics/#:~:text=Freeform%20optics%20is%20key%20to,few)%20special%20freeform%20lens.%E2%80%9D, 12.01.2022.

[48] https://ntrs.nasa.gov/citations/20130012648, 15.01.2022.

[49] https://ntrs.nasa.gov/api/citations/20170010419/downloads/20170010419.pdf, 12.01.2022.

[50] N. Tilly, D. Hoffmeister, Q. Cao, S. Huang, V. Lenz-Wiedemann, Y. Miao, G. Bareth, Multitemporal crop surface models: accurate plant height measurement and biomass estimation with terrestrial laser scanning in paddy rice, *J. Appl. Rem. Sens.* 8(1) (2014) 083671. https://doi.org/10.1117/1.JRS.8.083671.

[51] T. Agócs, R. Navarro, N. Tromp, R. Vink, Freeform mirror based optical systems for nanosatellites, Proc. SPIE 10562, International Conference on Space Optics- ICSO 2016, 1056228 (2017). https://doi.org/10.1117/12.2296187.

[52] https://www.nasa.gov/feature/goddard/2021/nasa-to-explore-divergent-fate-of-earth-s-mysterious-twin-with-goddard-s-davinci, 15.01.2022.

[53] T. Kyono, J. Lucas, M. Werth, B. Calef, I. McQuaid, J. Fletcher, Machine learning for quality assessment of ground-based optical images of satellites, *Opt. Eng.* 59(5) (2020) 051403. https://doi.org/10.1117/1.OE.59.5.051403.

[54] https://humanhealth.iaea.org/HHW/MedicalPhysics/e-learning/Nuclear_Medicine_Handbook_slides/Chapter_10._Non-Imaging_Detectors_and_Counters.pdf, 15.01.2022.

[55] J. O'Gallagher, *Nonimaging Optics in Solar Energy*, Morgan & Claypool (2008). https://doi.org/10.2200/S00120ED1V01Y200807EGY002.

[56] Y. Fang, D. Feuermann, J. M. Gordon, Maximum-performance fiber-optic irradiation with nonimaging designs, *Appl. Opt.* 36 (1997) 7107–7113.

[57] J. Li, C. Yuce, R. Klein, A. Yao, A two-streamed network for estimating fine-scaled depth maps from single RGB images, *Comput. Vision Image Understanding* 186 (2019) 25–36. https://doi.org/10.1016/j.cviu.2019.06.002.

[58] J. Jiang, S. To, W. B. Lee, B. Cheung, Optical design of a freeform TIR lens for LED streetlight, *Optik* 121(19) (2010) 1761–1765. https://doi.org/10.1016/j.ijleo.2009.04.009.

[59] X. Zhang, L. Zheng, X. He, L. Wang, F. Zhang, S. Yu, G. Shi, B. Zhang, Q. Liu, T. Wang, Design and fabrication of imaging optical systems with freeform surfaces, Proc. SPIE 8486, Current Developments in Lens Design and Optical Engineering XIII, 848607 (2012). https://doi.org/10.1117/12.928387.

[60] Z. Li, F. Fang, J. Chen, X. Zhang, Machining approach of freeform optics on infrared materials via ultra-precision turning, *Opt. Express* 25 (2017) 2051–2062. https://doi.org/10.1364/OE.25.002051.

[61] Z. Hong, R. Liang, IR-laser assisted additive freeform optics manufacturing. *Sci. Rep.* 7 (2017) 7145. https://doi.org/10.1038/s41598-017-07446-8.

[62] M. Brunelle, J. Yuan, K. Medicus, J. D. Nelson, Importance of fiducials on freeform optics, Proc. SPIE 9633, Optifab 2015, 963318 (2015). https://doi.org/10.1117/12.2195350.

[63] A. Bruneton, A. Bäuerle, R. Wester, J. Stollenwerk, P. Loosen, Limitations of the ray mapping approach in freeform optics design, *Opt. Lett.* 38 (2013) 1945–1947. https://doi.org/10.1364/OL.38.001945.

[64] J. D. Nelson, K. Medicus, M. Brophy, Fabricating and testing freeform optics: current capabilities, lessons learned and future opportunities, in Classical Optics 2014, OSA Technical Digest (online) (Optical Society of America, 2014, paper OW3B.2. https://doi.org/10.1364/OFT.2014.OW3B.2.

[65] M. C. Knauer, J. Kaminski, G. Hausler, Phase measuring deflectometry: a new approach to measure specular free-form surfaces, Proc. SPIE 5457, Optical Metrology in Production Engineering (2004). https://doi.org/10.1117/12.545704.

CHAPTER 3

Current Status of Light Pollution and Approaches for Its Prevention

Manish Srivastava, Anjali Yadav, Anamika Srivastava,
Anjali Banger, and Bharti Mehlawat

Banasthali Vidyapith

CONTENTS

DOI: 10.1201/9781003185109-3

3.1 INTRODUCTION

Light pollution involves the usage of sources of artificial lights for brightening the night sky, which is scattered into the environment. Artificial lighting is possibly the greatest success of humans and has to turn out to be an essential part of the urban surroundings [1,2]. Light pollution has been considered to be the most speedily growing variation to the surroundings. It was supposed that artificial light offers security and enhances visibility but a large section of lighting does neither. Lighting which is overused, misdirected, or else conspicuous is mainly pollution. Artificial lighting is considered to be the latest form of pollution. So, this issue is attracting extensive attention. During the last few decades, fast industrialization and urbanization lead to the extensive usage of artificial light. This has significantly altered the nightscapes, although their harmful impacts are over and over again ignored.

Globally, lighting is considered for the consumption of one-fourth of all energy, and about 19% of electricity was utilized to generate light at night. Greenhouse gases are released as a consequence of the burning of fossil fuels for the production of light. These greenhouse gases result in global warming and the extinction of non-renewable resources. The theory of light pollution has attracted attention in the 1970s to illustrate its adverse effects [3]. Hence, a reduction in excessive lighting could have a similar decrease in carbon dioxide as eliminating about 9.5 million cars from the path [4]. Light pollution results in negative consequences on plants, human beings, and animals [5]. Additionally, overexposure consumes energy that could be used elsewhere because it is an uneconomical use of energy.

Various effects are associated with light pollution are breast cancer in women [6,7], diabetes, depression, and fatigue to humans, interrupt sleeping, feeding, migration, and reproduction in animals, can alter plant's metabolism such as growth, sleeping, feeding, reproduction, and other behaviors, and hence affects food consumption, orientation, migration, growth, communication, storage, locomotion, and reproduction [8]. The increase in lightning usage and inappropriate design of lighting fixtures results in the enhancement of pollution. It is the latest type of environmental pollution and causes threats to living beings. Generally, there are five different types of light pollution which include- glare, over-illumination, clutter, trespass, and sky glow [9].

The increase in the usage of artificial light is because of the reality that human beings are diurnal animals that are attempting to expand actions into the dark. Shopping centers are extensively lighted to focus the attention of the public and form an atmosphere to encourage spending. Lighting in public areas is usually high to prevent criminal activity. Certainly, the urban centers are extensively lighted and the ensuing light pollution can expand out from the town's border. There is consistent proof that this light pollution causes a serious effect on human health [10]. Various studies demonstrated that light pollution results in a low level of melatonin production which consequence in the risk of breast cancer in females. Light pollution keeps humans from falling asleep, which leads to reducing the generation of melatonin. For instance, light pollution is responding to disturb the movement of nocturnal birds and can result in the eating of sea turtles by predators. Artificial light impacts the behavior, perceived photoperiod, and eventually fitness of birds.

The artificial lights not only influence the birds but also harm moths and bats. It can affect competitive interactions, animal behavior, reproductive behavior, navigation, predator-prey relations, and animal physiology [11]. Lights are responsible for the extinction of local species by attracting animals [11,12]. Also, it hurts astronomical research as light pollution disturbs both specialized and proletarian astronomy by decreasing the visibility of nebulae, galaxies, and various other outer space substances.

Therefore, in this book chapter, we will focus on light pollution. The book chapter is divided into five parts: -sources of light pollution, types of light pollution, various impacts associated with light pollution, some common effects, and methods to reduce light pollution.

3.2 SOURCES OF LIGHT POLLUTION

Various sources of light which affect animals and human beings are listed below:

3.2.1 Billboards

It was believed that the video Billboard resulted from the excessive exterior lighting. This excessive lightning can disturb animals and human beings.

3.2.2 Car Headlights

Car headlights majorly affect those peoples which live near roads, highways, streets, etc. Their life was affected due to overexposure to light.

3.2.3 Spotlights

Spotlights are also a major cause of light pollution and can affect every living being.

3.2.4 Street Lights

Street lights mainly affect those peoples who live nearby because these lights are on all night time and hence have a negative impact. These lights have more effect on insects and they don't differentiate between day and night. It can even lead to loss of life.

3.2.5 The Consciousness of Light Pollution

Regrettably, the majority of humans are not familiar with light pollution and its adverse effects on mankind and the ecosystem.

3.3 TYPES OF LIGHT POLLUTION

There are several types of light pollution and it has an effect of reduction of natural lights such as moonlight or starlight as well. The five types of light pollution are discussed below:

3.3.1 Over-Illumination

It is the chief cause of light pollution. This pollution involves the extreme use of light. Excessive lighting is not forever superior and can consequence in enhanced running costs and glare [12].

Over-illumination is Caused by Various Factors Which are Discussed Below:

- Light can be extinguished when not needed by not using occupancy sensors, timers, or other controls.
- A high level of light that is not required for certain tasks results from inappropriate design, particularly in office spaces.
- Inaccurate selection of lighting devices.
- Inappropriate choice of hardware to exploit large energy than desired.
- Imperfect teaching of building managers to utilize lighting systems capably.
- Enhancement in the energy costs and stray light due to insufficient lighting maintenance.
- "Daylight lighting" can be utilized by shop owners to catch the attention of the customers and by citizens to decrease crime.
- Replacement of mercury lamps with metal halide lamps utilizing similar electrical power.
- Indirect lighting methods, like lighting a wall to bounce photons on the floor [13].

3.3.2 Clutter

It involves the unnecessary groupings of lights. Clutter results in creating misunderstanding, divert from obstacles, and leads to accidents. This is predominantly visible on streets where intensely lit advertising surrounds the roadways and where the road lights are poorly designed. The design and placement of the lights which are installed by organizations or persons may result in a distraction or even lead to accidents. Clutter is also dangerous for the aviation sector as the non-relevant lighting attracts the pilot's attention. As aircraft crash evasion lights may be puzzled with lights on the grounds and airstrip lighting may be puzzled with an array of commercial lighting. The strip of Las Vegas is an ideal example of severe light cluttering [13].

3.3.3 Light Trespass

It is also a severe type of light pollution. This happens when light shines in the external region it is planned to light up. A general light trespass trouble happens when a powerful light penetrates the glass of the house from the exterior which results in blocking of an evening view or sleep deprivation [12].

3.3.4 Skyglow

It represents the orange glow that is seen over a highly populated region at night. It is the amalgamation of all light reflected from the enlightened surface and evasion up into the sky and is redirected again to the ground by the atmosphere. This type of scattering is associated with the wavelength of light as the atmosphere is extremely unambiguous. The sky is visible blue in the daytime as Rayleigh scattering dominates in such clear air. There is less dependence of scattered light on wavelength when there is considerable aerosol and creates a whiter daytime sky. Skyglow results in inconvenience to astronomers as it decreases distinction in the dark and it makes it difficult to see the intense stars. Light is mainly challenging for amateur astronomers, whose capability to watch the dark sky is probable to be repressed by any stray light from close by. The majority of optical astronomical laboratories are enclosed by zones of firmly enforced restrictions on light emissions [12].

3.3.5 Glare

It is a consequence of extreme contrast between areas of intense and dimness in the field of observation. It involves the direct viewing of the filament of a poorly shielded light. The shining of light in the eyes of drivers and pedestrians results in incomprehensible night image, and it results in intricacy for man's eye to adjust the dissimilarity in vividness. It is predominantly a concern in road protection since it to some extent blinds pedestrians or drivers and hence leads to accidents.

Glare is Characterized into Various Types Which are Discussed Below:

a. Blinding Glare

This type of glare involves the effects which are due to direct gazing to Sun. It results in short-term or everlasting sight deficiency.

b. Disability Glare

It involves various impacts like a scattering of light in the fog, blinded by approaching cars' lights, reflections from dark regions that make them bright, with a considerable decrease in vision abilities.

c. Discomfort Glare

This type of glare does not usually cause risky circumstances in itself. It results in tiredness when experienced over a large time interval [12].

3.4 IMPACTS OF LIGHT POLLUTION

The research in the field of light pollution is still in its beginning years and hence the effects are not entirely implicit. Nightglow is one of the major impacts of light pollution and several other disturbing features are still unknown: for instance, the fact that light pollution involves large energy wastage. There are various other effects of light pollution on human beings, the animal kingdom, and the plant kingdom [14].

3.4.1 Effects on Human Health and Psychology

The presence of enough light at right time is necessary for the health of humans. However, overexposure has negative effects. Studies suggested that the health of human is extensively affected by light pollution [12]. Fatigue, lack of sleep, increased headache occurrence, increased level of anxiety, loss of visual acuity, stress, decrease in sexual function, and hypertension are the various health-related effects due to over-illumination [14]. The presence of light and dark spots on the retina causes the interference of muscles that open the pupil in the dark with the muscles that wish to close the pupil in light. Lighting can affect road safety by blinding drivers and pedestrians and hence results in accidents.

Daylight is responsible to determine the sleep-wake cycle. The nocturnal illumination can disturb the hormone balance and hence the inner clock. Sleep disturbances are often the result. Falling asleep in the evening and waking up in the morning is hard due to the delay of sleep hormone melatonin by artificial light. Sleep plays a major role in the functioning of the immune system, learning, and memory formation. Also, a chronic sleep disorder can be a contributing aspect to obesity, diabetes, and hypertension. Also, studies have been done on the relation of light pollution with other disorders like the overall quality of sleep and asthma. But, the majority of the medical research on light has attracted his attention toward the benefits mainly for the cure of depression [15].

Research also suggested that nocturnal light results in depression and is also accountable for children coming into teenage years prior. Garcia-Saenz et al. [16] illustrated the connection between exposure to light and the danger of prostate and breast cancer which is due to the suppression of melatonin [17]. Studies demonstrated that the decreased generation of the melatonin hormone may improve the danger of breast cancer. The studies performed at the National Institute of Environmental Health Sciences and National Cancer Institute (NCI) have accomplished that artificial lighting is accountable for breast cancer. Various studies concluded that there is a relation between night shift work and the improved occurrence of breast cancer [14]. Disturbance of biological rhythms particularly melatonin generation resulted from overexposure to light pollution which leads to cancer [18]. Overexposure to light results in circadian disturbance and hence affects human health. Elevated morbidity related to light pollution is that night light can straightly influence the mood through unusual signals transmitted from essentially photosensitive retinal ganglion cells in the retina to brain areas used in the regulation of emotions. Fluorescent lighting is adequate to raise blood pressure by about eight points. Exposure to high lighting results in a decrease in sexual performance [14].

3.4.2 Effect on Plants

Light is needed by plants for their survival and maintenance and hence generates bio-chemical energy for the environment. It is thus not astonishing that usually, plants do not respond as robustly to light pollution as lots of animals act. But the light pollution results in disturbance in the development, growth, wintering, and flowering patterns. It was observed that field plants rely on natural light which cannot be modified while there is an increased use of artificial light in the greenhouse to enhance the production of plants which results in light pollution. Studies suggested that extra light for plants is considered to be hazardous and their life cycle can be affected. The metabolism of plants can be altered by light pollution as if they are not placed in a dark place for the duration of time; they do not at all produce flowers. Also, the studies demonstrate that light at night is responsible to stop the discharge of the photochrome hormone which leads to the extinction of the plants. Low-pressure sodium lamps can disturb the photoperiodic regulation of plant growth. Skvareninova et al. [19] displayed that light pollution results in the delay of 13–22 days at the beginning of autumn vegetative phenological phases. Exposure to light pollution results in the delay in the period of leaf fall by around 1 week and leaf coloring by 6–9 days.

French-Constant et al. [20] investigated the other effects of light pollution which demonstrated that in intensely lit regions, trees' budburst takes place 7.5 days prior.

Plants also react to the duration of the night. Due to this cause, short-day plants need extended nights. It was observed that light pollution in the region of lakes averts zooplankton from eating surface algae, causing algal blooms which can destroy the plants of the lake and lower the quality of water. Light pollution also influences the ecosystem. For instance, Entomologists and Lepidopterists have documented that the power of navigation of moths and other nocturnal insects is influenced by night-time light. Night light affects the night blossoming flowers which depend on moths for pollination [17].

3.4.3 Effect on Animals

Life has resulted from the usual patterns of light and dark, hence interruption in these patterns affects various characteristics of animal behavior. Light pollution can affect predator-prey relations, animal navigation, animal physiology, and competitive interactions.

3.4.3.1 Birds

Overexposure to light has extensive behavioral effects on birds. Light from airports, houses, parking, harbors, fireworks, car headlights, stadiums, and factories has attracted the attention of birds, particularly in the cloudy environment, and is used for hunting [14]. The reason behind the disorientation of birds by lighting is not well recognized. Research demonstrated that exposure to light disturbs the navigation of birds for the direction [21]. Also, the study suggested that black-tailed godwits (*Limosa Limosa*) favored breeding far away from artificial light. Light pollution also attracts the attention of seabird fledglings which results in high mortality. It was observed that in various songbird species, like great tit (*Parus major*), artificial light advanced the beginning of the action. As the activity pattern of birds is affected by the artificial light it also impacts their sleep behavior. Also, Blue tits (*Cyanistes caeruleus*) are found to regulate their arousing time as per local light situations. Therefore, illumination cause birds to awaken up earlier and fall asleep later with potentially less sleep. Kempenaers et al. [22], Dominion et al. [23], and Raap et al. [24] illustrated their research on Blackbirds and Tits that the female birds found to lay eggs, 1.5 days before [22], their reproductive system starts operational up to 1 month before [23], and the nestling body mass increases due to effect of streetlights.

Also, Dominion et al. [25] demonstrated that birds were found to lay eggs earlier on exposure to green or white light. Cabrera-Cruz et al. [26] and Sorte et al. [27] showed that during migration season, light pollution confounds the reproductive system of migrant birds.

Light pollution also has a sex-related impact on birds, according to which male birds are more sensitive to light exposure than female ones [24,28–30].

3.4.3.2 Sea Turtles

3.4.3.2.1 Impact on Adult Females

Lighting has a severe impact on female turtles searching sites for nests. The female turtles often choose less illuminated beaches for building nests over more illuminated beaches. This results in the selection of a suboptimal nesting habitat causing impacts on the higher hatchling mortality. The procedure can be discarded when turtles meet digging impediments, human disturbance, insufficient thermal cues, and large structures. Turtles go back to the sea after completing the nesting process. This procedure can be influenced by artificial light. Lightning from roads, housing developments, and car parking attracted the attention of the turtles [14].

3.4.3.2.2 Effect on Hatchling Sea Turtle Orientation

Skyglow can affect the hatchling. The means from which hatchling sea turtles discover the sea is related to the reality that the nocturnal horizon above the land is less dazzling than on the sea. Artificial lightning due to houses, street lights lead to misoriented hatchlings on their approach to the sea. These orientation troubles result in the crawl of hatchlings in an erroneous way wherever they are in danger by predators, high temperatures, and dehydration after daylight [14]. The studies have illustrated that various marine species are responsive to light pollution, the effect of increasing coastal illumination has been unexplored [14,31].

3.4.3.3 Fish

Response of fish to light depends on the species but influences their normal behavior. Various researches are available on the usage of light at fish farms and deep-sea fish. Research also suggested that fish stay away from white light sources. Nonetheless, species are present which are attracted by light and this is utilized to catch them by industrial fisheries [14].

3.4.3.3.1 Technique to Catch Mukene
Anglers used the light attraction technique to catch fish at the night. The FAO found that Mukene in Lake Victoria is caught by fishing with a floating lamp by utilizing nets pulled from canoes and shores. This method is utilized in shallow waters near the coastlines and hence can cause danger to nursery grounds for immature Mukene, Tilapia, and Nile perch [14].

3.4.3.3.2 Salmon Farms
Submerged light decreases the fish density of Atlantic salmon in production cages. These photoperiods are utilized to enhance growth and delay sexual maturation [14].

3.4.3.3.3 Halibut Farms
Their swimming behavior is affected by the light utilized in Halibut. There is the impact of light on the swimming action. Halibut grows more and swims less. It is due to the reason that the fish are mainly responsive to ultraviolet harm. Proof of harm includes skin lesions that have been seen in Halibut [14].

3.4.3.3.4 Deep-sea Fish
It was found from the studies that the normal behavior of the sea fish is disturbed by white light. The results showed that the "usual number of fish appearances on camera was appreciably lesser in white light than in comparison to red light". The cause behind this is the adaptation of the deep-sea fish's eyes in the dark and the potential harm to their eyes [14].

3.4.4 Effect on Astronomy
The contrast between galaxies and stars in the sky is decreased by sky glow and it is found to be hard to distinguish fainter objects. This is the reason for the building of new telescopes in remote regions. Several astronomers employ "nebula filters" which simply permit explicit wavelengths of light, or "light pollution filters" which are considered to decrease the impacts of light pollution to improve the contrasts and enhance the sight of nebulae and galaxies. Unluckily this influences color perception, hence no filter matches the efficiency of a dark sky for photographic reasons. The visibility of galaxies and nebulae is influenced by light pollution because of low surface brightness. An easy technique for determining the darkness of an area is to glance at the Milky Way.

The observations can be affected by the light trespass when the stray light from the off-axis passes through the telescope's tube and is then reflected from the surface except through the mirror of the telescope and finally reach the eyepiece [14].

3.5 COMMON IMPACTS
Various health effects such as depression and sleep disorder, weight gain and an eating disorder, tumors, reproductive health (including pollination), and locomotion and orientation changes are associated with light pollution and are discussed below:

3.5.1 Depression and Sleep Disorder

Depression and its general sign–sleep disorder are associated with light pollution exposure in vertebrates [32], humans [33], arthropods [34], and avian spices. Sleep is a significant animal action that is common across the animal kingdom. It was observed from the studies that sleep lets animals recuperate from everyday anxiety and that sleep deficiency has a key pessimistic impact. Sleep in birds allows conserving energy and also allows consolidating memory. Also, White-crowned sparrows (*Zonotrichialeucophrysgambelii*) can decrease sleep throughout migration with no adverse impacts but a reduction of sleep outside the migratory season results in the reduction of cognitive functioning. Blue tit males which sleep longer are more probable to sire extra-pair offspring however there was no prominent impact of deviation in sleep behavior on fitness. Pectoral sandpipers (*Calidris melanotos)* can remove sleep completely during the breeding season with no negative impacts and males that sleepless sire more offspring. Studies also suggested that differences in sleep may influence some aspects of fitness. The blue tits animals slept less due to earlier dawn song (because of light pollution) had increased male extra-pair paternity and an advanced laying date [17,33].

3.5.2 Weight Gain and Eating Disorder

Light pollution exposure in vertebrates [35], humans, aquatic organisms [36], avian species, and arthropods results in increased body mass and eating disorders. Light pollution is the major cause of obesity in human beings. It was observed from more than 80% of studies that other species have comparable relations between light pollution exposure and body mass. Laboratory male mice, sandy beach invertebrates, terrestrial arthropods, and moths are the various affected species. Out of the various studies carried out on Tammar wallabies [37], Wistar rats, and Zebra Finches did not discover such effects.

3.5.3 Tumors

Exposure to light pollution in humans and vertebrates results in the development of tumors. It was observed from the studies that people living in densely illuminated regions are prone to enhance the possibility of prostate and breast cancers, attributing this effect to improved melatonin suppression and circadian disturbance. Also, it was observed that exposure to light pollution in female mice results in increased tumor occurrence and a high number of malignant tumors.

3.5.4 Reproductive Output and Pollination

Exposure to light pollution in arthropods [38,39], aquatic organisms [40–42], vertebrates [43], and avian species [22,24] results in reproductive output. A negative effect was associated with exposure to light pollution in aquatic organisms, vertebrates, and arthropods, and in the case of plants, such exposure was observed to interrupt pollination by nocturnal insects [44]. However, the results of research about these consequences in avian species were incompatible [14].

3.5.5 Locomotion, Orientation, and Trajectory

Locomotion, orientation, and trajectory reveal diverse features of movement. In many studies, these characteristics were observed to be related to exposure to light pollution. For instance, light pollution was found to affect the orientation in migrating birds, locomotion in vertebrates, and aquatic organisms. A decrease in the effectiveness of energy usage and physiological reaction influences the trajectory of vegetation change in plants [45].

3.6 METHODS TO REDUCE LIGHT POLLUTION

Opportunely it is promising to reduce the effects of light pollution and, simultaneously, permit for the lighting that is typically apparent as necessitate by people. There are various ways which are present to reduce the light pollution in the environment: The preeminent technique for decreasing light pollution depends on unerringly what the difficulty is in any known cause. The present solutions concentrate on the reduction of the stated effects – over-illumination, glare, sky glow, clutter, and light trespass. The foremost focus of light reduction is to make sure that light is not emitted over the horizontal. Various methods which are used for the reduction of light pollution are listed below [12].

3.6.1 Light Shields

It involves the shielding of the street lights so that the pattern of illumination is under the horizontal plane of the light fixture and light is directed only where required to decrease the extent of light pollution. These shields help to prevent glare by decreasing the upward radiation and it is considerably utilized to decrease the light pollution when utilized in breeding colonies of shearwaters. These light shields reduced the attention of Newell's Shearwaters by 40% in Hawaii.

But, the fitting of shields is supposed to only be utilized as a final option. As these shields help in the decrease of light pollution, it leads to the wastage of a large amount of energy. Hence, its replacement by properly designed full cut-off lighting fixtures is an efficient way to decrease light pollution and also to decrease the wastage of energy.

3.6.2 Planning System

It requires the usage of light sources of the least power, essential and limiting needless or several lights. It needs legislation with enforcement from the preparation authority. A new lighting strategy is designed and the existing ones are re-evaluated. To support the initiatives, awareness should be raised between homeowners, architects, and builders. By altering the type of lights used, light pollution can be reduced. This is achieved by using high-pressure sodium vapor lamps in place of older high-pressure mercury vapor lights and joining them with "full cut-off luminaries" to decrease the glow and wastage of energy. Lights with different wavelengths have dissimilar abilities to attract animals. White lights are found to be the nastiest offender for the attraction of birds, with yellow lights being found to be superior in this feature. Blue and red lights are the least attractive. This may be imperative in the lighting of cranes and production of high constructions with lights on top of them etc [12].

3.6.3 Sports Arenas

The major lighting of the sports centers such as floodlighting or spotlighting should be turned off no more than 30 minutes after the ending of the day events [12].

3.6.4 Lighthouses

The presence of lighthouses causes danger to migrating species and there have been numerous reports all over the world of shearwaters and petrels being grounded by lighthouses. It was observed that steady rotating beams result in more bird loss as compared to intermittent lights at lighthouses. On the other hand, illuminating the lighthouse compounded the trouble.

3.6.5 Lights at Sea

Light-induced fishery sanctions require to be vigilantly believed in the context of their impact on seabirds. Vessels using bunkering zones, oil platforms, and other fishing vessels are too frequently intensely lighted and it is thus imperative that these lights be scaled back to decrease the attraction of birds [12].

3.6.6 Education

Various promotional messages like putting up notices in the hall, general educational displays, information in elevators, emailing tenants and staff during the critical period, and introducing information on light pollution in consumers information packs displayed by the individual building owners such as hoteliers for light pollution reduction. A communication strategy is required to create awareness among the society; this program could associate with government efforts to decrease global warming.

3.6.7 Light Restriction and Conservation During the Peak of Fledging

This involves the efforts to switch off street lighting and limit the utilization of non-valuable lights in the peak of fledging every time. Building lighting systems can be planned to attain a considerable diminution in night lighting. Rooftop floods, perimeter spots i.e. all exterior "vanity" lighting should be extinguished. Various options such as reprogramming timers, receptive lighting, adopting lower intensity lighting, using desk lamps, and task lighting should be used to decrease the light pollution in the areas where lights must be left on all night. Window covering by curtains also helps in decreasing light. There is a requirement for new planning applications to avoid an improvement in unsuitably lighted structures. Global awareness is essential to reach the goals [12].

3.6.8 Turning Lights off Using a Timer and Occupancy Sensor

Motion sensors can be used to activate the outdoor lights in comparison to switching. These forms of sensors assist in decreasing light pollution and also save energy. Various motion detectors can be set so that they are less responsive to the motion of small animals. Passive Infra-Red detectors (PIR) should be utilized in case of security lighting. It is sufficient to use150 W tungsten halogen lamp for domestic security lighting. Higher wattage lamps generate large glare, darker shadows, and intense lights. In general, lights can be switched off physically when not required. Attentiveness is important in achieving this objective.

3.6.9 Alternatives to Road Lighting

The first choice should be given to inert ways of forewarning drivers in rural regions. These warnings consist of fitting informational signs, reflector roadway markers, lines, and reflectors present at the sides of the path. These are utilized in various regions of the earth and majority of the cases, it reduces the usage of artificial lighting. Therefore, they should be fatally considered and reduction in light pollution helps in the decrease of cost. The surface of the road is completely illuminated as the design of full cut-off road lighting has been developed without reducing the spacing between streetlights and increasing the number of streetlights.

3.6.10 Full Cut-off Lighting Equipment

The 'Full Cut-Off' lighting equipment is the best option for lightning. It reverts their light output down, reduces the requirement of energy, and improves the visibility of the night sky. The light that would usually vector off into the sky is collected and reverted down. Currently, all trustworthy lighting producers make full cut-off lighting equipment [12].

3.7 CONCLUSION

Lighting is ethnically associated with the thoughts of modernity, growth, and advancement – the Enlightenment. Various religions employ light as a sign of integrity while darkness is associated with viciousness, disorder, and sin. Light pollution is a worldwide ecological topic, the environmental effects of which are only now beginning to be examined. Light pollution is a rising problem that has to be considered with an interdisciplinary approach. Light pollution is detrimental to several world's most geologically diverse and functionally vital living beings. Light pollution causes severe difficulty with implications for human health, wildlife, global warming, scientific study, and energy expenditure. Research into the various consequences of light pollution is rising. Over the decades, numerous efforts were made to evaluate the effects of light pollution on human and ecosystems health. There are lots of proof that humans have huge physiological harms. The study of light pollution requires further research. The increasing use of lights, like streetlights, advertising columns, greenhouses, and industrial services results in adverse effects on humans and wildlife species. Hence, this book chapter deals with the various types of light pollution, its effects on living beings, and ways to reduce the effect.

REFERENCES

[1] R.M. Lunn, D.E. Blask, A.N. Coogan, M.G. Figueiro, M.R. Gorman, J.E. Hall, J. Hansen, R.J. Nelson, S. Panda, M.H. Smolensky and R.G. Stevens. 2017. Health consequences of electric lighting practices in the modern world: a report on the National Toxicology Program's workshop on shift work at night, artificial light at night, and circadian disruption. *Sci. Total Environ.* 607: 1073–1084.

[2] J. Ngarambe and G. Kim. 2018. Sustainable lighting policies: the contribution of advertisement and decorative lighting to local light pollution in Seoul, South Korea. *Sustainability* 10: 1007.

[3] T. Stone. 2017. Light pollution: a case study in framing an environmental problem. *Ethics Policy Environ.* 20: 279–293.

[4] T. Stone, F.S. De Sio and P.E. Vermaas. 2020. Driving in the dark: designing autonomous vehicles for reducing light pollution. *Sci. Eng. Ethics* 26: 387–403.

[5] J.Q. Ouyang, M. de Jong, R.H. van Grunsven, K.D. Matson, M.F. Haussmann, P. Meerlo, M.E. Visser and K. Spoelstra. 2017. Restless roosts: light pollution affects behavior, sleep, and physiology in a free-living songbird. *Glob. Change Biol.* 23: 4987–4994.

[6] K.Y. Kim, E. Lee, Y.J. Kim and J. Kim. 2017. The association between artificial light at night and prostate cancer in Gwangju City and South Jeolla Province of South Korea. *Chronobiol. Int.* 34: 203–211.

[7] P. James, K.A. Bertrand, J.E. Hart, E.S. Schernhammer, R.M. Tamimi and F. Laden. 2017. Outdoor light at night and breast cancer incidence in the nurses' health study II. *Environ. Health Perspect.* 125: 1–11.

[8] Z. Hu, H. Hu and Y. Huang. 2018. Association between nighttime artificial light pollution and sea turtle nest density along Florida coast: a geospatial study using VIIRS remote sensing data. *Environ.Pollut.* 239: 30–42.

[9] J. Ngarambe, H.S. Lim and G. Kim. 2018. Light pollution: is there an environmental Kuznets curve? *Sustain. Cities Soc.* 42: 337–343.

[10] F. Falchi, P. Cinzano, C.D. Elvidge, D.M. Keith and A. Haim. 2011. Limiting the impact of light pollution on human health, environment, and stellar visibility. *J. Environ. Manag.* 92: 2714–2722.

[11] T. Longcore and C. Rich. 2004. Ecological light pollution. *Front. Ecol. Environ.* 2: 191–198.

[12] H. Raine, J.J. Borg, A. Raine, S. Bairner and M.B. Cardona. 2007. *Light pollution and its effect on Yelkouan Shearwaters in Malta; causes and solutions.* BirdLife Malta, Malta.

[13] S. Langenegger, F. Voll, R. Rodewald, S.L. Geschäftsleiter. *Actions to reduce light pollution in Swiss tourism destinations.* Bachelor Thesis, University of Applied Sciences HTW Chur. http://www.darksky.ch/dss/wp-content/uploads/2019/09/2019.08.09_IBT_Tou16_Langenegger_Simona.pdf

[14] R. Rajkhowa. 2014. Light pollution and impact of light pollution. *Int. J. Sci. Res.* 3: 861–867.

[15] J. Lyytimäki, P. Tapio and T. Assmuth. 2012. Unawareness in environmental protection: the case of light pollution from traffic. *Land Use Policy* 29: 598–604.

[16] A. Garcia-Saenz, A.S. De Miguel, A. Espinosa, A. Valentin, N. Aragonés, J. Llorca, P. Amiano, V. Martín Sánchez, M. Guevara, R. Capelo and A. Tardón. 2018. Evaluating the association between artificial light-at-night exposure and breast and prostate cancer risk in Spain (Mccspain study). *Environ. Health Perspect.* 126:047011.

[17] A. Svechkina, B.A. Portnov and T. Trop. 2020. The impact of artificial light at night on human and ecosystem health: a systematic literature review. *Landsc. Ecol.* 35: 1725–1742.

[18] R.J. Reiter, F. Gultekin, L.C. Manchester and D.X. Tan. 2006. Light pollution, melatonin suppression, and cancer growth. *J. Pineal Res.* 40: 357–358.

[19] J.S. ˇkvareninova´, M. Tuha´rska, J.S. ˇkvarenina, D. Baba´lova´, L. Slobodnı´kova´, B. Slobodnı´k, H. Strˇedova´ and J. Mind'asˇ. 2017. Effects of light pollution on tree phenology in the urban environment. *Morav. Geogr. Rep.* 25: 282–290

[20] R.H. Ffrench-Constant, R. Somers-Yeates, J. Bennie, T. Economou, D. Hodgson, A. Spalding and P.K. McGregor. 2016. Light pollution is associated with earlier tree budburst across the United Kingdom. *Proc. R Soc. B: Biol. Sci.* 283: 20160813.

[21] D.M. Dominoni. 2015. The effects of light pollution on biological rhythms of birds: an integrated, mechanistic perspective. *J. Ornithol.* 156: 409–418.

[22] B. Kempenaers, P. Borgstro¨m, P. Loe¨s, E. Schlicht and M. Valcu. 2010. Artificial night lighting affects dawn song, extra pair siring success, and lay date in songbirds. *Curr. Biol.* 20: 1735–1739.

[23] D. Dominoni, M. Quetting and J. Partecke. 2013. Artificial light at night advances avian reproductive physiology. *Proc. R Soc. B* 280: 1756.

[24] T. Raap, G. Casasole, R. Pinxten and M. Eens. 2016a. Early-life exposure to artificial light at night affects the physiological condition: an experimental study on the ecophysiology offree-living nestling songbirds. *Environ. Pollut.* 218: 909–914.

[25] D.M. Dominoni, J.K. Jensen, M. Jong, M.E. Visser and K. Spoelstra. 2020. Artificial light at night, in interaction with spring temperature, modulates timing of reproduction in a passerine bird. *Ecol. Appl.* 30: e02062.

[26] S.A. Cabrera-Cruz, J.A. Smolinsky and J.J. Buler. 2018. Light pollution is greatest within migration passage areas for nocturnally-migrating birds around the world. *Sci. Rep.* 18: 1–8.

[27] F.A. Sorte, D. Fink, J.J. Buler, A. Farnsworth, and S.A. Cabrera-Cruz. 2017. Seasonal associations with urban light pollution for nocturnally migrating bird populations. *Glob. Change Biol.* 23: 4609–4619.

[28] I. Malek and A. Haim. 2019. Bright artificial light at night is associated with increased body mass, poor reproductive success, and compromised disease tolerance in Australian budgerigars (Melopsittacus undulates). *Integr. Zool.* 14: 589–603.

[29] T. Batra, I. Malik and V. Kumar. 2019. Illuminated night alters behavior and negatively affects physiology and metabolism in diurnal zebra finches. *Environ. Pollut.* 254: 112916.

[30] S. Moaraf, Y. Vistoropskya, T. Poznera, R. Heibluma, M. Okuliarova´c, M. Zemanc and A. Barneaa. 2020. Artificial light at night affects brain plasticity and melatonin in birds. *Neurosci. Lett.* 716: 134639

[31] M. Brei, A. Pérez-Barahona and E. Strobl. 2016. Environmental pollution and biodiversity: light pollution and sea turtles in the Caribbean. *J. Environ. Econ. Manag.* 77: 95–116.

[32] W.H. Walker, J.C. Borniger, M.M. Gaudier-Diaz, O.H. Mele´ndez-Ferna´ndez, J.L. Pascoe, A.C. De Vries and R.J. Nelson. 2019. Acute exposure to low-level light at night is sufficient to induce neurological changes and depressive-like behavior. *Mol. Psychiatry* 25: 1080.

[33] J.Y. Min and K.B. Min. 2018. Outdoor artificial nighttime light and use of hypnotic medications in older adults: a population-based cohort study. *J. Clin. Sleep Med.* 14: 1903–1910.

[34] N.R. Rodrigues, G.E. Macedo, I.K. Martins, K.K. Gomes, N.R. de Carvalho, T. Posser and J.L. Franco. 2018. Short-term sleep deprivation with exposure to nocturnal light alters mitochondrial bioenergetics in Drosophila. *Free Radical Biol. Med.* 120: 395–406.

[35] M. Touzot, L. Teulier, T. Lengagne, J. Secondi, M. The´ry, P.A. Liboure, L. Guillard and N. Mond. 2019. Artificial light at night disturbs the activity and energy allocation of the common toad during the breeding period. *Conserv. Physiol.* 7: coz002.

[36] R.H.J.M. Kurvers, J. Dra¨gestein, F. Ho¨lker, A. Jechow, J. Krause and D. Bierbach. 2018. Artificial light at night affects emergence from a refuge and space use in guppies. *Sci. Rep.* 8: 14131.

[37] A.M. Dimovski and K.A. Robert. 2018. Artificial light pollution: shifting spectral wavelengths to mitigate physiological and health consequences in a nocturnal marsupial mammal. *J. Exp. Zool. Part A* 329: 497–505.

[38] L.K. McLay, V. Nagarajan-Radha, M.P. Green and T.M. Jones. 2018. Dim artificial light at night affects mating, reproductive +output, and reactive oxygen species in Drosophila melanogaster. *J. Exp. Zool. Part A* 329: 419–428.

[39] N.J. Willmott, J. Henneken, C.J. Selleck and T.M. Jones. 2018. Artificial light at night alters life history in a nocturnal orb-web spider. *PeerJ* 6: e5599

[40] Z.A. Weishampel, W.H. Cheng and J.F. Weishampel. 2016. Sea turtle nesting patterns in Florida vis-a`-vis satellite-derived measures of artificial lighting. *Remote Sens. Ecol. Conserv.* 2: 59–72.

[41] T.N. Simo˜es, A.C. da Silva and C.C. de Melo Moura. 2017. Influence of artificial lights on the orientation of hatchlings of Eretmochelys imbricata in Pernambuco Brazil. *Zool* 34: 1–6.

[42] P. Wilson, M. Thums, C. Pattiaratchi, M. Meekan, K. Pendoley, R. Fisher, and S. Whiting. 2018. Artificial light disrupts the nearshore dispersal of neonate flatback turtles Natator depressus. *Mar. Ecol. Prog. Ser.* 600: 179–192.

[43] V.A. Underhill and G. Ho¨bel. 2018. Mate choice behavior of female Eastern Gray Treefrogs (Hyla versicolor) is robust to anthropogenic light pollution. *Ethology* 124: 537–548.

[44] E. Knop, L. Zoller, R. Ryser, C. Gerpe, M. Ho¨rler and C. Fontaine. 2017. Artificial light at night as a new threat to pollination. *Nature* 548: 206–209.

[45] J. Bennie, T.W. Davies, D. Cruse, F. Bell and K.J. Gaston. 2017. Artificial light at night alters grassland vegetation species composition and phenology. *J. Appl. Ecol.* 55: 442–450.

CHAPTER 4

Source, Impact, and Perspective of Light Pollution

Essia Hannachi and Yassine Slimani

Imam Abdulrahman Bin Faisal University

CONTENTS

4.1 INTRODUCTION

The succession between the night and day cycle is the most central monitor of a large range of physiological patterns in living beings, involving humans [1]. Owing to the inclusion of electricity and non-natural light nearly a century ago, the type and period of light exposure has varied radically, hence night light becomes an upward and vital portion of the up-to-date lifestyle. Exposure to illumination at night appears to be correlated to serious behavioral and health complications, involving cardiovascular disease and cancer [2,3]. According to the daily perturbation proposition, the light at night may disorder the rhythm of self-circadian, and this particularly inhibits the production of the hormone indole melatonin and its excretion into the blood [4]. According to the International Agency for Research on

DOI: 10.1201/9781003185109-4

Cancer (IARC) [5], new cancer cases, 19.3 million have been recorded worldwide. Breast cancer was a major burden disease. Another case of more rapid cancer is colorectal cancer, and it has been estimated at 11.0% in 2021. The total cancerous cases are anticipated to reach, in 2040, 28 million due to the population growth. Some epidemiological proofs indicate that increased exposure to LP may be accountable for an amplified occurrence of breast and colon cancers in humans [6].

At first, it was supposed that light of at minimum 2,500 lux was enough to adjust the excretion of melatonin by the human pineal gland. However, later studies proved that low illumination of blue light can considerably prevent the production of melatonin [7]. Two experiments were conducted in real-life, home settings on 33 youth volunteers. Young volunteers were exposed to 96 hours of 1 lux artificial light or 48 hours of 5 lux artificial light. Two tests in normal conditions have been performed by K. Stebelova et al. [8]. The results showed that subjects were more delicate to dim light in the whole night, as previously anticipated. Artificial light up to 5 lux decreased significantly the biosynthesis of melatonin and varied the quality of sleep as proved by an augmented proportion of 1 minute inactivity and a tendency to increase the index of fragmentation.

Khodasevich and co-workers [9] showed that light at night has plentiful other effects. Interestingly, light plays a crucial role in circadian rhythm. Humans have largely replaced cycles of natural light with a diversity of artificial light sources and spend large amounts of time indoors. The response to artificial light leads to a change in temporal regulation in various organisms. The comprehensive observation that light-coerced intermediaries practice over various body systems, for instance, produces many targets upon which light-caused disturbances can act, inducing a large range of physiological variations and possibly dangerous medical effects. In a wider framework, the supporting physiological mechanisms adjust a diversity of activities, from generation to foraging, producing expanded targets for light disturbance. This chapter summarizes the impact of LP on health. The medical effects and related troubles in normal, seasonal, and biological clock functions are also discussed.

4.2 LIGHT POLLUTION

The key accomplishment of the last century in light may be called artificial light. However, artificial illumination is very useful for modern society. It also poses significant risks to human health in the form of pollution. LP can be clearly described as needless and wrongly directed artificial lighting [10]. LP is defined as a condition in which the wrong use of night light may impede people's comfort and health [10]. Among the known consequences, human health problems due to exposure to night light are the most urgent. Through the eye, light enters the human body. Most light intercedes other biological processes in humans, like cycles of light and darkness. According to the first atlas of artificial night sky brightness, around one-fifth of the world's populace, live under polluted skies [11].

Two sorts of LP can be existing; indoor artificial light (IAL) and outdoor artificial light (OAL). The sources of IAL are diverse. The increased evening exposure to short-wavelength indoor light from LEDs, television, tablet computers, and smartphones disrupt circadian rhythms. The electricity system has improved in the past century owing to lower electricity prices [12]. Moreover, artificial light has increased today through the widespread use of solar panels. This makes night light accessible in remote zones. Consequently, the residents' exposure to indoor night light has augmented dramatically [12]. In contrast to normal light which had stable and known emission spectra. Whereas OAL is the output of the modern world. However, most people now live under dim light all day without exploring the sunlight and muted light at night. OAL involves street lighting, vehicles, marketing, highways, architecture, railways, households, shopping malls, commercial buildings, sports facilities, industries, etc. The light at night released from these sources is received by the satellites through the sensors, orbiting the Earth and transmitting the captured information to the Meteorological Defense Satellite Program database (DMSP). The DMSP involves annual information after ignoring natural light sources like sun, moonlight, and others like fire and lightning. These images denote a small fraction of the light from the surface of ground light, and they epitomize the night light levels at ground level.

4.2.1 Light Pollution Caused by Urban Progress

Urbanization advances resulted in the requirement for non-natural illumination in roads, stadiums, malls, universities, parking areas, and houses. Some of this light is dispersed in the air, causing the natural sky to shine beyond background levels, and is called urban sky glow [12]. The increased urban growth over the past centuries is also the story of the widespread release of OL at night. Night illumination is one of the urban comforts of a range of sorts: functional, cultural, economic, and social. Night illumination is serious about finding the way, safety, and business success. Public lighting and the quality of substructure and related maneuvers are among the signs of good governance and the economic capacity of the city. Light at night also participates in identification as more cities incorporate lighting designs into tourism and leisure development strategies share affordable modern technologies such as LEDs (light-emitting diodes) in superior levels of night illumination, a welcome advance in backward areas but also augmenting worries about LP. Annually, artificial illumination grows globally at a rate of 2% [13]. Evidence is also growing on the risks raising the spotlight on human well-being and safety, environment protection, energy conservation, climate impacts, and costs.

Previous studies on night light using DMSP- Operational Linescan (OLS) (DMSP-OLS) system for cities, countries, and worldwide analyzes have revealed that the intensity of light is directly correlated to the density of residents and economic status [14,15]. They also surveyed the social and cultural influences that influenced levels of light release and the actions of people surrounding the use of light. Thus, urban ecological and cultural aspects are possible explicatory factors for dissimilarities in light resurrection levels among municipalities. Additionally, scientists examined human activities in urbanized areas of intense lighting. Hahn et al. [16] employed the data of DMSP-OLS to investigate the evolution in the light intensity and local differences in China in the period ranging between 1992 and 2012. A correlation between light intensity and urban areas/regional industrial features was demonstrated. Levin and Zhan [17] employed Visible Infrared Imaging Radiometer Suite day-night band (VIIRS-DNB) data to perform a universal analysis of zones with dense populations. Their investigation demonstrated that city illumination in the studied areas correlates with the economy. Kuechly and co-workers [18] used aerial scanning data (high spatial resolution) to analyze LP in Berlin. Their study, which recognized the main sources of LP based on urban terrestrial usage, presented a "brightness factor", that permits direct evaluations of illumination releases by urban land use. Earlier investigations have proposed numerous urban environmental indicators (for example, population, urban areas, and cultural variances) that can produce much use of lighting in urbanized zones. This proposal indicates that the night light intensity changes with different urban development patterns, terrestrial use characteristics, and built surroundings in the municipal and among cities [19].

4.2.2 Light Pollution Caused by Shift Work

Aside from the occasional exposure to light produced by night lighting, today's society suffers from 5 to 9 workdays cancellations in exchange for a larger number of night shifts and increased outputs and profits. For instance, fast-food restaurants in North America profit throughout the late night and at dawn. Additionally, in surveys conducted, in any urbanized society, around 20% of persons function in alternate shifts [20]. Shift-workers are living their lives outside of the stage with usual local time, but frequently cannot fully regulate their daily patterns owing to the varying shift calendars and the essential re-regulation of rest days [20]. Consequently, shift workers are subjected to intense light at night that can disorder the regular behavioral and physiological rhythms of the day.

4.3 IMPACTS OF LIGHT POLLUTION ON HEALTH

4.3.1 Light Pollution-Induced Eye Damage

The menace of exposure to LP to human health has attracted the interest of many scientists to conduct extensive research on this topic. Particularly, relatively high-energy blue illumination can cause harm to the tissue of the eye. Intense exposure to blue can induce many types of modifications, like oxidative stress, DNA injury, inflammatory apoptosis, mitochondrial apoptosis, and mitochondrial apoptosis, leading to dry eye disease, keratitis, and glaucoma. Blue light can affect circadian rhythm, causing prolonged cognition and improved alertness, which is linked with stimulation of ganglion cells of photosensitive retinal (ipRGCs) [21]. Light is believed to harm the retina through different mechanisms involving photomechanical photothermal, and photochemical (Figure 4.1). Some investigations showed that blue illumination with high energy can enter the cornea, lens and directly arrive at the retina, resulting in photochemical injury to the eye [22].

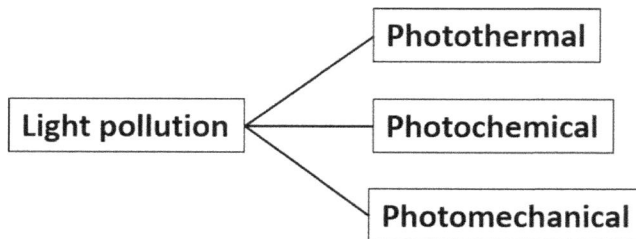

FIGURE 4.1   Schematic representation of the three main forms of light-induced eye injury.

Photochemical mechanisms have been identified, since 1966, cause damage to rod cells. Subsequently, several experiments have proved the danger of blue light and have demonstrated that the intensity of injury is associated with various parameters like the intensity of the illumination that the eye receives, the distance of the lighting, the direction of the sightline, and the light source spectrum. The morphology, as well as the role of the retina, could be protected by using blue-covering filters. The position report by the International Commission on Illumination (CIE) showed that the efficacy of blue light risks is meticulously linked to the supreme safe blue light radiation [23]. There is no clear indication to confirm that exposition to blue illumination rises the danger of photochemical damage through usual conditions of use. An evaluation of light with low energy, PCs, and tablets showed that exposure limits to blue light are not exceeded in these devices even under longstanding observing situations [24].

In 1978, it was reported for the first time that blue illumination causes lesions of the retina induced from photochemical harm rather than from thermal damages [25]. Intensifying proofs indicate that excessive exposition to blue light leads to a marked upsurge in reactive oxygen species (ROS) production, which causes apoptosis, lipid peroxidation, and photoreceptor loss [26]. The influence of blue illumination and photochemical reflection of blanching will exacerbate the photochemical injury and induce activation of DNA injury, inflammatory reactions, mitochondria curb, and function of the lysosome. Developing studies show that the risk of blue light is not confined to the retina only; it impairs also the ocular surface via inflammatory response and oxidative stress. Moreover, some reports indicate that the crystalline lens transparency decreases with age progress, resulting in a gradual upsurge in the absorption in the interior of the blue light spectrum [27].

Recent epidemiological investigations showed that exposure to blue light might be related to the progress of age-related maculopathy (ARM) [28]. Nevertheless, a meta-analysis performed in 2018 indicated that there is no association between sunlight subjection and the menace of developing age-related macular degeneration (AMD) [21], a result supported the CIE position report [23]. However, exposure to blue light can be a conducive parameter in age-related macular degeneration (ARMD). Exposure to blue light may upsurge the production of ROS. The accumulation of ROS can control CC chemokine receptor activity and enhance the liberation of inflammatory factors [29]. In developed countries, ARMD is the principal cause of irreparable blindness and frequently influences individuals with the range age between 50 and 65 years old. Ironically, the light we need to see severely damages our visual organisms. ARMD is induced by injury to the retinal macula (macula is the inner region of the retina and comprises the maximum number of photoreceptors) and is brought about fuzzy sight or, in severe situations, not seeing in the center of the visual field. Photoreceptors can be injured by light, especially short-wave illumination [30]. Light with a short wavelength is a hazard as photons at this range exhibit elevated energy and can harm the role of photoreceptors, and the cellular structure. Blue illumination was more efficient by 50–80 times than photoreceptor impairment [31]. The cornea blocks most UV rays below 295 nm, while the lenses absorb much ultraviolet (with a wavelength between 315 and 400 nm) and ultraviolet (280–315 nm) rays. Some radiations <400 nm can achieve the retina. In nature, the macular region has yellow color owing to the occurrence of lutein (one of the xanthophyll carotenoids), which may protect from blue light. For example, some experiments conducted on mice showed apoptosis of photoreceptors under exposure to light from 400 to 480 nm [32].

Lougheed and coworkers made experiments on albino mice exposed to illumination by white LEDs. Their results showed that, after just 9 days of exposure, harm of retinal and death of cell [33]. Krigel et al. [34] reported the effect of light on retinal injury using different LEDs with different colors (green, blue, cool white) in albino and pigmented mice with enlarged pupils. Their analyses do not consider the impacts of repetitive exposure, they are based principally on intense light exposure. The results showed that the blue constituent of the white LEDs can induce harmfulness of retina at professional domestic illuminance and not only in extremist experimental circumstances. Shang and co-workers [35] reported that LEDs (blue; 460 nm) are more harmful to the retina of the eye in freely operating mice compared to LEDs (green; 530 nm) or LEDs (red; 620 nm) for analogous levels of corneal radiation. Ryo Mukai's group controlled the functional and morphologic variations of the monkey retina after light exposure. Their findings showed that exposure to light produced an enlarged reflection of the photoreceptor operating system, which corresponds to an intracellular gap and irregularity of the laminar structure of the operating system [36]. Although most experiments have been done on animals, lately, an investigation performed by Margrain revealed that blue light is also hazardous for developing ARMD in humans [37].

Over the past 10 years, the selling of incandescent bulbs has been escaped progressively in the European Union and it is assessed that using white LEDs to substitute other lighting sources will lessen the emissions of carbon dioxide [37]. Nevertheless, replacing tungsten light with short-wavelength LEDs is a possible health danger. ARMD is among the serious menaces that can upsurge with augmented usage of light (particularly blue light) in buildings and workstations. Alim-Marvasti reported a study on two cases in which persons suffered from bouts of momentary "blindness" from watching a phone during 10–20 minutes with only one eye, whereas the other eye was masked with a pillow. The persons practiced momentary "blindness" that undertook some minutes to recuperate. The authors noted that such experiences of photo-bleaching are owing to the normal optical adaptation of lighting in one eye and obscurity in the other eye [38].

In 2017, James and co-workers [39] assessed the possible threats from LEDs projected for house usage. They observed that if rated at a level of luminance of 500 lux (as endorsed for luminaires proposed for General Lighting Service (GLS), all LED lights fall into the exemption class. Nevertheless, if rated at a measuring distance of 0.2 m, as endorsed for non-GLS bulbs, then one of the LEDs falls into the RG Class 2. This lamp was very bright and potentially very painful, which leads to an aversion reaction in a large category of persons. In 2019, the ANSES (French Agency for Food, Environmental, and Occupational Health & Safety) presented an update on the numerous health impacts that may be resulting from exposure to rich blue light and other features of LEDs. The ANSES assessment proves the results obtained in 2010 on the risk of blue light toxicity on the eye, leading to vision impairment. ANSES emphasizes the short-term impacts on the retina associated with high contact to blue light, and the long-term impacts associated with the start of

ARMD and strongly requests limiting the selling of LEDs devices that produce very high levels of blue light through amendments to precise regulations for devices other than spots and lighting fixtures.

Moreover, based on recent experimental studies on the phototoxicity mechanisms, ANSES highlights the need to revise the limit values of blue light exposure to consider the privacy of kids, for whom the efficiency of the lens of the eye filters out blue light is less than that of adults. Additionally, ANSES focuses on the troublemaking impacts of circadian rhythms and sleep-related exposition to dim blue light at night, especially via television or laptop and phones. The agency advises restraining residents' exposure to the high level of blue at night, particularly children [40]. Very recently, a group from the United Kingdom has published a paper in which the researchers highlighted the energy-saving-effectiveness of LED spots against the other conventional illumination system. Nevertheless, their study is incapable to deal with the harmful and uncontrollable impacts of LEDs on human health, particularly the eye [41].

Future studies are needed to address the indirect benefits of light and health impact. In the domain of ophthalmology, extra care is concerted on the deterrence of the blue light danger; accordingly, it is essential to expose the causal mechanisms included in the blue light-caused injury to the retina. It has been suggested that blue light only influences cells with a healthy mitochondrial respirational chain function, indicating that mitochondria are essential to damage light [42]. Indeed, mitochondria approximate multiple cell death pathways in mammalian cells [43]. In this awareness, mitochondria are prospective potential and proactive targets of blue light on the retina. Consequently, they are anticipated to become desirable targets for precautionary and curative treatments that save retinal injuries caused by blue light. Unlike this, the cogent usage of LEDs is the first stage in avoiding the danger of blue light.

4.3.2 Physiological Impacts of Light Pollution

4.3.2.1 Behavioral Effect

The daily pacemaker is accountable for regulating the scheduling of the whole body, which includes the body's multiple systems. As mentioned above, light is received by ipRGCs in the eye. A group of ipRGCs produces the phantom retinal area that protrudes and enters a set of neurons that form the daily oscillators in the suprachiasmatic nuclei (SCN), and that regulate melatonin secretion. Melatonin is an indoleamine that regulates variations in numerous physiological roles in reply to a change in the day extent. The nighttime period of melatonin is the serious factor accountable for shifting the impacts of light on both the individual body systems and the neuroendocrine axis [44]. Prolonged exposure to light modifies melatonin quantities in all organisms, including humans. Therefore, exposure to light in the evening results in a diversity of behavioral impacts, which are possibly intermediated by various amounts of melatonin. Additionally, direct compassionate monitoring of physiological processes following contrast in light conditions has been recognized independently of melatonin production. Thus, prolonged exposure to light can modify human behavior.

In a previous report that was conducted on the populations of the Czech Republic. Forejt et al. [45] discovered that 5% of the residents perceive undesirable light from the outside as one of the principal reasons for their sleeping difficulties. Seven percent also grumbled of light quantities that were not attenuated to tolerable levels, and another 20% decreased LP in bedrooms to dark levels deemed enough, nevertheless, 5% of the Czech population felt troubled by the loss of full natural morning light due to the blocks against the light at night. Most of the investigations regarding the influence of darkness on human behavior have been performed on shift workers. Since it is not known, the amount of improvement observed in shift work studies concerning daily maladjustment is due to bright light at work and the amount of sleep scheduled during the day in the dark. Numerous mechanisms for this adaptive mechanism can exist. Sleep itself might operate as non-optical behavioral synchrony, or darkness may behave as a synchronizer in itself (a "dark pulse"). It is possible that the strong phase shift effect observed may be entirely due to the sleep scheduling of subjects undergoing treatment, where bright light plays only a minor role. Contact to bright light affects sleep trouble, mental capacity, concentration, and attentiveness. Investigations also assumed that the risks of LP become a social matter and possibly will provoke social conflicts in the upcoming years.

Benfield et al. [46] reported the psychological influence of LP in three U.S. countrywide parks. The experiments were made on 138 participants from different origins, including 48 males and 89 females. The results showed that LP affects an array of psychological dimensions and act evaluation. LP affected individuals who are experimented in a simulated natural environment. Contributors who have experimented with a reasonable amount of LP informed lower humor scores and lower landscape quality compared to fewer LP acts. This indicates that managing light LP in these areas might become worthy to create agreeable visitor experiences and maintain positive outcomes related to landscapes in general or at night particularly.

4.3.2.2 Metabolic Disturbance

Effectual energy metabolism is critical to the whole physiological role. Disruptions or complications in the productivity of the metabolic process can lead to various disorders, including obesity, diabetes type 2, and heart disease. There is plenty of proof showing the impact of exposure to higher levels of LP on the metabolic process and numerous of these epidemiological endpoints. P. H. Manríquez and co-workers [47] examined the effect of light in the evening on the physiological features of Concholepas concholepas (C. concholepas). Authors utilized juveniles C gathered in tidal habitations that are not formally subjected to light in the evening and exposed them to dark and white light-emitting diodes (LEDs) to evaluate the light in the evening effects on metabolism. The experiments showed that juveniles C search for and select their target more effectively in dark zones.

LEDs illumination increased metabolism times. Long-standing exposure of the mice to continuous light had adjusting effects on metabolism, particularly on the liver's carbohydrate metabolism. Tests on chickens have shown that continuous light conveys metabolic productivity. Female chickens raised in a steady light atmosphere acquired a harshly higher proportion of lipids compared to controls raised in a cycle of 12 L: 12 D photosynthesis. Male broilers also gained substantially much more weight when subjected to steady light [48]. The interruptions of steady light in the nocturnal excretion of melatonin have also been demonstrated to lead to metabolic disturbance. Melatonin looks to disturb the adjustment of body mass, thermogenesis, bowel proficiency, and a metabolic degree in some mammalian types. Therefore, the fundamental processes linked to energy procurement and use are changed after prolonged exposure to artificial lighting.

Numerous investigations indicate that humans experienced analogous impacts when exposed to light in the evening. For instance, adverse impacts of shift work were detected on lipid metabolism and carbohydrate, High blood pressure and coronary heart disease (CHD), insulin resistance, etc. [49]. Such effects are resulting from the direct physiological impacts of exposition to light and/or indirect influences correlated with deficiency of sloper [50]. Sleep deficiency leads to a marked change in metabolic and endocrine parameters linked to obesity, diabetes, and other complaints. Furthermore, melatonin amounts, which directly mirror variations in the light, have been correlated to CHD. For instance, in an associated study, patients suffering from CHD had considerably lower melatonin levels in the evening compared to patients that did not suffer from CHD [48]. Melatonin decreases the activities of the sympathetic nervous system and considerably decreases the rate of circulation of norepinephrine in the heart, hence increasing norepinephrine and epinephrine which are responsible for accelerating the absorption of harmful cholesterol. Prolonged exposure to dim night lighting reduces the secretion of melatonin in humans [50].

4.3.2.3 Immunological Dysfunction

Exposure of individuals to long-lasting artificial night illumination can change immune response (IR), which is an essential agent for survival, through combinations of neurotransmitter endocrine pathways or oxidative. In a recent study, T. A. Bedrosian et al. [51] have reported the influence of light at night in the photoreceptive genus in their original residence. The authors have examined the immune response of Siberian hamsters. After 1 month of exposure to mild light at night, IR was evaluated according to diverse challenges: (i) Delayed hypersensitivity (DH), (ii) Fever caused by lipopolysaccharide (LPS), and (iii) Bactericidal activities (BA) in the blood. It was demonstrated that illumination at night repressed the DH response and lessened BA after treatment of lipopolysaccharides. Additionally, in addition, illumination at night changed quotidian patterns of locomotor activities, indicating that human violation of environments by night illumination may accidentally harm IR and eventually physical fitness.

Likewise, M. E. Kernbach et al. [52] showed that light in the evening expanded the infectious window into the vector of zoonotic pathogens in wild reservoir classes. Sparrows subjected to artificial light at night preserved infectious viral titers for 48 hours longer than controls but did not exhibit greater deaths due to the West Nile virus (WNV) in this window. Relatively, artificial light at night changed the expression of gene monitoring networks involving major axons (TRAP1, PLBD1, and OASL) and effector genes recognized to influence WNV propagation (SOCS). Although antiviral IR escalated earlier, transcriptional signatures indicated that individuals subjected to artificial light at night may have exposed to harm from immunopathology and pathogens, possibly due to elusion, of immune stimuli. Previous studies in laboratory rodents discovered that individuals subjected to different sources of light at night had overstated IR, many of which had the potential to cause collateral destruction [48].

Different mechanisms by which artificial light at night modifies immune defenses among them, hormones (like melatonin) could play a role. Injecting melatonin into Syrian hamsters or keeping hamsters in short light periods leading to increased levels of melatonin resulted in amplified total splenic lymphocyte count, splenic masses, and macrophage counts [53]. A lot of investigations proved the presence of melatonin receptors in lymphoid tissues and circulating cells in the immune system [48]. Though splenic melatonin receptor proliferation usually fluctuates such that numbers of the receptor are small at night once melatonin concentrations are elevated, the binding site level in light is still high. This corroborates the supposition of the regulatory function of melatonin on the immune system where melatonin reduces its bonding status. Melatonin has been informed to be a drug-resistant or hormone-based immunosuppressant and looks to have overall immunomodulatory properties. Inhibition of melatonin for shift workers or by subjection to LP can pit down these immune properties. Moreover, steady light usually impedes autoimmunity T cells by removing melatonin. The influences of melatonin on the immune system have been reviewed in detail in Ref. [54]. Investigations in both mice and humans showed that sleep deficiency may stimulate one of the main neuroendocrine stress systems (the hypothalamic-pituitary-adrenal (HPA) axis). The impact of 40 hours of sleeplessness on a diversity of immune variants was studied in the peripheral blood of ten normal males. Sleep deficiency enhanced the nocturnal interleukin-1-like activities and nocturnal plasma. The elevated nocturnal response of lymphocytes to mitogen provoked stimulation during the normal whole day sleep-wake cycle was overdue due to sleep scarcity, but the response to mitogen phytohemagglutinin was not affected. With the resumption of night sleep, there was an extended decrease in natural destroyer cell activities and an increased return of the mitogenic response. The changing patterns of immune function arose individualistically of the circadian rhythm of cortisol, which persisted unaffected. In a previous study in the rats, P. Meerlo et al. [55] have demonstrated that sleep deficiency stimulates the HPA axis in mice and modifies post-stress responses, which may lead to indirect influences on the immune system.

4.3.2.4 Testosterone Levels (T)

LP also affects the level of testosterone of various organisms. D. M. Dominoni et al. [56] have studied the impact of light in the evening exposure on the reproductive physiology of Turdus merula (European blackbirds). The birds were subjected to 0.3 lux at night and nights during 2 year cycles. The results showed that the birds who stayed on nights showed two cycles of testes and T during the experiment. Urban birds have evolved testes quicker than controlled rural species. On the other hand, while blackbirds that were subjected to light in the first year showed a usual but early gonad cycle in comparison with control birds, during the subsequent year the reproductive system did not show any development at all. The size of testicular, as well as amounts of T, were at a baseline level in the whole birds. These results demonstrated that long-lasting dim light in the evening can significantly disturb the reproductive system.

T. Le Tallec et al. [57] performed experiments on the LP impact on the Microcebus murinus (nocturnal mouse lemur). The authors arbitrarily subjected 12 males in the winter for sexual comfort to moonlight or night street light simulating light for 5 weeks and controlled plasma amounts of T, motor activity, etc. Their results showed that males of Microcebus murinus subjected to LP exhibit changes in activation of reproductive function. S. J. Schoech's group studied the outcome of lightning at night on the male and female western scrub-jays (Aphelocoma California). Their results disclosed that light in the evening tends to impede reproductive hormone secretion, although not in a gender-consistent manner [58].

4.3.2.5 Locomotor Activity

LP greatly alters the daily rhythm of LA. Though not adjusted for its entire extent, LA offered a postponement in both its start/end movement and decreased drastically at night under LP exposure. Such modifications were observed in various species. For instance, recent studies on bats exposed to light in the evening have been performed and showed that illumination has serious influences on the upkeep of bats [59,60]. The lightning exposed to home-dwelling bats delays their appearance or considerably extends the period of appearance and, in dangerous cases, destroys the entire settlement. Differences in forearm length and body mass may indicate that the time of birth begins later and/or that the rate of growth is lesser in bats that live in lightened buildings.

Other studies have been performed in nocturnal rodents subjected to light in the evening [61]. The impact of the intensity of illumination on activity patterns of Patagonian Leaf-Eared Mice (Phyllotis Xanthopygu) has been studied by K. M. Kramer et al. [61]. The findings showed a decrease in the activity of the mice exposed to high light intensities. The photoresist can bypass the internal biological clock of animals and thus modulate their activities. In nocturnal assortments, the illumination impulse destroys activity. Moreover, LP may be an issue when taking into consideration activities related to moonlight. Previous studies in various species, involving mammals, have revealed that moonlight has an impact on the rhythm of activity, reproductive synchronization, navigation, habitat use, communication, and plundering. Consequently, in addition to changing the natural light/darkness and the perception of it by living species, LP can also alter the observation of moonlight.

4.3.3 Light Pollution-Induced Cancer

Cancer is often fought by the endocrine glands, antioxidants, and immune function. All these factors may be affected by exposure to light. Consequently, the appearance of many types of cancer diseases in living species including animals and humans is swelling. According to the European Commission, artificial illumination might cause endometrial cancer, ovarian cancer, prostate cancer, breast cancer, skin cancer, and so on. Urbanized countries display a higher risk of progressing cancer than the least urbanized countries. Generally, approximately 50% of breast cancer cases cannot be explained by traditional menace factors [62]. Western countries are progressively becoming 24 hour civilizations as more people are exposed to a large quantity of artificial light during the evening at the house and especially in the workroom.

A report done by IARC in 2000 [5] indicated that breast cancer is a giant disease problem in industrialized nations. It is the most communal cancer in females with valued breast cancer new cases of 999,000 (around 21%) each year, giving rise to around 370,000 death cases. The majority of the cases have been recorded in industrialized nations: around 200,000 in North America and 330,000 in Europe. The syndrome is not yet widespread among females in uncivilized countries, though its prevalence is growing. In the beginning, it was supposed that illumination (at least 2,500 lux) was enough to adjust the secretion of melatonin by the human pineal gland. However, prudently considered investigations have exposed, that melatonin excretion represses when illumination is as low as 1.3 lux [1]. An intense decrease in the level of melatonin has been detected in humans who were subjected within 2 weeks to irregular exposure to light in the evening.

In mammals, the SCN stimulates the nocturnal production by the pineal gland of melatonin which can impact the activity of the clock gene and be included in inhibiting cancer progression (Figure 4.2). Some research papers report that exposure to light-induced daily turbulences promoted cancer. Several research papers showed that circadian rhythm disturbances are correlated with many types of cancer, involving hormone-linked cancers that recur in shift workers since they developed irregular work hours able of disrupting the daily rhythm. In the past few years, numerous researchers have demonstrated that there is a relationship between control loops in the modulation of the nervous immune system and the onset and progression of cancer. Various results in humans and animals have shown

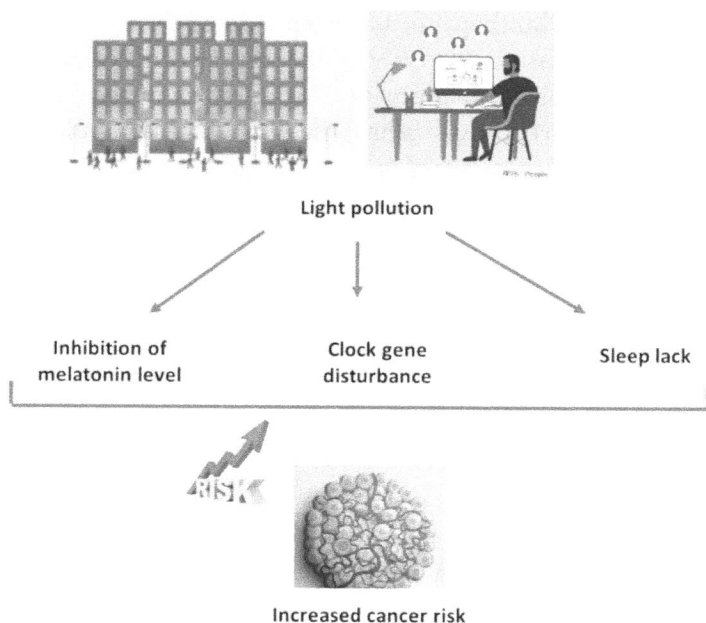

FIGURE 4.2 Summary of the effects of light pollution on an increased risk of cancer diseases.

that light can inhibit melatonin secretion and may upsurge estrogen, cortisol, and androgen levels [63].

Stevens' group hypothesized that exposure to light in the evening may be dangerous and cause breast cancer in developed civilizations through its capacity to destroy the nocturnal creation of melatonin [62]. This supposition is founded on investigations indicating that melatonin impedes the development of breast cancer, while surgical resection of the pineal gland or exposition to steady light promotes the formation of breast tumors in rodents [64]. This hypothesis was further supported by the validation that nocturnal levels of melatonin directly hinder the proliferation of human MCF-7 breast cancer cells in culture [65].

Following the suppositions of Stevens, Kerenyi and his coworkers suggested that LP may have a possibly significant pathology effect on the origins of other cancers [66]. In a previous epidemiological study, Davis and coworkers showed that females who are shift workers have a higher risk of mounting breast cancer owing to their increased likelihood of exposition to light in the evening providing extra evidence for this supposition [67]. Davis et al. [67] assessed the reaction of rats carrying either hepatocellular or hepatic grafts to human breast cancer, to upsurge the light exposure intensity during obscurity on tumor growth. Importantly, the authors identified signal transduction of hepatocellular carcinomas and xenografts of breast cancer grafts to in situ perfusion with blood gathered from human premenopausal volunteers whose signal of physiological melatonin was blocked by exposure to bright white fluorescent light in the evening.

The only justification for introducing hepatocellular carcinomas in tissue-isolated mice relied on the fact that they display a marked sensitivity to melatonin. Via this study, the authors aim to assess molecular/biological responses to light in the evening in a novel isolated tissue- human breast cancer xenograft model with that in a rat liver cancer model hypersensitive to melatonin which acts as a positive control. Contrasting to breast cancer, human liver cancer is more frequent in the unindustrialized world (having less LP exposure) than in urbanized countries (with more LP exposure), and its occurrence declines as civilizations are more western.

Another study has been performed by J. Hansen et al. [68] on the evolution of breast cancer among Danish females who work most of the time at night. The ages of the women who were tested are in the range of 30–54 years old. The likelihood of breast cancer among shift workers women for at least half a year at night was great, and there has been a probability to upsurge the odds ratio by rising the period of night work. A very recent investigation done by Garcia-Saenz in 2020 showed that there has been a correlation between OL exposure and colorectal cancer in Spain [69]. The results showed that exposition to blue light has an impact on the emergence of colorectal cancer (odds ratio=1.6; 95%). Odds ratios were analogous (odds ratio=1.7; 95%) upon further adjustment of status of the area's socioeconomic, family history, smoking, diet patterns, and sleep. Reduced sleep duration is associated with exposure to light in the evening. The obvious example of troubled sleep habitats is those who suffer from jet lag or shift workers, associating light exposure.

Experimental investigations in animals and humans have confirmed the possible role of LP in postponed sleep start, early wake time, and plunge in the sleep period. Reduced duration of sleep is also linked with augmented cancer risk. Though the precise mechanism of the relationship between duration of sleep and menace of cancer remains a mystery, scarce mechanisms have been not compulsory that, firstly, shorter sleep leads to reduced melatonin concentrations and thus the growth of cancers diseases such as the breast, endometrium, ovary, and prostate [70]. Second, deficiency of sleep may be a result of chronic stress that causes the occurrence of cancer. Chen et al. [71] studied the correlation between duration of sleep and the cancer appearance using categorical, dose, and response meta-analysis. Conclusive meta-analyses showed that a shorter duration of sleep augmented the risk of cancer in Asians and that longer duration of sleep augmented colorectal cancer risk, nonetheless, these results were not reliable in a dose and response meta-analyses. The authors suggested further upcoming studies are mandatory to precisely assess causation and clarify the underlying mechanisms of the relationship between the extent of sleep and cancer risk.

In a recent study published in 2020, a scientific group from Denmark, in collaboration with the US, Australia, and the Republic of Korea researchers reported that there is no association between light in the evening and cancer in the breast in general. The authors have exploited information on around 17,000 Danish nurses who were tracked from the cohort baseline in 1999 till 2012 in the Danish Breast Cancer Collaborative Group for ER Breast Cancer, the Danish Cancer Registry of Breast Cancer Incidence, and Public Relations Status. The data of light at night exposure were gotten from the US DMSP and were allocated to the study contributors' accommodation addresses during monitoring. Among the all followed-up Danish nurses, 745 suffered from breast cancer. The authors concluded a suggestive relationship between light at night at breast cancer development [72].

Another study done by J. Ritonja et al. [73] proved that small or no effect of illumination at night on cancer emergence. Residents-based case-control investigations were conducted in British, Vancouver, Columbia Ontario, Kingston, and Canada with the occurrence of breast cancer cases, and age-matched frequency controls in the identical region. The examination was limited to 844 statues and 905 controls who offered a residential history of life. Through the time-weighted mean duration in each house from 5 to 20 years preceding the entry of the analysis, two measurements of the accumulative middling of the light at night were estimated using two data sources satellite. For the estimation of the association between light at night at cancer emergence logistic regression was exploited by taking into account interactions related to menopausal status and night shift work. These results support the conclusion accomplished by Clarke's group [72]. Despite a large number of studies on LP, there is an urgent need for in-depth studies on its effects on cancer risk.

4.4 CONCLUSIONS AND PERSPECTIVES

Exposure to LP is customary in the developed world. LP causes damages to health with investigation dating back to the nineties but people are still unconscious of its effect and threats. First, light can enact risk on the eye (retina and underlying structures). Light can result in injury to the retina via photothermal, photochemical, and photomechanical mechanisms. In addition, LP hinders the secretion of melatonin, a decisive monitor of daily rhythms. The disturbance of melatonin induces several negative effects on physiological functions including behavior, metabolism disruption, immunological dysfunction, oxidative stress, and increase epidemiological diseases.

Evening shift work has been categorized by the IARC as a possible human hazard based on enough proof of cancer and robust mechanistic indication in animals, and restricted epidemiological proof, involving prostate, colorectal, and breast cancerous diseases. LP is receiving far less attention than any other universal issue. There is no doubt that the increase in lighting helps people sense safe, but there should be a way less damaging to lighting our towns. Bright light is commonly seen as a blessing, not a curse. Light is described as a sign of knowledge, progress, and finesses in all civilizations while dark is associated with poorness, badness, and unconsciousness.

The possible harmful impacts of light are not identified to all people, and the dangers of hidden light have not yet been demonstrated in full public awareness. Time is needed to focus on consciousness campaigns so that individuals are attentive to the menace of LP and can be effectively handled with night light. The analysis showed that the level evidenced by obtaining information from the media and the recall of media criticism of LP were the main influences that stimulated the increased perception of risk. This indicates that the government and controlling establishments need to enhance communication with the populace through unceasing use of the Internet and social media. When offering LP information to the populace, the transparency and the accuracy of the information must be ensured.

REFERENCES

[1] S. M. Reppert, D. R. Weaver, Coordination of circadian timing in mammals. *Nature*, 418 (2002) 935–941.

[2] R. J. Reiter, Potential biological consequences of excessive light exposure: melatonin suppression, DNA damage, cancer and neurodegenerative diseases. *Activitas Nervosa Superior*, 49 (2007) 33–37.

[3] V. N. Anisimov, The light-dark regimen and cancer development. *Neuroendocrinology Letters*, 23 (2002) 28–36.

[4] R. G. Stevens, Circadian disruption and breast cancer: from melatonin to clock genes. *Epidemiology*, 16 (2005) 254–258.

[5] H. Sung, J. Ferlay, R. L. Siegel, M., Laversanne, I. Soerjomataram, A. Jemal, F.Bray, Global cancer statistics 2020: GLOBOCAN estimates of incidence and mortality worldwide for 36 cancers in 185 countries. *CA: A Cancer Journal for Clinicians*, 71 (2021) 209–249.

[6] E. S. Schernhammer, F. Laden, F. E. Speizer, W. C.Willett, D. J. Hunter, I. Kawachi, G. A. Colditz, Rotating night shifts and risk of breast cancer in women participating in the nurses' health study. *Journal of the National Cancer Institute*, 93 (2001) 1563–1568.

[7] G. C. Brainard, J. P. Hanifin, Photons, clocks, and consciousness. *Journal of Biological Rhythms*, 20 (2005) 314–325.

[8] K. Stebelova, J. Roska, M. Zeman, Impact of dim light at night on urinary 6-sulphatoxymelatonin concentrations and sleep in healthy humans. *International Journal of Molecular Sciences*, 21 (2020) 7736.

[9] D. Khodasevich, S. Tsui, D. Keung, D. Skene, M. E. Martinez, The influence of light pollution and light-at-night on the circadian clock. *MedRxiv* (2021) 1–15.

[10] M. Liu, B. G. Zhang, W. S. Li, X. W. Guo, X. H. Pan, Measurement and distribution of urban light pollution as day changes to night. *Lighting Research & Technology*, 50 (2018) 616–630.

[11] P. Cinzano, F. Falchi, C. D. Elvidge, The first world atlas of the artificial night sky brightness. *Monthly Notices of the Royal Astronomical Society*, 328 (2001) 689–707.

[12] A. Haim, B. A. Portnov, Light-at-Night (LAN) as a General Stressor. In *Light Pollution as a New Risk Factor for Human Breast and Prostate Cancers* (2013) pp. 35–40, Dordrecht: Springer.

[13] C. C. M. Kyba, T. Kuester, A. S. De Miguel, K. Baugh, A. Jechow, F. Hölker, J. Bennie, D. Elvidgekevin, J. Gaston, L. Guanter, Artificially lit surface of Earth at night increasing in radiance and extent. *Science Advances*, 3 (2017) e1701528.

[14] Z. Chen, B. Yu, Y. Hu, C. Huang, K. Shi, J. Wu, Estimating house vacancy rate in metropolitan areas using NPP-VIIRS nighttime light composite data. *IEEE Journal of Selected Topics in Applied Earth Observations and Remote Sensing*, 8 (2015) 2188–2197.

[15] C. Kyba, S. Garz, H. Kuechly, A. S. De Miguel, J. Zamorano, J. Fischer, F. Hölker, High-resolution imagery of earth at night: new sources, opportunities and challenges. *Remote Sensing*, 7 (2015) 1–23.

[16] P. Han, J. Huang, R. Li, L. Wang, Y. Hu, J. Wang, W. Huang, Monitoring trends in light pollution in China based on nighttime satellite imagery. *Remote Sensing*, 6 (2014) 5541–5558.

[17] N. Levin, Q. Zhang, A global analysis of factors controlling VIIRS nighttime light levels from densely populated areas. *Remote Sensing of Environment*, 190 (2017) 366–382.

[18] H. U. Kuechly, C. C.M. Kyba, T. Ruhtz, C. Lindemann, C. Wolter, J. Fischer, F. Hölker, Aerial survey and spatial analysis of sources of light pollution in Berlin, Germany. *Remote Sensing of Environment*, 126 (2012) 39–50.

[19] S. Cheon, J. A. Kim, Quantifying the influence of urban sources on night light emissions. *Landscape and Urban Planning*, 204 (2020) 103936.

[20] S. M. W. Rajaratnam, J. Arendt, Health in a 24-h society. *The Lancet*, 358 (2001) 999–1005.

[21] P. L. Yang, S. I. Tsujimura, A. Matsumoto, W. Yamashita, S. L. Yeh, Subjective time expansion with increased stimulation of intrinsically photosensitive retinal ganglion cells. *Scientific Reports*, 8 (2018) 1–9.

[22] D. Van Norren, J. J. Vos, Light damage to the retina: an historical approach. *Eye*, 30 (2016) 169–172.

[23] M. Spitschan, G. K. Aguirre, D. H. Brainard, A. M. Sweeney, Variation of outdoor illumination as a function of solar elevation and light pollution. *Scientific Reports*, 6 (2016) 1–14.

[24] J. B. O'hagan, M. Khazova, L. L. A. Price, low-energy light bulbs, computers, tablets and the blue light hazard. *Eye*, 30 (2016) 230–233.

[25] W. T. Ham, J. J. Ruffolo, H. A. Mueller, A. M. Clarke, M. E. Moon, Histologic analysis of photochemical lesions produced in rhesus retina by short-wave-length light. *Investigative Ophthalmology & Visual Science*, 17 (1978) 1029–1035.

[26] M. Marie, K. Bigot, C. Angebault, C. Barrau, P. Gondouin, D. Pagan, S. Fouquet, T. Villette, J. A. Sahel, G. Lenaers, S. Picaud, Light action spectrum on oxidative stress and mitochondrial damage in A2E-loaded retinal pigment epithelium cells. *Cell Death & Disease*, 9 (2018) 1–13.

[27] H. S. Lee, L. Cui, Y. Li, J. S. Choi, J. H. Choi, Z. Li, G. E. Kim, W. Choi, K. C. Yoon, Influence of light emitting diode-derived blue light overexposure on mouse ocular surface. *PLoS One*, 11 (2016) e0161041.

[28] P. V. Algvere, J. Marshall, S. Seregard, Age-related maculopathy and the impact of blue light hazard. *Acta Ophthalmologica Scandinavica*, 84 (2006) 4–15.

[29] Y. Kuse, K. Tsuruma, Y. Kanno, M. Shimazawa, H. Hara, CCR3 is associated with the death of a photoreceptor cell-line induced by light exposure. *Frontiers in Pharmacology*, 8 (2017) 207.

[30] M. Boulton, M. Różanowska, B. Różanowski, Retinal photodamage. *Journal of Photochemistry and Photobiology B: Biology*, 64 (2001) 144–161.

[31] L. M. Rapp, S. C. Smith, Morphologic comparisons between rhodopsin-mediated and short-wavelength classes of retinal light damage. *Investigative Ophthalmology & Visual Science*, 33 (1992) 3367–3377.

[32] J. Wu, S. Seregard, B. Spångberg, M. Oskarsson, E. Chen, Blue light induced apoptosis in rat retina. Eye, 13 (1999) 577–583.

[33] T. Lougheed, Hidden blue hazard? LED lighting and retinal damage in rats. *Environmental Health Perspectives*, 122 (2014) A81–A81.

[34] A. Krigel, M., Berdugo, E. Picard, Levy-Boukris, R., I. Jaadane, L. Jonet, M. Dernigoghossian, C. Andrieu-Soler, A. Torriglia, F. Behar-Cohen, Light-induced retinal damage using different light sources, protocols and rat strains reveals LED phototoxicity. *Neuroscience*, 339 (2016) 296–307.

[35] Y. M. Shang, G. S. Wang, D. H. Sliney, C. H. Yang, L. L. Lee, Light-emitting-diode induced retinal damage and its wavelength dependency in vivo. *International Journal of Ophthalmology*, 10 (2017) 191.

[36] R. Mukai, H. Akiyama, Y. Tajika, Y. Shimoda, H. Yorifuji, S. Kishi, Functional and morphologic consequences of light exposure in primate eyes. *Investigative Ophthalmology & Visual Science*, 53 (2012) 6035–6044.

[37] T. H. Margrain, M., Boulton, J. Marshall, D. H. Sliney, Do blue light filters confer protection against age-related macular degeneration? *Progress in Retinal and Eye Research*, 23 (2004) 523–553.

[38] A. Alim-Marvasti, W. Bi, O. A. Mahroo J. L. Barbur, G. T. Plant, Transient smartphone "Blindness". *The New England Journal of Medicine*, 374 (2016) 2502–2504.

[39] R. H. James, R. J. Landry, B. N. Walker, I. K. Ilev, Evaluation of the potential optical radiation hazards with LED lamps intended for home use. *Health Physics*, 112 (2017) 11–17.

[40] C. Martinsons, D. Attia, F. Behar-Cohen, S. Carré, O. Enouf, J. Falcón, C. Gronfier, D. Hicks, A. Metlaine, L. Tahkamo, A. Torriglia, Correspondence: an appraisal of the effects on human health and the environment of using light-emitting diodes. *Lighting Research & Technology*, 51 (2019) 1275–1276.

[41] M. Pagden, K. Ngahane, M. S. R. Amin, Changing the colour of night on urban streets-LED vs. part-night lighting system. *Socio-Economic Planning Sciences*, 69 (2020) 100692.

[42] G. Lascaratos, D. Ji, J. P. Wood, N. N. Osborne, Visible light affects mitochondrial function and induces neuronal death in retinal cell cultures. *Vision Research*, 47 (2007) 1191–1201.

[43] R. Khosravi-Far, Death receptor signals to the mitochondria. *Cancer Biology & Therapy*, 3 (2004) 1051–1057.

[44] B. J. Prendergast, R. J. Nelson, I. Zucker, Mammalian seasonal rhythms: behavior and neuroendocrine substrates, *Hormones, Brain and Behavior*, 2 (2002) 93–156.

[45] M. Forejt, K. Skočovský, R. Skotnica, J. Hollan, Sleep disturbances by light at night: two queries made in 2003 in Czechia, *Cancer and Rhythm Conference* (2004) Graz, Austria.

[46] J. A. Benfield, R. J. Nutt, B. D. Taff, Z. D. Miller, H. Costigan, P. Newman, A laboratory study of the psychological impact of light pollution in national parks. *Journal of Environmental Psychology*, 57 (2018) 67–72.

[47] P. H. Manríquez, M. E. Jara, M. I. Diaz, P. A. Quijón, S. Widdicombe, J. Pulgar, K. Manríquez, D. Quintanilla-Ahumada, C. Duarte, Artificial light pollution influences behavioral and physiological traits in a keystone predator species, Concholepas concholepas. *Science of the Total Environment*, 661 (2019) 543–552.

[48] K. J. Navara, R. J. Nelson, The dark side of light at night: physiological, epidemiological, and ecological consequences. *Journal of Pineal Research*, 43 (2007) 215–224.

[49] E. Haus, M. Smolensky, Biological clocks and shift work: circadian dysregulation and potential long-term effects. *Cancer Causes & Control*, 17 (2006) 489–500.

[50] S. M. Pauley, Lighting for the human circadian clock: recent research indicates that lighting has become a public health issue. *Medical Hypotheses*, 63 (2004) 588–596.

[51] T. A. Bedrosian, L. K. Fonken, J. C. Walton, R. J. Nelson, Chronic exposure to dim light at night suppresses immune responses in Siberian hamsters. *Biology Letters*, 7 (2011) 468–471.

[52] M. E. Kernbach, D. J. Newhouse, J. M. Miller, R. J. Hall, J. Gibbons, J. Oberstaller, D. Selechnik, R.H. Jiang, T.R. Unnasch, C.N. Balakrishnan, L.B. Martin, Light pollution increases West Nile virus competence of a ubiquitous passerine reservoir species. *Proceedings of the Royal Society B*, 286 (2019) 1051.

[53] M. K. Vaughan, G. B. Hubbard, T. H. Champney, G. M. Vaughan, J. C. Little, R.J. Reiter, Splenic hypertrophy and extramedullary hematopoiesis induced in male Syrian hamsters by short photoperiod or melatonin injections and reversed by melatonin pellets or pinealectomy. *American Journal of Anatomy*, 179 (1987) 131–136.

[54] A. Carrillo-Vico, J. M. Guerrero, P. J. Lardone, R. J. Reiter, A review of the multiple actions of melatonin on the immune system. *Endocrine*, 27 (2005) 189–200.

[55] P. Meerlo, M. Koehl, K. Van der Borght, F. W. Turek, Sleep restriction alters the hypothalamic-pituitary-adrenal response to stress. *Journal of Neuroendocrinology*, 14 (2002) 397–402.

[56] D. M. Dominoni, M. Quetting, J. Partecke, Long-term effects of chronic light pollution on seasonal functions of European blackbirds (Turdus merula). *PLoS One*, 8 (2013) e85069.

[57] T. Le Tallec, M. Théry, M. Perret, Melatonin concentrations and timing of seasonal reproduction in male mouse lemurs (Microcebus murinus) exposed to light pollution. *Journal of Mammalogy*, 97 (2016) 753–760.

[58] S. J. Schoech, R. Bowman, T. P. Hahn, W. Goymann, I. Schwabl, E. S. Bridge, The effects of low levels of light at night upon the endocrine physiology of western scrub-jays (Aphelocoma californica). *Journal of Experimental Zoology Part A: Ecological Genetics and Physiology*, 319 (2013) 527–538.

[59] S. Boldogh, D. Dobrosi, P. Samu, The effects of the illumination of buildings on house-dwelling bats and its conservation consequences. *Acta Chiropterologica*, 9 (2007) 527–534.

[60] E. L. Stone, G. Jones, S. Harris, Conserving energy at a cost to biodiversity? Impacts of LED lighting on bats. *Global Change Biology*, 18 (2012) 2458–2465.

[61] K. M. Kramer, E. C. Birney, Effect of light intensity on activity patterns of Patagonian leaf-eared mice, Phyllotis xanthopygus. *Journal of Mammalogy*, 82 (2001) 535–544.

[62] R. G. Stevens, M. S. Rea, Light in the built environment: potential role of circadian disruption in endocrine disruption and breast cancer. *Cancer Causes & Control*, 12 (2001) 279–287.

[63] R. J. Reiter, Potential biological consequences of excessive light exposure: melatonin suppression, DNA damage, cancer and neurodegenerative diseases. *Neuroendocrinology Letters*, 23 (2002) 9–13.

[64] D. E. Blask, L. A. Sauer, R. T. Dauchy, Melatonin as a chronobiotic/anticancer agent: cellular, biochemical, and molecular mechanisms of action and their implications for circadian-based cancer therapy. *Current Topics in Medicinal Chemistry*, 2 (2002) 113–132.

[65] D. E. Blask, S. M. Hill, Effects of melatonin on cancer: studies on MCF-7 human breast cancer cells in culture. *Journal of Neural Transmission*, Supplementum 21 (1986) 433–449.

[66] N. A. Kerenyi, E. Pandula, G. Feuer, Why the incidence of cancer is increasing: the role of "light pollution". *Medical Hypotheses*, 33 (1990) 75–78.

[67] S. Davis, D. K. Mirick, R. G. Stevens, Night shift work, light at night, and risk of breast cancer. *Journal of the National Cancer Institute*, 93 (2001) 1557–1562.

[68] J. Hansen, Increased breast cancer risk among women who work predominantly at night. *Epidemiology*, 12 (2001) 74–77.

[69] A. Garcia-Saenz, A. S. de Miguel, A. Espinosa, L. Costas, N. Aragonés, C. Tonne, V. Moreno, B. Pérez-Gómez, A. Valentin, M. Pollán, M. Castaño-Vinyals, Association between outdoor light-at-night exposure and colorectal cancer in Spain. *Epidemiology*, 31 (2020) 718–727.

[70] M. Derwahl, D. Nicula, Estrogen and its role in thyroid cancer. *Endocrine-Related Cancer*, 21 (2014) T273–T283.

[71] Y. Chen, F. Tan, L. Wei, X, Li, Z. Lyu, X. Feng, Y. Wen, L. Guo, J. He, M. Dai, N. Li, Sleep duration and the risk of cancer: a systematic review and meta-analysis including dose–response relationship. *BMC Cancer*, 18 (2018) 1–13.

[72] R. B. Clarke, H. Amini, P. James, M. von Euler-Chelpin, J. T. Jørgensen, A. Mehta, T. Cole-Hunter, R. Westendorp, L.H. Mortensen, S. Loft, J. Brandt, Outdoor light at night and breast cancer incidence in the Danish Nurse Cohort. *Environmental Research*, 194 (2021) 110631.

[73] J. Ritonja, M. A. McIsaac, E. Sanders, C. C. Kyba, A. Grundy, E. Cordina-Duverger, J.J. Spinelli, K. J. Aronson, Outdoor light at night at residences and breast cancer risk in Canada. *European Journal of Epidemiology*, 35 (2020) 579–589.

CHAPTER 5

Health Impacts/Risks of Light Pollution

Humira Assad and Ishrat Fatma

Lovely Professional University

Ashish Kumar

Nalanda College of Engineering

CONTENTS

DOI: 10.1201/9781003185109-5

5.1 INTRODUCTION

The invention of the light bulb by Thomas Edison in 1897 marked a watershed moment in the history of mankind. Humans were only exposed to low-intensity lights from specific sources at the period, such as candles, petroleum lanterns, and so on. As technology and invention progressed, so did the extent to which humans were exposed to artificial light. The incorporation of artificial light directly or indirectly into surroundings either by humans or by any other source is called light pollution (LP). This kind of pollution is developing persistently all over the world. The growth of the global community is matched by the overall economy and basic infrastructure. However, as a consequence of this inevitable economic expansion and industrialization, the light output from street lamps and huge building lamps, which are the primary sources of LP, has increased [1]. This practice continues to grow, even though few people realize it is harming the environment and the health of living creatures. Undoubtedly, by 2001, the portion of land beneath skies that were brightened falsely surpassed 10% in 66 nations, and presently manufactured lighting dispersed extensively and is spreading at a rate of 6% per year all around the world. According to the "International Dark-Sky Association", LP refers to any unwanted impact of artificial light such as "skyglow, light clutter, glare, energy waste", etc. LP is responsible for diminishing the visibility of stars, disrupting the entire ecosystems as well as food webs.

According to the research, LP is a major worry for astronomers since it interferes with nocturnal astronomical awareness, preventing celestial investigation. Recently it was explored that LP not only shows its impact on astronomy but also causes major disturbances in the ecological balance of wildlife and serious problems in human health. Many astronomers across the world are concerned about skylight over large towns because it blackens starlight, even on crystal clear evenings. According to the experts, the sky luminance in broad portions of urban areas in Europe and North America is at least 2–4 times higher than normal. LP can have a detrimental effect on one's health such as recurrent migraines, weariness, expanded stretch, and expanded uneasiness. There was also a few more research that indicated there is a link between LP and breast cancer because the usual nightly production of melatonin is suppressed. LP is also thought to be a contributor to smog. LP, according to the "American Geophysical Union", smashes nitrate radicals, interfering with the regular nocturnal smog decomposition. Because the ambient environment is not completely dark, estimating the overall light intensity contamination in a specific zone is a tough and complicated process. LP's detrimental effects were first acknowledged in the late 1800s, but efforts to mitigate them did not emerge until the 1950s. With the formation of the International Dark-Sky Association (IDA) in the 1980s, a worldwide dark-sky movement arose. There are now such educational and support organizations in many countries throughout the world [2–7]. Hence, this chapter covers the types, causes of LP as well as the impact of light contamination on the internal rhythms of humans, animals, and plants, as well as the negative physiological effects that imbalance ecological sustainability, and so on. In addition, the chapter briefly covers the economics of LP and the efforts that must be done to mitigate its harmful effects.

5.2 CLASSIFICATION OF LP

The concept of LP is very deep, which mainly occurs due to abnormal or inefficient utilization of artificial light. Generally, LP is of five types, as shown in Figure 5.1, which includes light trespass, over-illumination, glare, light clutter, and sky glow, all of these are discussed below:

5.2.1 Light Trespass

When undesirable light penetrates one's estate, for example, by glinting off a compatriot's barrier, it is known as light trespass. Laws have also been drafted to limit the total illumination at the domain border and in the background, but they may be unreasonable or questionable. Reasonable restrictions and lucidity in estimation require to be given. Expressing "zero light at the property line" is too imprecise. Blatant zero suggests that if a lamp placement is a mile away and the source of light is visible, it is in noncompliance and would necessitate the use of coverings across each lighting system. Depending on whether

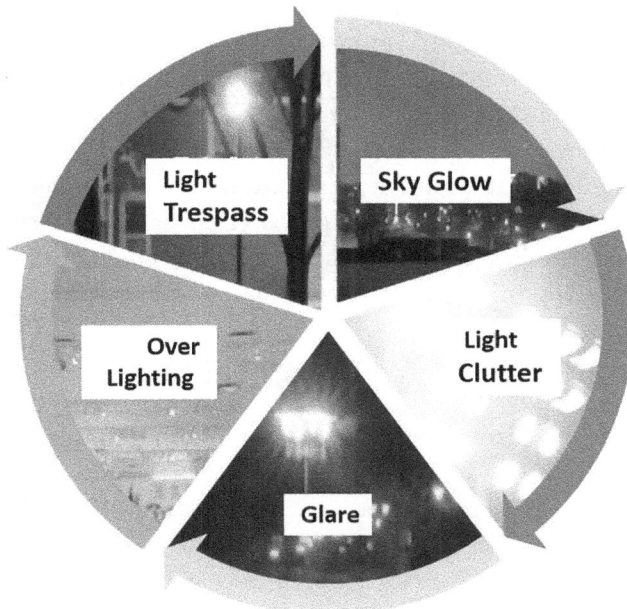

FIGURE 5.1 Types of light pollution.

a region is commercial or industrial, urbanized, provincial, or nation, what is acceptable may alter Regulations have also been drafted to limit the amount of light near the estate border and in the past, however, these may be unrealistic or questionable. Reasonable limits and estimations that are clear must be provided. The LEED rating stipulates boundaries at the site border and 10–15 feet beyond it. LEED restricts light to 0.01 fc at a distance of 10–15 feet (full moon offers 0.03 fc, while a moonless night delivers 0.004 fc). Generally, a challenging constraint to adhere to while providing illumination on a parking lot or basement. For both indoor and exterior illumination estimates, level assessments are typical. However, when it comes to light trespass, the issue is how much light gets into a person's eye. Estimates can be made at eye level of the perpendicular light intensity facing into the site, or by pointing at the sharpest illumination. Whenever drivers approach the roadway, exemptions may be made. This would allow road lights to be installed at the drive entry, making automobiles highly apparent as they enter the action zone. Another typical legal technique for reducing light trespass is to limit the height of posts. When the statute additionally includes maximum: minimum portions for security considerations, this becomes counterproductive. Dim patches on a place will increase as pole elevation is reduced. Due to the breadth of the passageways and stops, expanding the number of shafts was feasible up to a limit. To ensure constant illumination, additional poles would need to be placed within the stopping zones and passages.

5.2.2 Over-Illumination

The use of light in excess is known as over-illumination, which is responsible for nearly 2 million barrels of oil per day in energy waste in the United States alone. This is frequently based on the United States' use of petroleum equivalents. According to the same United States Department of Vitality source, the industrial, technological, and private firms utilize nearly 30% of total vigor. According to energy audits of existing structures, the illuminating portion of personal, professional, and commercial processes consumes 20%–40% of those arriving utilizes, depending on location and come usage. Illumination energy contributes to around 4 or 5 million gallons of oil (comparable) each day in this manner. According to new energy review data, roughly 30%–60% of the energy spent in illumination is unwanted or undesirable. The datum that marketable property illumination devours a surplus of 81.68 terawatts of power (1999 data) is the starting point for an optional estimate. As a result, according to the other justification for assessing U.S. lighting energy usage, industrial illumination devours around 4 and 5 million barrels of petroleum lonely each day.

Over-illumination is Caused by Several Reasons as Follows:

- Not using timers, occupancy sensors, or other switches to turn off lights when they aren't in use.
- Immoral planning, especially in occupational settings, by setting higher lighting conditions than are required for a specific task.
- Imprecise setup or light bulb selection that does not distribute light into categories as required.

- The embarrassing decision by devices to use more energy than is necessary to finish the lighting operation.
- Housing administrators and tenants receive sporadic training on how to effectively use lighting frameworks.
- Increased stray illumination and vigor costs due to inadequate lighting compliance.
- unshine lighting" may be mandated by people to minimize crime or by shop owners to attract customers, therefore excessive illumination is a plan decision rather than a fault. In both circumstances, aim achievement is debatable.
- Using the same light-generating devices, replace old mercury lights with even more effective sodium or metal halide lighting.

The majority of these issues can be quickly resolved using widely accessible, low-cost innovation; nevertheless, there is tremendous stagnation in the sector of illumination projects and in proprietor/tenant relationships that obstructs hasty resolution of these issues. For industrialized nations to reap the expansive reward in reducing over-illumination, public awareness would have to improve significantly.

5.2.3 Glare

Glare usually occurs due to excessive differences between bright and dim regions within the sector of see. For instance, glare can be defined as observing the fiber of an unfortified or severely guarded light directly. Pedestrians and vehicles can have their night visualization obliterated for up to an hour subsequently being exposed to bright light. Glare, which is generated by a large dissimilarity between light and dark regions, can also formulate it harder for the human eye to regulate to brilliance contrasts. Glare is a particular problem in street security since vivid and/or poorly covered lighting on the street can blind vehicles and cause accidents. Owing to light dispersion within the eye caused by extreme luminosity, or emission of light from dark bands within the visual regions with a brightness comparable to the baseline luminance, glare can cause reduced contrast. This type of glare is known as veiling glare, and it is a type of incapacity glare. As illustrated in Figure 5.2, glare is further divided into three types: blinding glare, disability glare, and discomfort glare.

5.2.3.1 Blinding Glare
It depicts consequences such as those induced by gazing directly at the Sun. It entirely blinds you and departs you with either temporary or permanent visual problems.

5.2.3.2 Disability Glare
It displays effects such as being dazzled by oncoming car lights, or light dispersing in mists or the eye reducing differentiation, as well as reflections from print and other dark ranges that make them shine, resulting in a major reduction in locating abilities.

5.2.3.3 Discomfort Glare
It rarely causes a dangerous scenario in and of itself, and is at best annoying and uncomfortable. It has the potential to make you tired if you use it for a lengthy period.

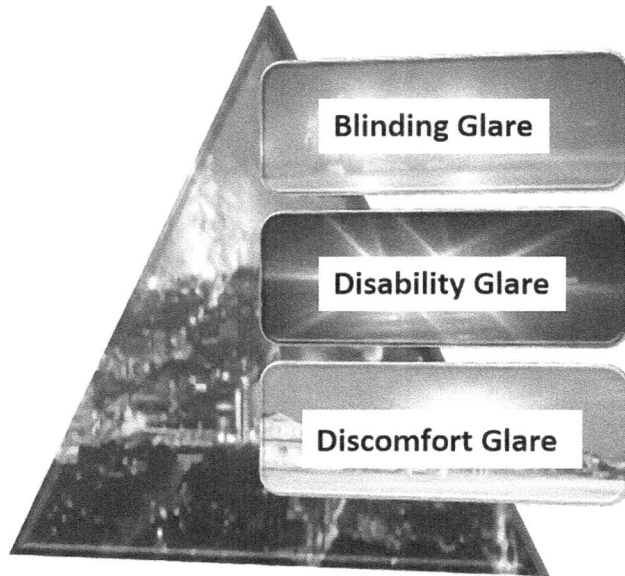

FIGURE 5.2   Types of glare.

5.2.4 Light Clutter

Clutter refers to light clusters that are excessively large. Clusters of lights can be perplexing, obstruct barriers (including those they're supposed to enlighten), and even create mishaps. Clutter is particularly apparent on streets with poorly planned street lights or where brilliantly lighted advertising blankets the lanes. The arrangement and construction of the lights, relying on the mental processes of the individual or organization who introduced them, may indeed be envisioned to engage drivers and can exacerbate catastrophes. Clutter could pose a concern in the air travel context if flying protection lighting has to contend with non-relevant lights for pilot attention. Runway lighting, for example, might be confused with a grouping of suburban business illumination, while airship smash evasion lights could be chaotic with landscape lights.

5.2.5 SkyGlow

The term "skyglow" states the "gleam" influence that can be perceived above densely inhabited zones. It's the result of all the light reflected off what it's irradiated absconding into the sky, as well as all the synchronized light in that zone escaping into the sky and being dispersed (diverted) returning toward the surface by the atmosphere. When the discussion is very vivid, this dispersing is very tightly connected to the wavelength of the light. Rayleigh invades in such clean air, causing the sky to seem blue during the day. When there is a lot of vaporized matter, the dispersed light is less wavelength-dependent, resulting in a whiter midday sky. Because of the increased susceptibility of the eye to white or blue-rich lighting systems when accustomed to extremely low light levels, white or blue-rich light contributes substantially more to sky glow than a breakdown even with the same quantity of yellow light. Astronomers are particularly irritated by sky shine because it reduces the distinction between stars in the night sky [8–13].

5.3 CAUSES OF LP

LP is extraordinary as only human beings can do it. Some of the important causes of LP include poor planning, irresponsible use, overpopulation, smog and clouds, light from vehicles and houses, street lights, etc as shown in Figure 5.3.

5.3.1 Poor Planning

Engineers arrange the situation of signage and road lights, and if they do not take into consideration the impact arrangement on the encompassing environment, which is responsible for making glare, light trespass, and light clutter.

5.3.2 Irresponsible Use

You will adore Christmas lights, but clearing out them all night may be a frame of contamination, as is clearing out a room with the lights still on. Not effectively choosing to play down vitality squander could be a tremendous source of LP.

5.3.3 Overpopulation

This is a big issue. As well numerous businesses or as well numerous homes, gathered in one range can cause LP of numerous sorts.

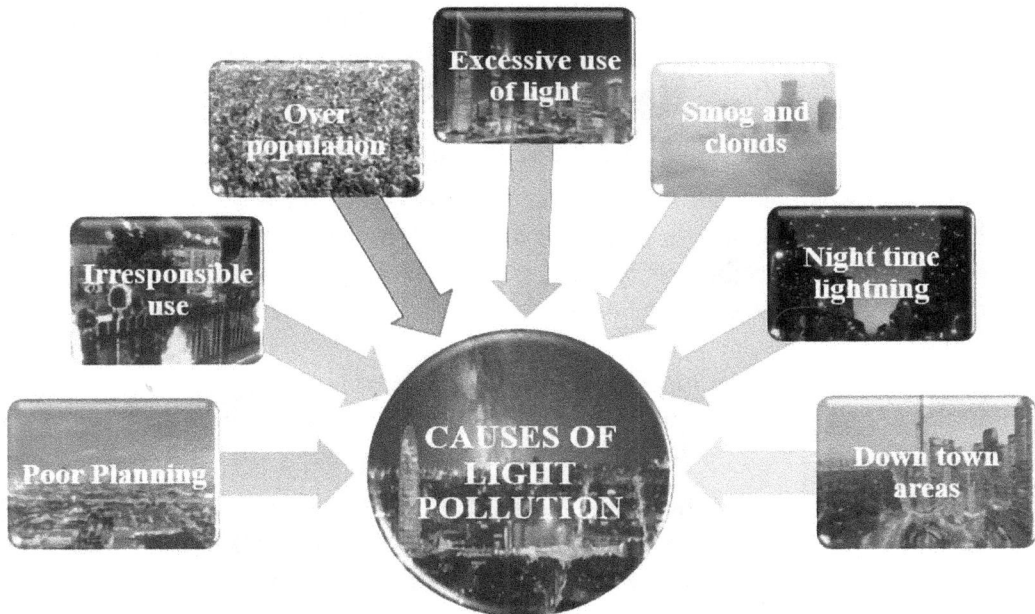

FIGURE 5.3   Causes of light pollution.

5.3.4 Excessive Use of Light

One critical cause of LP is the overutilization of electric light. Power has ended up very cheap over time; in this manner, individuals do not care much about power utilization.

5.3.5 Smog and Clouds

Smog and clouds can reflect the light radiated by cities and hence make the encompassing environment see much brighter, causing LP.

5.3.6 Nighttime Lightning

All sorts of nighttime lighting can influence individuals in an antagonistic way, and thus cause light contamination. A great illustration would be the advertising spaces at nighttime, which will cause light contamination to the encompassing houses.

5.3.7 Downtown Areas

Because many of the lights will be on 24 hours a day, 7 days a week for engagement but also commercial purposes, they will contribute significantly to LP.

5.4 GROWTH OF LP

Around the world, LP is steadily increasing [14,15]. Indeed, by 2001, the percentage of area under artificially lit sky had surpassed 10% in 66 nations, and artificial lighting is now extensively used and expanding at a pace of 6% per year worldwide [16]. Around half of the world's population lives in cities, according to estimates [17]. Around 20% of the world's sector is pretentious by LP. Additionally, such light-polluted regions are home to 63% of the world's population and 99% of the populations of the United States and Europe. Furthermore, sky brightness is experienced by more than 80% of the US inhabitants even more than nights with a full moon [18]. North America, parts of the United Kingdom, and India have severe LP, whereas the continents of South America, Africa, and Australia have comparatively little LP. In developing countries like India, the primacy of fundamental problems frequently overshadows the emergence of new urban challenges that have the potential to negatively impact the ecosystem on a broader gage. LP is one such issue that requires attention [19]. Based on the current "world atlas of artificial night sky brightness" which positions the G-20 countries by populace and sector under LP, India is home to 58.5% of the global population, encompassing 24.7% of the land.

Fabio Falchi reported that LP is expected to be developing at a rapid rate, and with an outpouring of growth and no stringent rules in place, it may become a problem in the future [20]. As per current publications on LP in India,

> the states of West Bengal, Telangana, Tamil Nadu, Maharashtra, Karnataka, Uttar Pradesh, and Gujarat had an increase in light pollution from 2003 to 2013. Due to a lack of legislation, no decline was recorded. As a result, operative administration is required to lower the intensities. [21]

5.5 ADVERSE EFFECTS OF LP

Although LP research is yet in its early stages, the consequences of this phenomenon are not completely comprehended. Although the most well-known effect is the rising illumination of the night sky of LP's various effects, many other concerning elements remain unknown. Despite the necessity of nighttime illumination, LP would be a serious concern for the environment and humans, with negative consequences for human and wildlife health. LP has been reported to increase the occurrence of headaches, anxiety, depression, and uneasiness in humans, as well as diminish melatonin content in the skin that leads to sarcoma, according to studies [17,18]. Lights consume one-fourth of all worldwide energy, and systematic reviews have shown that several sorts of over-illumination waste energy, including the non-beneficial upward inclination of evening lights. On a global basis, almost 19% of electricity is used to provide illumination at night. Electric lighting, which is produced by the burning of fossil fuels, emits greenhouse gases as a by-product. These gases are to blame for global warming, carbon dioxide emissions, and nonrenewable resource exhaustion. In 2005, around 30% of the electricity utilized for outdoor lighting in the United States was squandered as LP. The annual price that accounts for energy waste in the United States is $6.9 billion [22].

Many other environmental effects are also caused by LP. The animal kingdom, the vegetable kingdom, and humanity are all affected negatively. While LP is especially hazardous to nighttime and itinerant species, animals in a journey, as well as harms plants, as shown below in Figure 5.4.

5.5.1 Impact on Humans

In the previous several years, the standard of living and health maintenance in many parts of the globe has significantly upgraded, especially in cities. These urban and industrial

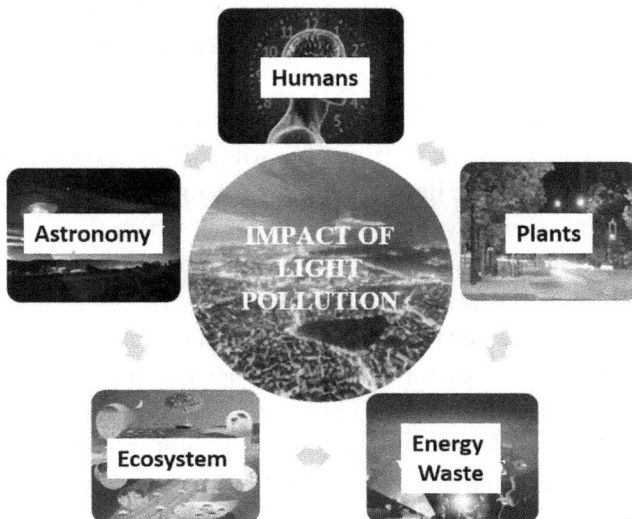

FIGURE 5.4 Impacts of light pollution.

environments, however, do not appear to be the healthiest places to live. Aside from the "normal" health concerns, such as air pollution and lifestyle-related stress, a slew of new health risks has cropped up regularly. Humans, by nature, require a consistent rhythm of day and night hours during their lives. According to medical studies on the influence of extreme light on the human body, LP or extreme contact to light can produce several negative health impacts and even some lighting design textbooks employ public health as a specific requirement for good interior illumination [23]. Artificial light mimics the brilliance of the Sun, disrupting melatonin's efficiency in the phase-shifting circadian clock. This interruption can result in widespread disruptions of numerous body systems, resulting in serious medical repercussions for individuals. Enhanced headache occurrence, worker weariness, medically defined distress, decreased sexual performance, and nervousness are some of the health repercussions of over-radiance or inappropriate spectral arrangement of light [24,25]. The average amount of fluorescent lighting in an office raises blood pressure by around eight points. Long-term daily stimulation to moderately bright illumination has been linked to decreased sexual function. Visible light is a portion of the (violet-red) spectrum that human eyes can perceive. A blue region of the visible light spectrum has been identified as having the ability to control human time rhythms in a recent study [26].

Several published research also implies a relationship between nighttime light exposure and the risk of breast cancer, owing to the inhibition of melatonin's natural nocturnal generation. "Researchers from the National Cancer Institute (NCI) and the National Institute of Environmental Health Sciences" have discovered that manmade light throughout the dark phase may be a risk parameter for breast sarcoma. Breast sarcoma is a danger for women, particularly night shift workers. According to Blask et al., intense light exposure during the night can cause breast cell proliferation [27]. The brightness of bedroom illumination does not protect the female from breast sarcoma menace, mainly short-wavelength light (460 nm), was revealed to be nearly 95% closely linked to the chance of breast tumor [28]. This explains why the risk of acquiring breast cancer in industrialized nations is up to five times more than in poor countries [29].

Kloog et al. discovered that people who were exposed to severe radiance in dark had a higher chance of prostate cancer than people who were exposed to less LP [30]. Due to increased exposure to variable light changes, rotating shift workers were included. A trial in Iceland was carried out by Sigurdardottir et al., where they found that men with a greater amount of 6-sulfatoxymelatonin, a major melatonin decomposition component, had a 75% lower incidence of prostate cancer [31].

Kidney disease can also be caused by shift employment or exposure to light at night. There is no direct experimental documentary evidence relating to artificial light at night, melatonin levels, and kidney disorders. Cukor et al. looked at the link between sleep disturbances and renal disease. Sleep disturbances can disrupt the renin-angiotensin system, resulting in diminished overnight blood pressure lowering, which is a major risk factor for renal disease progression [32]. Moreover, Mehta and Drawz found that when 30 end-stage hepatic encephalopathy patients on hemodialysis were contrasted to 20 healthy volunteers in a cross-sectional investigation, the sufferers with the end-stage renal disease had substantially lower evening melatonin levels [33].

This demonstrated that the dangers of nighttime lighting surpassed the advantages. As a result, both exterior and in-house lighting should have their configuration and exposure time reconsidered at night.

5.5.2 Impact on Astronomy

The luminance disparity between the night sky and the celestial bodies determines the visibility of celestial bodies in the sky in astronomy. When identifying a dim celestial object in the sky, human vision has a particular luminance threshold [34,35]. The Bortle scale is used to quantify the recognition of an astronomic entity in a specific night sky.

Sky brightness in heavily populated regions can be up to 100 times brighter than in natural settings [36]. Because the disparity between the brilliance amplitude of the astronomic entity and the sky in dark would be tinier in a severely populated sky with a lesser amount of sky brightness, celestial objects would be less visible to both naked and optically aided observations [37,38]. This illustrates why, due to its impact on ground-based investigations, LP is fitting a chief basis of apprehension for stargazers. As a result, examining the quantity of LP at the observation site and its effects on astronomical data is crucial.

Moreover, the comparison between stars and galaxies in the sky and the sky itself is reduced by sky glow, rendering it more challenging to see hazier entities. This is one of the reasons that modern telescopes are being constructed in progressively distant locations. Some astrophysicists utilize narrow-band "nebula filters" , which permit only light at precise wavelengths frequently observed in nebulae, or wide-ranging-band "light pollution filters" , which are aimed to minimize the impacts of LP by screening out spectral lines prevalently transmitted by "sodium and mercury-vapor lamps", augmenting distinction and continuing to improve the perspective of blur artifacts such as galaxies and nebulae. Unluckily, because this impacts color view, these screens can't be used to objectively evaluate changeable star sunniness, and no screens can equal the optical or photographic efficacy of a dark sky. Thus, LP has a greater impact on the discernibility of diffuse sky entities like nebulae and galaxies than it does on the visibility of stars. Based on the assumption that light travels directly from the source to the "optic" – the observer's eye or telescope – the consequence of stray light is designated as "optical pollution" in one Italian provincial lighting regulation.

5.5.3 Impact on Plants

Darkness is used by plant species in a lot of formats. The regulation of their physiology, maturation and life plans are all impacted. Night length, or the period of darkness, is measured and reacted to by plants. Additional light for plants can be harmful to them, according to earlier findings, and light can change their life cycle. Many plants use the length of darkness to determine their rate of development and flowering and fruiting times. Plant metabolism can be altered by LP, and if they are not exposed to darkness for an extended period, they will never be able to produce flowers or colors. Lights at night can block the production of photochromic hormones, culminating in plant extinction. Low-pressure sodium lighting can interfere with plant growth and development by disrupting photoperiodic control. Plants with short days need long nights to complete their metabolism. When a plant is unexpectedly exposed during a long night, it reacts and perceives as if it had two short nights instead of one long night with a respite. Its blossoming and developmental patterns may be drastically disturbed as a consequence. As the days get shorter, short-day plants usually bloom in the autumn. They take advantage of the elongated evenings to begin flowering and, as the nights are longer, to begin dormancy, which allows them to withstand the rigors of winter [39].

Moreover, LP harms trees that support entire ecosystems for a variety of animal species. Because artificial light makes it difficult for trees to react to seasonal changes: LP prevents many trees from shedding their leaves. This has ramifications for creatures who rely on trees as a habitat. Birds, for example, are unable to nest on trees as a consequence of LP in the area. Additionally, LP near lakes hinders zooplankton such as Daphnia from consuming superficial algae, contributing to algal blooms that can harm lake plants and degrade the quality of the water. Ecosystems may be harmed by LP in other ways as well. Nighttime light has been shown to impede the capacity of moths and other nighttime microbes to traverse, according to lepidopterists and entomologists. Because no substitute pollinator is unaffected by artificial light, night-blooming flowers that rely on moths for pollination may be harmed by dark illumination. This can result in the extinction of plants that are unable to reproduce, as well as a disruption in the long-term ecology of a region.

5.5.4 Impact on Animals

Aside from humans, LP harmed animal survival. The luminosity of the night was previously governed by the rotation of a solar body, viz. the moon, throughout the synodic month. The night sky's luminosity ranged from full moon to new moon, with the full moon being the brightest. Faunas, microbes, and all other organisms, particularly nocturnal organisms, adapted their behavior to the synodic month cycle. Because life began with natural light and dark rhythms, disruption of those regularities has an impact on a variety of animal behaviors, which impeded their otherwise perfect life cycle. Animal navigation can be hampered by LP, which can also modify competitive dynamics, alter predator-prey relationships, and affect animal physiology, in the following manner.

- **Disorientation**: LP can cause the orientation to shift in the wrong direction.
- **Mortality**: LP can kill animals, especially insects and birds because they don't comprehend the time or how to behave properly.
- **Nesting Instinct**: Too much light can alter the nesting behavior patterns of almost all birds.
- **Attraction to Illumination**: Birds and insects might be attracted to artificial lighting. Because they cannot comprehend the directions and smash into items around them, mortality rises as a result of their attraction to lights.
- **Insomnia**: Most specialists agree that both people and animals can experience sleeplessness. Artificial light has been shown to have an impact on the environment and can alter childbirth, life, and mortality, according to studies. It can cause a spike in algae in the pools and a decline in water quality, both of which can be harmful to marine life. It can alter their reproductive patterns by degrading their eggs and sperm. Thus, LP can have a direct impact on ecosystems.

Since the 19th century, the influence of light in the guise of fire or lamps enticing migratory and non-migratory birds at night, particularly when misty or cloudy, has been known and is still employed as a method of hunting [40]. As a result, hundreds, if not thousands, of birds can be hurt or killed in a single night in a single building. The reasons for avian confusion caused by artificial night lights are unknown. Lighting and sky glow, according to experts, disturb the navigation of birds who use the horizon as a direction indicator. The number of birds killed at lighthouses is determined by the sort of signal used. More birds are drawn to fixed white lights than to flashing or colored lights [41].

Bruderer et al. [42] investigated the attitude of birds subjected to light and X-Band radiations. The light beam induced a 15° alteration in flight direction as well as a 3 m/sec drop in velocity. On October 6–8, 1954, a cold front passed through the Southeast, killing around 50,000 migratory birds (the biggest kill ever recorded by a ceilometer) at Warner Robins Air Force Base in Georgia [40].

Moreover, telecommunication and broadcasting towers are becoming more numerous and taller, increasing fatal collisions with migratory birds. These constructions obstruct songbird migration pathways. When they hit with other approaching birds or swoop into the scaffolding and its man cables, as the relatively small, illuminated space becomes increasingly crowded, they die. Recent research shows that using spinning or flickering lights, as well as white strobe lights, can help to reduce the threat of birds being trapped in illuminated turrets, although more research is needed to better understand the overall effect on migration [43].

Moth populations are declining dramatically over the world [44]. Artificial lights (AL) at night are one of the possible causes of this reduction. The behavior of moths flying about streetlamps in dark, which enhanced their poaching by bats and other hunters who reap the benefits of these artificial feeding spots, could explain this reduction. This reduction could have an impact on insect pollination [45].

LP can also pose a threat to the growth and development of fish. The manner fish behave to artificial light (attraction or avoidance) varies by species, but it has an impact on their natural behavior in both directions. Fish, according to the majority of studies, shun white light sources. However, some species are drawn to light, which is utilized by sport fishermen and industrial fisheries to trap them. In Atlantic salmon rearing enclosures, for example, submerged light increases swimming amplitude while lowering fish abundance. These artificial photoperiods are used to enhance development and prolong sexual development. According to research conducted in these farms, salmon retain schooling behavior by positioning themselves in respect to the artificial light gradient [46].

New light control solutions are required to reduce the detrimental impacts of artificial lighting. The utilization of light must be more accurate. To be less bothersome to species, it should be less intense and broadcast over longer wavelengths. The requirements must be enacted through legislation, as most counties in Florida have already done [47,48].

5.5.5 Impact on Air Pollution

Pollution is exacerbated by excessive light in the darkness, according to research from "Colorado University's Cooperative Institute for Research in Environmental Sciences

(CIRES) and the National Oceanic and Atmospheric Administration (NOAA)". Uplight from the outdoors, which leads to the radiance over areas, also competes with spontaneous chemical effects that aid purifying the air throughout the nighttime, according to findings presented "American Geophysical Union meeting in San Francisco". Each night, naturally produced nitrate radicals reduce chemicals and pollutants from a variety of car exhausts and other human-made sources, preventing them from turning into air pollution, ozone, or various irritants. Because daylight eliminates the presence of these nitrate radicals, this technique is only performed during hours of complete darkness. LP from cities, according to measurements, is responsible for stifling this process. Given that some cities are approaching their allowed limits, this information is likely to pique the public's curiosity in light reducing emissions as a way to improve the quality of the air among municipalities, nations, and federal agencies, as well as the "Environmental Protection Agency (EPA)". Bob Parks of the "International Dark-Sky Association (IDA)" is optimistic that the findings will inspire cities to adopt night sky lighting that is ecologically friendly, including low radiance intensities, and only turn on the lights when/where they're needed. However, this alone will not enhance city air quality. Under the Clean Air Act, the EPA was pleaded in 2008 to evaluate LP to maintain a close eye on and reduce region discoloration of the night sky. The EPA has yet to respond to the dark sky petition in any formal way [49].

5.5.6 Impact on Economic Sustainability

The massive use of artificial lighting in inconvenient settings is a ravage of reserves. Energy waste has significant monetary and ecological consequences. "In the United States alone, 3,600 billion kilowatt-hours of energy were created for electrical consumption in 2018, equating to a cost of USD362 million (EIA) [50]". Lighting accounted for 19% of the total, with outdoor lighting accounting for 16.85% of the total, primarily to illumine roadways and car parks. Because 30% of lighting systems are insufficiently illuminated, a total of 35 billion kilowatt-hours of energy is lost annually for street lamps, costing USD3.5 billion [51]. "According to the Department of Physics and Florida Atlantic University, human electrical energy consumption was equivalent to about 968 million gallons of gasoline or 3,664.28 million liters" [52].

Hunter and Crawford (1991), revealed that the cost of producing electricity was only USD0.7 billion which was ten times lower [53]. Despite the vast number of LP studies, the tenfold surge in excessive electrical consumption of artificial lightning suggested a stagnated concern about LP. This perfectly illustrates that LP is not only a threat to extraterrestrial investigations, public health, and ecosystem services, but it is one of the most serious threats to sustainable development.

5.6 FUNDAMENTALS AND ECONOMICS OF LP

Light has been addressed as an environmental issue by economists extensively throughout the years [54]. Earlier, population designs were used to generate qualitative and quantitative statistics on LP. Indeed, there is a substantial correlation between artificial LP and populace density in a given location. LP is also influenced by the amount of economic improvement in the sector as well as native ordinances. Few have also stated that the descriptive statistics on demography are less reliable because most models include the overall population of any nation but not the demography in the city [54,55]. Researchers also utilized and examined remote sensing photos and statistics from satellites such as the "Defense Metrological Satellite Program (DMSP)" and the "Operational Line Scan System (OLS)" to acquire a more accurate assessment of LP on the planet. For instance, statistics from the DMSP-OLS were formerly utilized to construct a cloud-free composite picture of the "Earth at Night" [56–58].

Moreover, Cinzano et al. compiled satellite observations during "the darkest nights of the lunar cycles" in 1996 and 1997 approximately 30 nights were part of a study. The wavelength frequencies of the most popular exterior lighting: mercury vapor, high-pressure sodium, and low-pressure sodium were (545 and 575 nm) (540–630 nm) and (589 nm) respectively that were included in the statistics [59]. Lights that did not reappear in a similar location three whiles were removed from the records. Importantly, assessments based grounded on terrain were utilized to certify that the information was accurately converted into a metric of a specific form of LP. The spread of LP was then calculated using modeling techniques that took into consideration light scattering and diffraction. Streamlining hypotheses resulted in conclusions that focused on the dispersion of LP rather than demonstrating how environmental conditions and altitude significantly affect sky brightness [59].

Gallaway et al. analyzed distant sensing techniques and monetary statistics from the world bank to assess the commercial repercussions of LP internationally in pioneered recorded evidence. The study's proportional logit regression model revealed that demographics and per capita GDP are major variables in determining worldwide degrees of LP. LP and other commercial issues such as "Foreign Direct Investment (FDI)" and terrestrial management practices have also been suggested as essential elements to examine and determine [55].

5.7 PREVENTION AND POLICIES OF LP

The many effects of LP on humans and other species have been discussed extensively. The expansion of AL in a typically dark environment, as well as changes in the form of light, length of illumination, and employment shifts, have all contributed to ecological LP. As a result, all of the above-mentioned factors should be minimized to lessen the effects of ecological LP. Minimizing LP entails a variety of factors, including lowering sky glow, lowering glare, minimization of artificially illuminating native dark areas, limitation of illumination endurance, minimization of lighting trespass in sectors not supposed to be lighted, fluctuation in ambient light, modification in the light spectrum, lowering clutter and so on [60,61]. As a result, the optimal strategy for minimizing LP is dependent on the specific problem at hand. Among the proposed approaches are:

- Using light sources with the smallest possible intensity to achieve the light's goal.
- It can be beneficial to provide automatic control devices and sensors, as well as configurable lighting control systems, that automatically turn off all outdoor lighting and decrease lighting levels when sufficient daylight is detected, or turn off the setup after a certain number of hours at night.

- Increasing the precision with which lighting installations direct light with fewer complications where it is desired.
- Changing the types of lights utilized such that light rays produced are less prone to instigate major LP glitches. Assessing current illuminating schemes and re-scheming part or all of them relying on whether or not the current light is required.

The findings suggest that the most efficient strategy to reduce LP is to establish an inherently dark environment, nonetheless it constantly clashes with other strategies in terms of social and economic goals. These strategies include:

- Decreased artificial lighting period will save money on energy bills. Not only will this result in carbon emissions, but it will also have epidemiological implications for nocturnal species.
- The uniformity in the well-lit area will be reduced if lighting trespass is lessened.
- Reduced lighting brightness reduces energy consumption while also limiting sky-glow and the area influenced by high-intensity direct light.
- As light is released across a greater range of wavelengths, adjustments to "whiter" light are anticipated to enhance the possible range of environmental repercussions.
- Additionally, radiometry of an area's night sky might provide valuable knowledge about the city's light-out distribution.
- Several investigators also propose using radiometry to create a "City Emission Functions (CFEs)" database, which would usher in a novel era in LP exploration with substantial price and improved night sky illumination predictive performance [62].

5.8 RECOMMENDATIONS TO REDUCE LPs

- Only turn on the lights when they're needed.
- Don't overlight the room
- Don't squander your brightness
- Use screens and reflectors to direct light downwards.
- When possible, employ sensors to turn on the lights just when they're needed.
- There has to be a considerable amount of work done on the effects of LP.
- To emphasize the need for LP prevention, avoidance, and reduction, public and government consciousness must be raised. Light trespass and "second hand" light, the extravagance of extensive dark illumination, and the significance of employing the proper illumination for correct circumstance would all require a shift in public attitude.
- By-laws amended technical standards, and construction codes should be developed to encourage and demand dark sky compatible illumination.

The diversification of ecological parameters are significant since it helps with resource partitioning and biodiversity. Numerous biological processes can only occur at night in the dark. Relaxing, mending, celestial navigation, commemorating, and energizing equipment are some examples. As a result, darkness serves the same and complementary functions as daylight. It is required for the proper development of biological systems.

5.9 CONCLUSION AND OUTLOOK

Across most regions of the world, LP is still not considered a problem worth addressing. No doubt, every property's lighting can be appealing and can increase the building's protection and personal security, but it must not harm the living or work environment. As previously documented, LP affects humans, wildlife, and even levels of air pollution. The biological clock of almost all living species receives incorrect information due to the lengthening of the light phase in the light: dark cycle, causing physiological adaptations that were not necessary for that duration of the day. Prolonged contact to such settings tops to a variety of bodily changes in humans as well as in animals. So, it is necessary to regulate the duration and intensity of illumination to save energy and to minimize its negative influences. According to the function of places, the light should be kept to a minimum; it should be used in the correct place at the correct time, and government agencies should encourage better ambient lighting. Officials participating in illumination system management must determine whether lighting is required and whether alternatives are accessible. The implications of LP, as well as the tactics and guidelines that are necessary for the prevention of LP, are briefly examined in this chapter. The study's findings may be useful in developing policies to control LP. Nevertheless, more extensive exploration particular to LP bases is needed in this sector, particularly in those areas, which have a high pace of growth and enhanced degrees of LP, and could become concern regions in the years ahead. Artificial light is undoubtedly necessary for public growth, but now is the moment for everyone to consider "how much is enough." The state government should create standard methods to regulate current and projected outdoor lighting consumption so that the competent experts can take appropriate action against individuals accountable to alleviate light inconveniences.

REFERENCES

[1] Walker, M. F. (1977). The effects of urban lighting on the brightness of the night sky. *Publications of the Astronomical Society of the Pacific, 89*(529), 405.

[2] Schernhammer, E. S., & Schulmeister, K. (2004). Melatonin and cancer risk: does light at night compromise physiologic cancer protection by lowering serum melatonin levels? *British Journal of Cancer, 90*(5), 941–943.

[3] Elsahragty, M., & Kim, J. L. (2015). Assessment and strategies to reduce light pollution using geographic information systems. *Procedia Engineering, 118*, 479–488.

[4] Owens, A. C., Cochard, P., Durrant, J., Farnworth, B., Perkin, E. K., & Seymoure, B. (2020). Light pollution is a driver of insect declines. *Biological Conservation, 241*, 108259.

[5] Hölker, F., Wolter, C., Perkin, E. K., & Tockner, K. (2010). Light pollution as a biodiversity threat. *Trends in Ecology & Evolution, 25*(12), 681–682.

[6] Narisada, K., & Schreuder, D. (2013). *Light pollution handbook* (Vol. 322). Springer Science & Business Media.

[7] Portree, D. S. (2002). Flagstaff's battle for dark skies. *Griffith Observer, 66*(10), 2–16.

[8] Kyba, C., Garz, S., Kuechly, H., De Miguel, A. S., Zamorano, J., Fischer, J., & Hölker, F. (2015). High-resolution imagery of earth at night: new sources, opportunities and challenges. *Remote Sensing, 7*(1), 1–23.

[9] Fotios, S., & Gibbons, R. (2018). Road lighting research for drivers and pedestrians: the basis of luminance and illuminance recommendations. *Lighting Research & Technology, 50*(1), 154–186.

[10] Mizon, B. (2012). Light pollution: penetrating the veil. *Journal of the British Astronomical Association, 122*(4), 204–207.

[11] Motta, M. (2009). US physicians join light-pollution fight. News. Sky & Telescope. Archived from the original on, 06–24.

[12] Gaston, K. J., Bennie, J., Davies, T. W., & Hopkins, J. (2013). The ecological impacts of night-time light pollution: a mechanistic appraisal. *Biological Reviews, 88*(4), 912–927.

[13] Rajkhowa, R. (2014). Light pollution and impact of light pollution. *International Journal of Science and Research (IJSR), 3*(10), 861–867.

[14] Falchi, F., Cinzano, P., Elvidge, C. D., Keith, D. M., & Haim, A. (2011). Limiting the impact of light pollution on human health, environment and stellar visibility. *Journal of Environmental Management, 92*(10), 2714–2722.

[15] Le Tallec, T., Perret, M., & Théry, M. (2013). Light pollution modifies the expression of daily rhythms and behavior patterns in a nocturnal primate. *PLoS One, 8*(11), e79250.

[16] Davies, T. W., Bennie, J., Inger, R., De Ibarra, N. H., & Gaston, K. J. (2013). Artificial light pollution: are shifting spectral signatures changing the balance of species interactions?. *Global Change Biology, 19*(5), 1417–1423.

[17] Khorram, A., Yusefi, M., & Fardad, M. (2014). Assessment of light pollution in Bojnord city using remote sensing data. *International Journal of Environmental Health Engineering, 3*(1), 19.

[18] Navara, K. J., & Nelson, R. J. (2007). The dark side of light at night: physiological, epidemiological, and ecological consequences. *Journal of Pineal Research, 43*(3), 215–224.

[19] Rakibul Shogib, J. E. Spinney (2018). *Types of light pollution why is it important?* Department of Geography, South Dakota State University: South Dakota State Geography Convention.

[20] Falchi, F., Cinzano, P., Duriscoe, D., Kyba, C. C., Elvidge, C. D., Baugh, K.,... & Furgoni, R. (2016). The new world atlas of artificial night sky brightness. *Science Advances, 2*(6), e1600377.

[21] Sufa Rehman, H. S. (2019). Analyzing the trend in artifcial light pollution pattern in India using NTL sensor's data. *Urban Climate, 27*, 272–283.

[22] Gallaway, T., Olsen, R. N., & Mitchell, D. M. (2010). The economics of global light pollution. *Ecological Economics, 69*(3), 658–665.

[23] Steffy, G. (2002). *Architectural lighting design*. New York: John Wiley & Sons.

[24] Burks, Susan L. (1994). *Managing your migraine*. New Jersey: Humana Press.

[25] Baum, A., Newman, S., Weinman, J., West, R., & McManus, C. (Eds.). (1997). *Cambridge handbook of psychology, health and medicine*. New York: Cambridge University Press.

[26] Abraham Haim, B. A. (2013). *Light pollution as a new risk factor for human breast and prostate cancer*. Dordrecht Heidelberg: Springer.

[27] Blask, D. E., Brainard, G. C., Dauchy, R. T., Hanifin, J. P., Davidson, L. K., Krause, J. A.,... & Zalatan, F. (2005). Melatonin-depleted blood from premenopausal women exposed to light at night stimulates growth of human breast cancer xenografts in nude rats. *Cancer Research, 65*(23), 11174–11184.

[28] Kloog, I., Portnov, B. A., Rennert, H. S., & Haim, A. (2011). Does the modern urbanized sleeping habitat pose a breast cancer risk? *Chronobiology International, 28*(1), 76–80.

[29] Stevens, RG. (2006) Artificial lighting in the industrialized world: circadian disruption and breast cancer. *Cancer Causes and Control, 17*, 501–507.

[30] Kloog, I., Haim, A., Stevens, R. G., & Portnov, B. A. (2009). Global co-distribution of light at night (LAN) and cancers of prostate, colon, and lung in men. *Chronobiology International*, *26*(1), 108–125.

[31] Gandaglia, G., Karakiewicz, P. I., Briganti, A., Passoni, N. M., Schiffmann, J., Trudeau, V.,... & Sun, M. (2015). Impact of the site of metastases on survival in patients with metastatic prostate cancer. *European Urology*, *68*(2), 325–334.

[32] Maung SC, Sara AE, Chapman C, et al. (2016). Sleep disorders and chronic kidney disease. *World Journal of Nephrology*, *5*(3): 224–232.

[33] Mehta, R., & Drawz, P. E. (2011). Is nocturnal blood pressure reduction the secret to reducing the rate of progression of hypertensive chronic kidney disease? *Current Hypertension Reports*, *13*(5), 378.

[34] Crumey, A. (2014). Human contrast threshold and astronomical visibility. *Monthly Notices of the Royal Astronomical Society*, *442*(3), 2600–2619.

[35] Shariff, N. N. M., Hamidi, Z. S., & Faid, M. S. (2017). The impact of light pollution on Islamic New Moon (hilal) observation. *International Journal of Sustainable Lighting*, *19*(1), 10–14.

[36] Faid, M. S., Shariff, N. N. M., & Hamidi, Z. S. (2019). The risk of light pollution on sustainability. *ASM Science Journal*, *12*, 134–142.

[37] Faid, M. S., Shariff, N. N. M., Hamidi, Z. S., Kadir, N., Ahmad, N., & Wahab, R. A. (2018). Semi empirical modelling of light polluted twilight sky brightness. *Jurnal Fizik Malaysia*, *39*(2), 30059–30067.

[38] Hölker, F., Moss, T., Griefahn, B., Kloas, W., Voigt, C. C., Henckel, D., ... & Tockner, K. (2010). The dark side of light: a transdisciplinary research agenda for light pollution policy. *Ecology and Society*, *15*(4), 13.

[39] Bidwell, Tony. (2003). Scotobiology of Plants. Conference material for the Dark Sky Symposium held in Muskoka, Canada, 22–24.

[40] Longcore, T., & Rich, C. (2004). Ecological light pollution. *Frontiers in Ecology and the Environment*, *2*(4), 191–198.

[41] Jones, J., & Francis, C. M. (2003). The effects of light characteristics on avian mortality at lighthouses. *Journal of Avian Biology*, *34*(4), 328–333.

[42] Bruderer, B., Peter, D., & Steuri, T. (1999). Behaviour of migrating birds exposed to X-band radar and a bright light beam. *Journal of Experimental Biology*, *202*(9), 1015–1022.

[43] Frank, K. D., Rich, C., & Longcore, T. (2006). Effects of artificial night lighting on moths. In *Ecological Consequences of Artificial Night Lighting*, edited by C. Rich & T. Longcore, pp. 305–344. Washington, D.C.: Island Press.

[44] Carvalheiro, L. G., Kunin, W. E., Keil, P., Aguirre-Gutiérrez, J., Ellis, W. N., Fox, R.,... & Biesmeijer, J. C. (2013). Species richness declines and biotic homogenisation have slowed down for NW-European pollinators and plants. *Ecology Letters*, *16*(7), 870–878.

[45] Macgregor, C. J., Pocock, M. J., Fox, R., & Evans, D. M. (2015). Pollination by nocturnal L epidoptera, and the effects of light pollution: a review. *Ecological Entomology*, *40*(3), 187–198.

[46] Juell, J. E., Oppedal, F., Boxaspen, K., & Taranger, G. L. (2003). Submerged light increases swimming depth and reduces fish density of Atlantic salmon Salmo salar L. in production cages. *Aquaculture Research*, *34*(6), 469–478.

[47] Salmon, M. (2003). Artificial night lighting and sea turtles. *Biologist*, *50*(4), 163–168.

[48] Montevecchi, W.A. (2006). Influences of artificial light on marine birds. *Ecological Consequences of Artificial Night Lighting*, 94–113.

[49] Bedi, T. K., Puntambekar, K., & Singh, S. (2021). Light pollution in India: appraisal of artificial night sky brightness of cities. *Environment, Development and Sustainability*, *23*, 18582–18597.

[50] EIA, U. (2019). *Electric sales, revenue, and average price*. https://www.eia.gov/electricity/sales_revenue_price

[51] Ashe, M., Chwastyk, D., de Monasterio, C., Gupta, M., & Pegors, M. (2010). *US lighting market characterization*. Washington DC: US Department of Energy, Office of Energy Efficiency and Renewable Energy.

[52] Department of Physics, & Florida Atlantic University. (2014). *Light pollution hurts the environment, hides the night sky*. https://cescos.fau.edu/observatory/lightpol.html

[53] Hunter, T, & Crawford, D. (1991) The Economics of Light Pollution. Light Pollution, Radio Interference, and Space Debris. In ASP Conference Series, 17(IAU Colloquium 112), Vol. 89, Flagstaff, AZ.

[54] Cinzano, P., Falchi, F., & Elvidge, C. D. (2001). The first world atlas of the artificial night sky brightness. *Monthly Notices of the Royal Astronomical Society*, *328*(3), 689–707.

[55] Chepesiuk, R. (2009). Missing the dark: health effects of light pollution. *Environmental Health Perspectives*. 2009 Jan; *117*(1), A20–A27.

[56] Elvidge, C. D., Baugh, K. E., Kihn, E. A., Kroehl, H. W., & Davis, E. R. (1997). Mapping city lights with nighttime data from the DMSP Operational Linescan System. *Photogrammetric Engineering and Remote Sensing*, *63*(6), 727–734.

[57] Elvidge, C. D., Imhoff, M. L., Baugh, K. E., Hobson, V. R., Nelson, I., Safran, J., ... & Tuttle, B. T. (2001). Night-time lights of the world: 1994–1995. *ISPRS Journal of Photogrammetry and Remote Sensing*, *56*(2), 81–99.

[58] Mayhew, C., & Simmon, R. (2000). NASA GSFC."Earth's City Lights". In *Data courtesy Marc Imhoff of NASA GSFC and Christopher Elvidge of NOAA NGDCManfred Schrenk*, edited by V. V. Popovich, D. Engelke, & P. Elisei . ISBN: 978-39502139-5-9 (Print)

[59] Cinzano, P., Falchi, F., E lvidge, C. D., & Baugh, K. E. (2000). The artificial night sky brightness mapped from DMSP satellite Operational Linescan System measurements. *Monthly Notices of the Royal Astronomical Society*, *318*(3), 641–657.

[60] Gaston, K. J., Davies, T. W., Bennie, J., & Hopkins, J. (2012). Reducing the ecological consequences of night-time light pollution: options and developments. *Journal of Applied Ecology*, *49*(6), 1256–1266.

[61] Kumar, P., Ashawat, M. S., Pandit, V., & Sharma, D. K. (2019). Artificial light pollution at night: a risk for normal circadian rhythm and physiological functions in humans. *Current Environmental Engineering*, *6*(2), 111–125.

[62] Kocifaj, M., Solano-Lamphar, H. A., & Videen, G. (2019). Night-sky radiometry can revolutionize the characterization of light-pollution sources globally. *Proceedings of the National Academy of Sciences*, *116*(16), 7712–7717.

CHAPTER 6

Light Pollution: Adverse Health Impacts

Manish Srivastava, Anjali Banger,
Anjali Yadav, and Anamika Srivastava

Banasthali Vidyapith

CONTENTS

DOI: 10.1201/9781003185109-6

6.1 INTRODUCTION

One of the enormous catastrophes of urbanization in the commotion of the biosphere includes alteration in light. This alteration refers to the man-made light or artificial light which results in the production of sky glow; hence it brightens the sky and distorts the brightness caused by the nature and finally the color of the night sky. This process of alteration of light is often referred to as pollution toward the environment – light pollution [1]. With the proliferation of population, there is immense evolution in terms of economy and social infrastructure. This inevitable development had led to increased light output via artificial light which includes spotlights, street lights, etc., and is considered as the major cause of light pollution [2]. The bitter truth resides in the fact that still, people are unaware of its consequences.

So, the question arises what is light pollution?

The natural nighttime light sources are the moon, stars, clouds coverage, and reflective surfaces like snow and ice. But with the evolution of homo sapiens who first made use of fire for heating, preparing their food, as a shield from wild animals, and later for lighting to outspread their work during the night [3]. Now, artificial lights have become of immense importance to humankind. Although its major drawback includes natural nightscape atmospheric pollution. The general definition of light pollution is the alteration in the levels of the nightscape as a result of extreme anthropogenic sources of light [4–6] (as shown in Figure 6.1). In terms of astronomy, light pollution serves as a barrier in observing astronomical bodies because of atmospherically scattered artificial light. Ecologically, light pollution is defined as the changes in the patterns of natural light that result in the degradation of the biosphere [7]. Figure 6.1 shows the mechanism of light pollution.

In excess

Light pollution

FIGURE 6.1   Mechanism of light pollution.

6.2 CLASSIFICATION OF LIGHT POLLUTION

Light pollution is caused by anthropogenic sky glow which includes well-lit skyscrapers, buildings, street lights, lamps of vehicles, etc., [8] over lighting, glare, and light trespass caused by industries and residential areas. The progression of manufacturing cost-effective and more operative lamps played a major role in the massive use of artificial lights.

Based on this information, some scientists have categorized light pollution into four discrete forms [9] relative to directional, quantitative, and spectral attributes of light, later light clutter was also included, namely:

1. Anthropogenic sky glow;

2. Glare;

3. Light trespass;

4. Over-illumination;

5. Light clutter.

6.2.1 Anthropogenic Sky Glow

The enhanced skylight over areas where the population resides is due to anthropogenic sources of radiation such as artificial outdoor lighting. It involves radiations that are emitted upward direction and the radiations which get reflected from the earth's surface. The poor installation of street lights, harbor lights, and other artificial lights serves as a barrier in astronomical annotations, and detonation of these radiations directly impacts the atmosphere of the planet. Other sources involve the lighting system of the buildings in residential and industrial areas, billboards, hoarders placed unnecessarily on various paths which creates unnecessary light pollution during nighttime. From small LED lights used for decoration in houses to sports lights on the grounds, all contribute to this. Bright light with different shades of colors radiating through the windows of buildings, industries, fronts of the greenhouse, and other such places directly into the sky during night creates an atmosphere that acts as a barrier during the astronomical studies, hence negatively affecting the procedure. Similarly, the light glow observed above the glasses of the greenhouse can infuriate the inhabitants residing nearby. This issue can be resolved by the installation of blinds in respective greenhouses which creates a negative impact on the life of inhabitants in that particular area. This form of light pollution produced by the city life can also affect the life of the organisms living at a farther distance from the source whether they are living on land or water [10,11]. The unnecessary light which emits from the anthropogenic sources of light (which includes light emitted from industries, residential buildings, skyscrapers, vehicular lights, beach lights, street lights, and other sources too) (Figure 6.2). The emission of such type of ineffective light disperses the particles present in the atmosphere and then gets reflected on the surface of the earth.

FIGURE 6.2   Artificial sources of light contributing to anthropogenic sky glow.

6.2.2 Glare

The excess amount of illumination that causes uneasiness to the eyes is known as glare. In other words, it is an unappealing light observed in the areas where there is unnecessary emission of artificial light and it causes an undesirable negative effect on the optical insight of living beings. It can further be classified into its types: blinding glare, disability glare, discomfort glare. When you stare for long into the sun, this causes blindness and it could lead to temporary or long-lasting eyesight deficiencies. The disability glare is kind of a glaze faced during fog or dark via car headlights, some bright light, reflections, and it could weaken the efficiency of your vision to see properly. The latter one i.e., discomfort glare is kind of a temporary glare that does not have many harmful effects. It could be exasperating, frustrating, irritating, and does not lasts longer. Distressing illumination occurs because of poorly lit sources of light. It results in irritation, exhaustion, and even discomfort to inhabitants and visitors. It weakens the observation skills of a person in absence of any distress or uneasiness. But later this visual impairment can result in aggravation, annoyance, irritation, and the inability of vision during riding or driving can directly lead to accidents, wounds, and grievance.

6.2.3 Light Trespass

The undesirable shine of light in areas where it is not required is known as light spill or intrusion. After the sun sets, this type of source of artificial light can undesirably affect the life of residents residing in the areas or the trespassers and the visitors along with them since it is an exterior source of light so it intervenes with the privacy of the residents whether it be their living area or cooking area. The undesirable invasion of light into the living room can badly affect the sleep period and the quality of sleep hence it directly impacts the circadian cycle of human beings. This disturbance can also be caused by the lights on street poles situated too close to the houses or buildings or from billboards, sports academies, etc. It not only stands for causing a reduction in visibility but also disturbs the environment of that particular area. In addition, since these lighting contrasts disrupt while driving so, they are responsible for a few traffic accidents too. The light spill is acknowledged to hurt wildlife since here the light comes from an anthropogenic source (illuminated structures, billboards, street lights, etc.), which attracts the migratory birds during their flight at night time [12]. In addition to migratory birds, other animals get affected too such as insects, beetles, rabbits, etc. causing bewilderment to their biological cycle and interrupting their habitats.

6.2.4 Over-illumination

Over-illumination can be defined as the extreme and unnecessary usage of anthropogenic light which is not required during a particular movement or action well-defined by lighting standards and norms. It can be caused by poor light maintenance, not turning off the switch-boards when there is no use, or dimming it off during nighttime. It affects sustainability since the source is getting misused or unexploited. It can be prevented via the use of sensors that turn off light when not in use or via setting timers. In this way, the energy remains conserved too. And due to dynamic changes in the light it can negatively impact astronomical studies and can prove to be life-threatening to humans as it can cause accidents or injury. It further directly or indirectly harms the circadian cycle of humans and the metabolism of biota.

6.2.5 Light Clutter

Light clutter stands for a group of excessive light sources in an area. They can be confusing and distracting for pedestrians, passer-by people, and car drivers. These distractions can cause serious injuries to humans and animals too. It can be observed in areas with poor

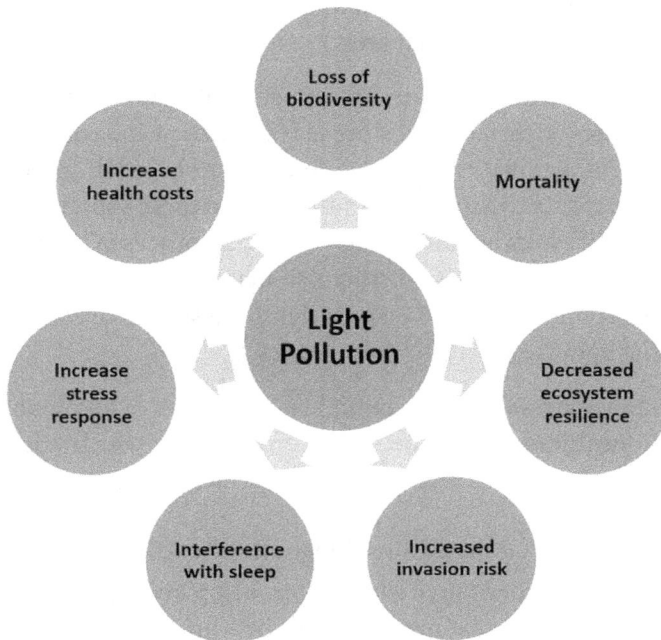

FIGURE 6.3 Hypothetical impacts of exposure to artificial light at night.

street light designs, overly lit billboards, or advertising boards and it can potentially lead to some serious accidents.

In a nutshell, some unwanted, uncomfortable glaze to eyes that harms the biotic veracity of the ecology, encroaches into homes and buildings, and produce sky glow in the atmosphere which annihilates the view of the above sky (stars, moon), this anthropogenic source of light or the artificial light is responsible for altering the world-wide nighttime ambiance, which had led to prevalent consequences to human beings, flora and fauna, together with a negative impact on their health and welfare [13–15] (Figure 6.3). For the past few years, light pollution has been in limelight as a major environmental issue.

6.3 IMPACT ON THE HEALTH OF LIVING BEINGS

Light pollution or in other words luminous/photo pollution plays a very crucial role in detreating our environment since it is associated with extreme artificial light. The source of extreme artificial light is usually exterior ones. It impacts the biodiversity and biosphere in many possible ways. It is responsible for eminent detrimental effects on the health and well-being of living beings in terms of humans, plant, and animal kingdom.

In general, the light spill or the undesirable light that intervenes from outside is responsible for causing sleeplessness, alteration in their mood, extreme fatigue, asthenopia (eye fatigue) in humans. Many areas have good exterior lighting system which protects the residents from such issues. Other outdoor sources of light whose extreme lighting affects the people include:

- **Billboards:** One of the sources of extreme outdoor lighting is billboards. The advertisements played on them or the picture radiating on their surface disturbs the humans and animals crossing that area.

- **Street Lights:** Residents living nearby street lights take more time falling asleep compared to those living far away since it is turned on the whole night. Along with humans, it affects the quality and the life duration of small insects which involves beetles, bugs, moths, etc. they easily get attracted to these sources of light. Eventually, they have a short life span but due to these artificial sources of light, their life span gets even shorter. They are unable to distinguish between day and night time. Residents should install a completely isolated street lightning system to completely eradicate or minimalize the light trespass into their houses.

- While the minor sources include spotlights, headlights of automobiles, bright lights on sea dorks, luminescence in the sports ground, etc. which affect human health, plants, and animals similarly but on a smaller scale.

6.3.1 Consequences of Excessive Artificial Light on Plants

The darkness of the night has been used in many ways by plants whether we talk about their metabolic processes, photoperiodic regulation of their growth and development, or their functions. Plants functions according to the brightness in the daytime and the darkness of the nighttime. Their activities and functions vary with the period and duration of the day

FIGURE 6.4   Effect of long-term exposure of street light on trees.

and the night. Continuous exposure to artificial light could lead to adverse consequences which directly or indirectly affect the growth, metabolism, and development of the plant (Figure 6.4). Accordingly, short-day plants need longer nights for their growth while long-day plants need a shorter night duration. If a short-day plant gets illuminated or irradiated with an artificial source of light during the night even for a short duration, it will react accordingly and construe it as a short night. Its functions, growth will get completely episodic. In normal conditions, these plants usually blossom during the autumn season when the length of the days is shorter. They make use of the longer nights to begin the onset of flowering and with time as the night lengthens, they get into the dormant state [16].

According to some studies performed by researchers, zooplanktons like Daphnia are inhibited from eating algae present on the surface of water bodies because of light pollution which resulted in the formation of algal blooms. These algal blooms lower the quality of water and kill the aerobic life forms of both plants and animals present in that water body. It affects the ecology in different ways. Another study recognized by Lepidopterists and entomologists shows that artificial light intervenes with the navigation process of nocturnal insects during nighttime. Insects like moths which act as pollinators help in the onset of the short-day flowering plants but since their navigation gets disrupted due to these artificial sources of light, the process gets inhibited which directly affects the growth of the flowering plant. This results in declination of the reproduction process of that particular plant species and which finally leads to alteration of the ecology of that particular area. Trees act as a natural habitat to various animals and mankind too. The increment in light pollution had adversely affected their development and density. It even hampers the seasonal changes in plants hence leading to their permanent dormancy condition. In short, their life cycle gets completely disrupted. Artificial light at night inhibits the secretion of photo chrome hormone in plants, leading to their termination and the passage of time to their extinction.

6.3.2 Consequences of Excessive Artificial Light on Animals

There is a natural biological cycle of day and night shifts. Any alteration to their natural being will directly or indirectly affect the life forms on earth. So, any sort of commotion to these shifts will stimulate the behavior of the animal kingdom. Artificial light can misperceive the navigation system of animals, their physiology, sleeplessness, their competitive connections, and most importantly the predator-prey relationship. According to some researchers, long-term exposure to artificial lighting systems could lead to alteration in birth and death patterns too. It can inhibit organisms from acting on the algal bloom leading to poor quality of water source and fatality of the organisms present in the water body. Because of the interference of artificial light at night, the organisms fail to respond naturally to the natural light during the daytime. It can lead to the degradation of the eggs of birds. Other than birds or terrestrial animals, marine life too gets affected by artificial light in many ways. It can lead to premature hatching of eggs. Other behavioral changes include their body structure, size, how they respond to natural light. Some have been briefly discussed below:

6.3.2.1 Threat to Birds

At night, during fog or on cloudy days, the illumination from artificial sources of light attracts the migrant and non-migrant birds toward it. This process has been used for hunting purposes since the 19th century [17]. As discussed earlier in the case of plants, their navigation gets disrupted because of interference of unwanted and uncomfortable glaze/ sky glow [18]. Since light pollution interrupts the timing of circadian and circannual rhythm, it adversely affects the body functions of the organism [19]. According to research birds living in well-lit areas work for longer hours as they outspread their work during night hours in comparison to those who live in darkness [20,21]. Further research in this area concluded that when birds are exposed to artificial light at night, they display agitated behavior [22,23] while the birds living in the uninhabited environment do not show such behavior because of the absence of any source of artificial light in that area [24].

- **Nesting:** One of the altered behavioral changes includes the nesting pattern. The artificial lighting had adversely affected the behavior [25] apparent photoperiod [26] and eventually the fitness [27,28] of the birds and other organisms as well.

- **Disorientation:** Due to the illumination or the disturbance caused by artificial light at night, the migratory and even the non-migratory birds get disoriented from direction and this ends up in altered navigation.

- **Mortality:** Artificial light at night not only affects the navigation of the birds but long-term exposure could be fatal. Since they are not adapted to such alteration of light in the environment. They perform their functions according to the light in the daytime and darkness during the nighttime. It leads to mortality in a way since they are not adapted to such sources of light at night so they easily get disoriented and crash into things that sometimes prove to be fatal or lead to injury or wounds.

- **Sleepless Nights:** Artificial lighting can cause sleeplessness to birds as well.

6.3.2.1.1 Lighthouses

Lighthouses serve as a source of attraction to many migratory and non-migratory sea birds. It was first observed in the mid-19th century and from that time onwards huge fatality rate has been observed in their case. Their fatality rate depends on the type and the amount of signal used in lighthouses during nighttime. It has been observed that white light attracts more birds compared to dynamic colorful lights.

6.3.2.1.2 Light Beams/Ceilometers

In the early 1940s when meteorologists mounted ceilometers to measure the height of the clouds at airports to prevent collision and other consequences. According to a report released in 1954, in October from 6th to 8th, ~50,000 deaths of migratory birds were reported at Warner Robins Air Force Base in Georgia and that was the largest fatality report ever recorded on ceilometer [17]. Later to overcome such types of incidents longer wavelength of the light sources was filtered and fixed beams were converted into rotating ones. Bruderer and his co-workers in 1999 reported that behavioral changes were observed in birds on the installation of these or when they came in contact with light beams and X-Band radars [29]. This disoriented the pattern of their flight and their direction too changed with a declination in their velocity and magnitude.

6.3.2.1.3 Light-induced Fisheries

Since the land has more sources of artificial light as compared to seashores, beaches, etc. So, the range and the intensity of the beam from a single source are much higher. Due to the absence of several artificial light sources, the marine birds easily get attracted to them and easily gets wounded and fatal, and in some cases, they get trapped and rotate around the sources of light lowering their reservoir of energy which inhibits them to reach to nearby places and ultimately, they lose their capability to survive through winters and perform reproduction. Light-induced fisheries adversely affect marine birds.

6.3.2.1.4 City Lights/Horizon Glow

The everlasting urbanization and the increment in the artificial sources of light along with it in form of streetlights, billboards, headlights of vehicles, glazing industries, etc. have adversely affected the life of migratory birds in several ways. These birds easily get trapped and attracted to the illuminated buildings and other nighttime sources of light. This affects their behavior, orientation, and reproducing ability. Their fatality rate increase during foggy winters and rainy days due to excessive use of artificial light.

6.3.2.1.5 Towers

The increasing number and altitude of telecommunication and broadcasting towers create a large sum of deadly collisions with migratory birds. These migration routes are severed by these structures, typically of songbirds. For collisions with towers, two reasons are given. The first reason is when the birds which fly in poor visibility do not look at the structure early enough to evade it, also called a blind collision. The other mechanism for mortality rises when there is a low cloud ceiling or when there are nebulous conditions, and the water particles on the tower are refracted off in the air, which creates a lit-up array about the tower. Birds generally misplace their stellar cues for nocturnal navigation under this kind of weather conditions. Also, they lead to losing all the wide orientation perspective which they might have had on the landscape since they are flying underneath rather a low cloud ceiling. While the illuminated area is passed, it can be like this, that the amplified visibility around the tower turned out to be the toughest cue the birds have for the navigation, and therefore they have a tendency to stay in the well-lit space nearby the tower. Mortality happens once they hover into the structure and its guy cables, or when a collision with other birds takes place, since, a large number of passing birds overcrowd the fairly minor, well-lit space [30]. Latest studies reveal that using spinning or flashing red lights and white strobe light can lessen the consequence of trapping birds at well-lit towers, but there is still work to do to expand the thoughtfulness of the entire consequence on the migration process [18].

6.3.2.2 Threat to Sea Turtles

6.3.2.2.1 Effect on Adult Females   Illuminance adversely affects adult female turtles when they search for places to lay eggs and for hatchlings. They search for locations where there is less interference of artificial light so that it does not affect their hatchings and the number of eggs [31,32]. Since artificial lighting could be led to fatality of the new hatchlings and could even change the behavior of the new ones. The nesting behavior of the female turtles depends on some factors. They abandon the process when they are intervened by human activities or insufficient thermal conditions. After completion of the nesting process, they return to sea but in some cases, they easily get attracted by artificial sources of light like the street lights, beach lights, car headlights, etc. and this leads to disorientation.

6.3.2.2.2 Effect on Hatchlings   The hatchlings get adversely affected by the artificial sources of light. Due to long-term exposure to illuminance, sky glow they easily get disoriented, they cannot differentiate between natural light and artificial light and work accordingly. Because of disorientation, the newborns crawl in the wrong direction where they become susceptible to high temperatures, predators.

6.3.2.3 Threat to Fish

The response of a fish to unwanted illumination varies from species to species. The artificial source of light generates behavioral changes in fishes whether they attract or do not react to it. Several studies have been performed on the behavior of the fishes reacting to an artificial source of light. There are several studies on the use of artificial light at fish farms and deep-sea fish. Most of the studies show that fish avoid white light sources. Nevertheless, some species are attracted by light and this is used to catch them by sport anglers or industrial fisheries.

6.3.2.3.1 Mukene   Artificial light has been extensively used to trap fish during the night since they easily get attracted to the artificial source of light. Fishes usually fall prey to white light as compared to other dynamic colors. Fishers use lamps, torches, and other sources of artificial light to trap fishes, since they are not adapted to such illuminance so they easily fell prey to these sources. This criterion has led to their extinction nearly.

6.3.2.3.2 Salmon and Halibut Farms   The depth level of swimming can be increased with the help of underwater light. It influences the behavior pattern of swimming. These artificial photoperiods are often cast-off to delay sexual development and upsurge evolution. Compared to salmon, halibut tend to swim less but their development ratio is more. These types of fishes can experience skin damage since they are prone to artificial light. This directly affects their natural behavior.

6.3.2.3.3 Deep-sea Fish   Researchers have reported that artificial sources of light have disrupted the natural behavior of fishes living underwater. Unlike the fishes which are found exactly below the surface of the sea, these fish respond to red light more often as compared to white or other dynamic colorful light sources [33]. The main reason behind this lies in the fact that since they are adapted to darkness so other bright colored lights possibly cause more damage to their eyes.

6.3.3 Consequences of Excessive Artificial Light on Human Health

Researchers have found that artificial light not only affects the life of plants and animals but humans too. With the growth in urbanization, there is an utter need for more sources of light along with other basic necessity requirements. Naturally, like other living beings' humans require a sturdy day and night time cycle. It can affect human health in many possible ways starting from their mental peace to physical strains. Long-term exposure to artificial illuminance replicated the brightness of the sun which disturbs the efficiency of the secretion of melatonin into a body which ultimately leads to disturbed circadian rhythm [14]. Long-term exposure to artificial sources of light could lead to migraine issues that begin with headaches experienced more often, fatigue, anxiety, stress, obesity, and other major diseases as bad as cancer, etc. [34]. According to a report, the luminescence observed in offices elevates the blood pressure of a healthy human being up to eight levels. Long hours of exposure to bright light could even lead to reduced sexual activity.

6.3.3.1 Cancer

Researchers from different parts of the world have reported different types of cancers associated with artificial lighting. Long-term exposure to bright light at night could lead to cancer ranging from prostate cancer to breast cancer. Exposure to excessive illumination during night hours disturbs the production and secretion of melatonin hormone in humans. Melatonin is an anti-carcinogenic agent which is secreted by the pineal gland. Usually, it is secreted during night hours in between 2 a.m. and 4 a.m. Several studies have suggested a direct link between the over illuminance and risk of breast cancer due to reduced production of melatonin. Earlier in 1982, Cohen and his co-workers had reported that suppression of melatonin hormone in lower levels could enhance the growth of cancer cells specifically breast cancer in females, and also cited environmental lightning as a factor [35]. Other associated reports include, scientists at the National Cancer Institute (NCI) and National Institute of Environmental Health Sciences have reported that nocturnal artificial light plays a vital role in increasing the cases of breast cancer in females. It was concluded from numerous other reports that the workers working night shifts have probably higher chances of facing breast cancer compared to those who work in day shifts [36]. A similar case study was found in the case of prostate cancer, the people having more hours of exposure to artificial light tend to have higher risks of prostate cancer compared to others.

The suppression of melatonin is more interrelated to breast cancer as compared to other cancer types which include prostate, liver, lung, colorectal, and larynx cancer [37]. It was discovered from another study that not only shift workers even women living at home have higher risks of breast cancer since they are more prone to bedroom lights with a lower wavelength which tends to enhance the tumor cells in the body [38]. This clarified the fact why developed nations have higher cases of breast cancer than under-developed nations.

The link between the artificial night at night and breast cancer is additionally reinforced by a laboratory experiment performed on rats. Black and his co-workers implanted human MCF-7 tumor cell (breast) xenografts into the groins of rats and further analyzed the rate of development of tumor cells, melatonin levels, and linoleic acid uptake [39]. The implanted rats were divided into two groups, out of which one was kept under light-dark conditions which involved 12 hours of light accompanied by 12 hours of darkness while the other group was kept under light-light conditions which involved 24 hours of illumination. Continuous exposure to light for long hours decreased the secretion of melatonin hormone and showed an inclination in the number of tumor cells and linoleic acid uptake. While its reverse results were observed in the first group of rats that resided under light-dark conditions. This whole experiment concluded that exposure to artificial light at night suppressed the melatonin hormone secretion and initiated the uptake of linoleic acid which enhanced the growth of tumor cells or the carcinogenic cells. Melatonin is an anti-carcinogenic hormone having defensive nature with reducing nature having progressed in all flora and fauna over the ages. While in vertebral organisms and nocturnal animals, melatonin is generally formed during the early hours at night which gets blocked due to exposure to artificial light at night. Hence proper use of artificial light is supposed to be maintained to lead a stress-free and healthy life. They are supposed to be designed in a way that they do not affect the circadian rhythms of flora, fauna, and mankind.

6.3.3.2 Retina of the Eyes/Visual Impairment

Biological circadian rhythm is found in all life forms ranging from algae to the largest mammal on the planet. Any charges incurred in the biological clock could lead to severe issues like lousy job performance, sleeplessness, lack of activity, metabolic disorders. The normal production of melatonin hormone is intricately delicate to day-time and nighttime theme and in the case of humans and animals, this system is activated with the help of the photons of light that strike the specific cells in the retina of the eye.

An excessive amount of exposure to artificial light could lead to the disintegration of the retina, death of retinal pigment epithelium cells, or quicken the genetic diseases associated with the retina of the eye. From the past few decades, humans have amplified the use of artificial light which has altered the light-dark rhythmic cycle, ranging from higher to lower wavelength and intensity of light.

The Retinal Damage Could be Caused by Two Main Reasons:

(a) Blue light

(b) Visual pigment-mediated procedure

So, when there is extreme exposure to the light, it could instigate some reactions which necessitate the stimulation of a photo-pigment "rhodopsin" [40]. It could lead to visual impairment and proliferates the risk of evolving cataracts.

Light plays an important role in the synchronization of the biological cycle. In the same way, the retina detects and senses the different wavelengths and intensities of light, which alters according to the light-dark hours, to regulate the endogenous clock system situated within the hypothalamus in our brain. Exposure to a blue night at night via the use of mobile phones, computers, and other screens directly impacts the functionality of the eye's retina. Since the retina got adapted to such physical changes of alteration of light-dark cycles, so to accomplish the necessities of our body, it consists of inherently photo-sensitive ganglion cells in the retina and facilitates its linkage to the brain cells.

6.3.3.3 Sleeplessness, Mood, and Fatigue

The duration, intensity, and wavelength of exposure to the artificial source of light are the key factors responsible for the degradation of the health of living beings. Long-term exposure to artificial sources of light could modify the neural functioning in cortical and sub-cortical regions. It could further lead to sleeplessness, vigor, fatigue, low performance, behavioral and physiological changes [41]. The intensity and the spectral conformation of the illuminance regulate the sleep of the living beings. The lower wavelength of the artificial light in bedrooms too contributes to sleeplessness. Exposure to white light and blue light at night tends to affect the sleep cycle compared to other dynamic bright colors. It could even lead to metabolic disorders.

6.3.3.4 Endocrine System

The artificial sources of light adversely affect the endocrine system. The endocrine system is a collection of glands that is responsible for the production of various types of hormones that regulate the different functions and activities performed by different organs of the human body from mood changes to sleeplessness. Artificial light at night could lead to hormonal imbalance in the body. This hormonal imbalance could further lead to worsening effects and weaken the efficiency of the body to function properly and effectively.

6.3.3.5 Fitness and Physique

Sleeplessness and irregular sleep-wake cycle routine could lead to various diseases. It directly affects the fitness of the individual since the circadian rhythm got disrupted. In worst cases, it could lead to depression and weaken the mental health of an individual affecting their performance at work. Long-term exposure to bright illuminance could lower sexual performance and the worst-case scenario includes infertility in females.

6.3.3.6 Fatality

The excessive use of artificial light could be fatal too. Due to continuous exposure to bright light, over-illumination could visually impair the drivers of the automobiles and ocular fatigue, or in scientific terms, it is known as asthenopia. It reduces their activity and performance at work. In the worst case, it could lead to road accidents. The bright headlights of the vehicles could distract the drivers and lead to the collision of the cars, which results in fatal injuries and lifetime wounds. The pedestrians too are at a high risk of getting attacked and injured and it could be fatal in case of worst accidents. Some of the hypothetical aspects which lead to the above-mentioned consequences are discussed below [42] (Figure 6.5):

In general, all living forms are adversely affected by light pollution. Earlier the nocturnal animals and birds reside on the natural brightness caused by celestial objects such as the light from the moon and the stars. The brightness of the moon varies with days, it seems to be brightest with the full moon and fades away with the new moon. The natural nocturnal light of the sky does not affect the functions and the activities of the living forms of the earth. They used to acclimatize their functions based on the synodic month cycle. But with the growth of industries and urbanization, the natural light has been replaced by artificial sources of light, which hindered the normal functioning routine and adversely affected their life cycle. The life expectancy rate of the different living forms has been shortened. Its most common example includes the moths, who has shown a rapid decline in their population throughout the world in the last few years [43] they get easily attracted from the artificial sources of light and congregate near street lights, lamps, bulbs, headlights, billboards, bright white lights in the sports ground, etc., they get trapped in their illuminance for longer hours and ultimately die. The exposure to artificial light reduced their life expectancy rate from days to a few hours. Other nocturnal mammals like bats suffered a lot from the poor lighting system. It disrupts their mode of navigation and redirects their direction in the wrong way. It inhibits the growth of plants that depends on small birds and different types of insects for the dispersal of seeds and pollination [44]. Usually, the pollination of some species of plants is carried out during night hours in complete darkness. Due to the artificial sources of light at

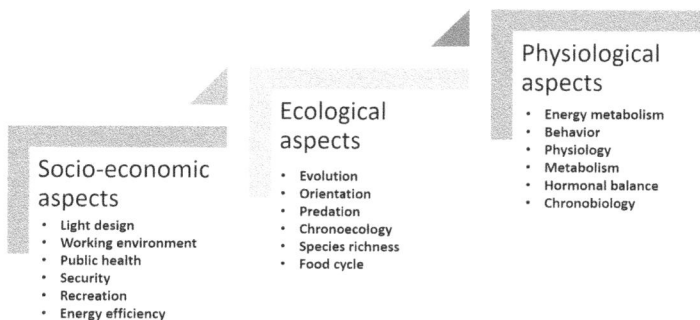

Socio-economic aspects
• Light design
• Working environment
• Public health
• Security
• Recreation
• Energy efficiency

Ecological aspects
• Evolution
• Orientation
• Predation
• Chronoecology
• Species richness
• Food cycle

Physiological aspects
• Energy metabolism
• Behavior
• Physiology
• Metabolism
• Hormonal balance
• Chronobiology

FIGURE 6.5   Hypothetical aspects of light pollution.

night these insects get attracted and hover around the illuminance and redirect their routes. Excessive illuminance is not only responsible for the behavioral changes in birds. Scientists have discovered that due to the exposure of bright white light at night, few songbird species perform their activities at dawn rather than in the early morning hours [45]. They get misled because of these artificial sources of light. It also affects the sleep-wake routine of the birds [46]. Because of disturbance in their sleep cycle, they tend to wake up earlier than their supposed time and hypothetically take fewer hours of sleep than the required duration of the sleep cycle. This delays their activity and reduces performance during day hours. Regarding marine life, as discussed previously, it impacts their navigation, life expectancy ratio, behavioral and physiological changes. Alteration in the breeding time of mammals, fishes, whales, octopuses, and other aquatic species. Light pollution further acts as a major factor in the extinction of the various species of flora and fauna.

This implies how over-illumination, glaze, artificial light sources, or in simple terms how light pollution unfavorably affects biodiversity. In this chapter, we have only discussed how it affects health and alters behavioral changes in different life forms present on the planet. It is a threat to our environment, sustainability. It hampers the astronomical observations; it disrupts the ecological balance [47]. It has degraded the quality of water and soil on the planet via different possible factors. All these things ultimately add up to conclude the extinction of species, in simple terms mortality.

6.4 CAUSES OF LIGHT POLLUTION

6.4.1 Poor Structural Designs

Poor structural designs of light sources could contribute to pollution. Engineers tend to do more research on the effective and efficient designing patterns of different sources of light so that there could be less illumination. The lightning sources having a higher wavelength need to be preferred over the sources which consist of lower wavelength. A lighting system with suitable intensities that are more sustainable and reliable needs to be processed in the markets.

6.4.2 Unsystematic Lightning

Engineers should be appointed to target the area where street lights need to be placed. There should be some proper space for billboards and advertising ads billboards that do not distract the people. Unsystematic lightning could lead to severe distractions and can result in injuries, wounds, and serious accidents. It can even cause visual impairment or asthenopia or other minor and long-lasting symptoms like occasional double vision, blurred vision, headaches, fatigue, irritation, pain around the eyes.

6.4.3 Overpopulation

With the developing era and the increasing population, there is more requirement of artificial sources of lights whether for terrestrial, aquatic, or atmospheric purposes. People prefer to live in areas that are well-lit as compared to the darker ones. To survive they work for longer hours and in day shifts and night shifts. So continuous exposure to artificial light for long hours immensely affects their well-being and mental peace.

6.4.4 Urbanization

With urbanization, there is more requirement for dynamic and brightly colored lights with different wavelengths and intensities. Irrespective of their designs, structure, maintenance people prefer to use more upgraded, decorative, brightly lit lights for different purposes. They use it on occasions, for security purposes, to decorate their houses and buildings and all of this ultimately contributes to light pollution in one way or another and degrade the environment, and is responsible for the ecological imbalance.

6.4.5 Excessive Use of Light Sources

The unnecessary and unwanted use of illuminance is bad for health and the environment. Overly lit buildings glaze, street lights, billboards, sky glow because of car headlights, automobiles, workshops, garage lamps, etc. degrades the biodiversity. The excessive use of white light and blue light with shorter wavelengths at night could increase the risks for obesity, depression, sleep disorders, cancer, frustration, anxiety, etc. as compared to dynamic colorful decorative lights.

6.4.6 Smog, Fog, and Clouds

At night when there is complete darkness all around, the artificial sources of light could lead to sky glow. The glaze, over illuminance from the buildings, billboards, car headlights, beach lamps go above in the sky. Smog and clouds reflect the light emitted from these artificial sources of light and make the sky glow. The sky and the surroundings now seem brighter leading to light pollution. Similarly in winters, the fog reflects the light emitted from the headlights of motor vehicles, etc., and contributes to light pollution.

6.4.7 Lack of Awareness

There is a dire need to educate the public regarding the advantages and disadvantages of artificial anthropogenic sources of light. Lack of awareness about light pollution has restricted the promotion of environment-friendly campaigns. People need to educate themselves and the younger generation how light pollution can deteriorate human health and biodiversity.

The other causes of pollution include lack of restrictions, use of good quality material, standard quality supervision. With the advancement in science and technology, people seek more utilization of electronic products [48]. Use of mobile phones, electronic gadgets, video games, laptops, and computers for entertainment and study purposes. Long-term exposure to a light stimulus leads to dizziness, blurred vision, anxiety, swelling of the hypothalamus (brain), and other health-related ailments.

The most effective way to defend biodiversity is the implementation of rules and regulations. Various social organizations and NGOs are working their best in this field and contributing efforts toward awareness of such environment degrading pollution. A strict prohibition of excessive use of artificial light needs to be implemented. The light sources must be used more precisely and ineffective manner that creates less environmental pollution. Light sources with longer wavelengths and intensities need to be promoted to save wildlife. Some countries have already implemented certain rules and regulations and strict actions are taken against those who do not abide by the rule.

6.5 RECOMMENDATIONS FOR REDUCTION OF LIGHT POLLUTION

The main objective behind the implementation of artificial light is to provide convenience to mankind. The street lights across streets were implanted to avoid crimes and ensure the safety of visitors, pedestrians, and nearby residents. People prefer to take paths that are lighted compared to the darker ones. Although it lacks to fulfill the norms required to ensure the safety of the ecosystem. The extreme use of artificial lights overweighed its aids. Figure 6.6 briefly depicts its causes, impacts, and recommendations. So, to make them more eco-friendly, their structural designs and time of exposure need to be reconsidered so that they could withstand the parameters of sustainability.

The light pollution can only be reduced by controlling the exposure of sky glows, glare, light trespass, over-illumination at night. Small steps should be implemented first to overcome this problem which can be initiated by self. A few of them are listed below:

- Switch off the lights when not in use or light only when required.

- Turn off the lights with the help of advanced technology like sensors, putting on timers.

- Improve light fixtures, to personalize their direction of exposure with minimal drawbacks.

- Adjust the wavelength of the lightning system according to requirement since a lower wavelength tends to cause more light pollution and health-related issues.

- Evaluate the pre-installed lightning pattern and make necessary structural changes wherever required.

- Re-constructing the lightning sources according to their place of requirement.

- Utilize the artificial sources of light which are less intensive and whenever necessary.

- Try to replace artificial sources of light with natural ones if possible.

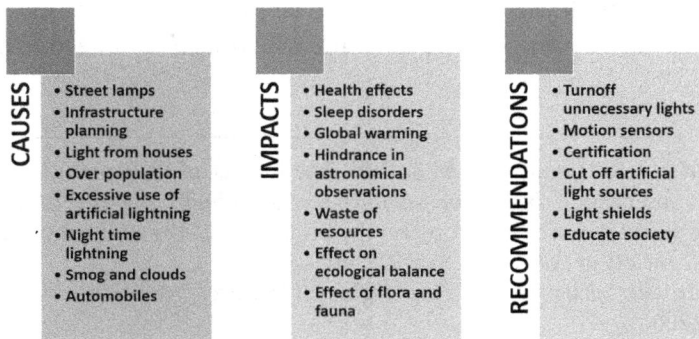

FIGURE 6.6   Causes, impacts, and recommendations of light pollution.

- Use shields and reflectors to give direction to the source of light.

- Make use of inventive/effective energy sources like LED and compact fluorescents.

- Promote dark sky initiatives.

- Modification in engineering ethics and manufacturing encryptions.

- Organize campaigns to educate people about light pollution.

- Use of certified lightning sources.

- Implementation of effective rules and regulations

The variation in ecological circumstances is significant because it subsidizes the partition of resources and better biodiversity. Innumerable natural progressions can only take place at night under complete darkness without the interference of any artificial source of light such as sleeping, refurbishing, restoration, navigation of celestial objects, and predating. And other such activities can occur only in night hours. The human body too carries out various functions at night like secretion and production of certain hormones, fluids, etc. For a similar purpose, darkness at night is as important as daylight. It is requisite for the well-being of humans, flora, and fauna. Subsequently, there is a dire need for planning to guarantee the well-being and fortification of the biota.

6.6 CONCLUSION

Light pollution is a topic of concern since it has some serious allegations for all forms of life existing on the planet Earth. Artificial light affects organisms from as simple as algae to mammals to aquatic birds and covers almost all living beings. It has altered the navigation, orientation, nocturnal activities, and other functions of the body. Even the natural functioning of the biological circadian clock is interrupted by exposure to extreme artificial light which directly affects the functionality of an organism. This has increased the mortality rates of certain species of insects, birds, and animals and even reduced their life expectancy ratio. Certain laws are meant to be implanted to inhibit the excessive use of artificial light. In this chapter, we have mainly emphasized the health impacts which are caused by light pollution. It is a vast field that involves various other aspects of study too. The design and proper planning for the structure of artificial light need to be focused on.

REFERENCES

[1] N.N.M. Shariff, Z.S. Hamidi and M.S. Faid. The impact of light pollution on Islamic New Moon (hilal) observation. *International Journal of Sustainable Lighting.* 19(1), 2017, 10–14.

[2] M.F. Walker. Effects of urban lighting on the brightness of the night sky. *Publications of the Astronomical Society of the Pacific.* 89, 1977, 405–409.

[3] W. Patterson. *Electricity vs fire: The Fight for Future.* Walt Patterson, 2015. ISBN-13: 978-0993261206.

[4] C.S.J. Pun, C.W. So, W.Y. Leung and C.F. Wong. Contribution of artificial lighting sources on light pollution in Hong Kong measured through a night sky brightness monitoring network. *Journal of Quantitative Spectroscopy and Radiative Transfer.* 139, 2014, 90–108.

[5] F. Falchi, P. Cinzano, C.D. Elvidge, D.M. Keith and A. Haim. Limiting the impact of light pollution on human health, environment and stellar visibility. *Journal of Environmental Management*. 92, 2011, 2714–2722.

[6] F. Falchi, P. Cinzano, D. Duriscoe, C.C.M. Kyba, C.D. Elvidge, K. Baugh, B.A. Portnov, N.A. Rybnikova and R. Furgoni. The new world atlas of artificial night sky brightness. *Science Advances*. 2, 2016, 1–25.

[7] Y. Katz and N. Levin. Quantifying urban light pollution—A comparison between field measurement and EROS-B imagery. *Remote Sensing of Environment*. 177, 2016, 65–77.

[8] T. Longcore and C. Rich. Ecological light pollution. *Frontiers in Ecology and the Environment*. 2(4), 2004, 191–198.

[9] K.M. Zielinska-Dabkowska and K. Xavia. Global approaches to reduce light pollution from media architecture and non-static, self-luminous LED displays for mixed-use urban developments. *Sustainability*. 11, 2019, 3446.

[10] J. Berge, M. Geoffroy, M. Daase, F. Cottier, P. Priou, J.H. Cohen, G. Johnsen, D. McKee, I. Kostakis, P.E. Renaud, D. Vogedes, P. Anderson, K.S. Last and S. Gauthier. Artificial light during the polar night disrupts Arctic fish and zooplankton behaviour down to 200 m depth. *Communications Biology*. 3, 2020, 102–108.

[11] C.A. Adams, A. Blumenthal, E. Fernández-Juricic, E. Bayne and C.C.S. Clair. Effect of anthropogenic light on bird movement, habitat selection, and distribution: A systematic map protocol. *Environmental Evidences*. 8, 2019, 13.

[12] S.A. Cabrera-Cruz, J.A. Smolinsky and J.J. Buler. Light pollution is greatest within migration passage areas for nocturnally-migrating birds around the world. *Scientific Reports*. 8, 2018, 3261.

[13] D.M. Dominoni, J.C. Borniger and R.J. Nelson. Light at night, clocks and health: From humans to wild organisms. *Biology Letters*. 12(2), 2016, 1–4.

[14] K.J. Navara and R.J. Nelson. The dark side of light at night: Physiological, epidemiological, and ecological consequences. *Journal of Pineal Research*. 43(3), 2007, 215–224.

[15] J. Zinsstag, E. Schelling, D. Waltner-Toews and M. Tanner. From "one medicine" to "one health" and systemic approaches to health and well-being. *Preventive Veterinary Medicine*. 101(3–4), 2011, 148–156.

[16] T. Bidwell. Scotobiology of plants. Conference material for the Dark Sky Symposium. Muskoka, Canada. 2003.

[17] T. Longcore and C. Rich. Ecological light pollution. *Frontiers in Ecology and the Environment*. 2(4), 2004, 191–198.

[18] P. Bogard. Ecological consequences of artificial night lighting. *Interdisciplinary Studies in Literature and Environmnt*. 13(1), 2006, 245–246.

[19] D.M. Dominoni, B. Helm, M. Lehmann, H.B. Dowse, and J. Partecke. Clocks for the city: Circadian differences between forest and city songbirds. *Proceedings of the Royal Society B: Biological Sciences*. 280(1763), 2013, 20130593–20130593.

[20] D.M. Dominoni and J. Partecke. Does light pollution alter daylength? A test using light loggers on free-ranging European blackbirds (Turdus merula). *Philosophical Transactions of the Royal Society of London B: Biological Sciences*. 370, 2015, 20140118.

[21] A. Russ, A. Ruger and R. Klenke. Seize the night: European Blackbirds (Turdus merula) extend their foraging activity under artificial illumination. *Journal of Ornithology*. 156, 2014, 123–131.

[22] M. De Jong, L. Jeninga, J.Q. Ouyang, K. Van Oers, K. Spoelstra and M.E. Visser. Dose-dependent responses of avian daily rhythms to artificial light at night. *Physiology & Behavior*. 155, 2016, 172–179.

[23] J.L. Yorzinski, S. Chisholm, S.D. Byerley, J.R. Coy, A. Aziz, J.A. Wolf and A.C. Gnerlich. Artificial light pollution increases nocturnal vigilance in peahens. *Peer Journal*. 3, 2015, e1174.

[24] M. de Jong, J.Q. Ouyang, R.H.A. van Grunsven, M.E. Visser K. and Spoelstra. Do wild great tits avoid exposure to light at night? *PLoS One*. 11(6), 2016, e0157357.

[25] A. Dawson. Photoperiodic control of the termination of breeding and the induction of moult in house sparrows Passer domesticus. *Ibis.* 140, 1998, 35–40.

[26] D.S. Farner. Photoperiodic control of reproductive cycles in birds. *American Scientist.* 52, 1964, 137–156.

[27] M. De Jong, J.Q. Ouyang, A. Da Silva, R.H.A. van Grunsven, B. Kempenaers, M.E. Visser and K. Spoelstra. Effects of nocturnal illumination on life-history decisions and fitness in two wild songbird species. *Philosophical Transactions of the Royal Society of London B: Biological Sciences.* 370(1667), 2015, 20140128–20140128.

[28] K. Spoelstra and M.E. Visser. The impact of artificial light on avian ecology. In *Avian Urban Ecology.* Edited by D. Gil and H. Brumm, Oxford University Press, 2014.

[29] B. Bruderer, D. Peter and T. Steuri. Behaviour of migratory birds exposed to X-band radar and a bright light beam. *Journal of Experimental Biology.* 202(9), 1999, 1015–1022.

[30] R. Rajkhowa. Light Pollution and impact of light pollution. *International Journal of Science & Research.* 2012, 2319–7064.

[31] M. Salmon. Artificial night lighting and sea turtles. *Biologist.* 50(4), 2003, 163–168.

[32] B.E. Witherington and R.E. Martin. Understanding, assessing, and resolving light pollution problems on sea turtle nesting beaches. *Florida Department of Environmental Protection.* 2, 1996, 1–71.

[33] E.A. Widder, B.H. Robison, K.R. Reisenbichler and S.H.D. Haddock. Using red light for in situ observations of deep-sea fishes. *Deep Sea Research Part I: Oceanographic Research Papers.* 52(11), 2005, 2077–2085.

[34] M. Smolensky. Review of Light pollution as a new risk factor for human breast and prostate cancers. *Chronobiology International.* 30, 2013, 1203–1204.

[35] L. Tamarkin, D. Danforth, A. Lichter, E. DeMoss, M. Cohen, B. Chabner and M. Lippman. Deceased nocturnal plasma melatonin peak in patients with estrogen receptor positive breast cancer. *Science.* 216(4549), 1982, 1003–1005.

[36] I. Kloog, A. Haim, R.G. Stevens and B.A. Portnov. Global co-distribution of light at night (LAN) and cancers of prostate, colon, and lung in men. *Chronobiology International.* 26(1), 2009, 108–125.

[37] I. Kloog, R.G. Stevens, A. Haim and B.A. Portnov. Nighttime light level co-distributes with breast cancer incidence worldwide. *Cancer Causes and Control.* 21(12), 2010, 2059–2068.

[38] I. Kloog, B.A. Portnov, H.S. Rennert and A. Haim. 2010. Does the modern urbanized sleeping habitat pose a breast cancer risk? *Chronobiology International.* 28(1), 2010, 76–80.

[39] D.E. Blask, G.C. Brainard, R.T. Dauchy, J.P. Hanifin, L.K. Davidson, J.A. Krause, L.A. Sauer, M.A.R. Bermudez, M.L. Dubocovich, S.A. Jasser, D.T. Lynch, M.D. Rollag and F. Zalatan. Melatonin-depleted blood from premenopausal women exposed to light at night stimulates growth of human breast cancer xenografts in nude rats. *Cancer Research.* 65(23), 2005, 11174–11184.

[40] M.A. Contin, M.M. Benedetto, M.L. Quinteros-Quintana and M.E. Guido. Light Pollution: The Possible Consequences of excessive illumination on retina. *Eye.* 30(2), 2015, 255–263.

[41] G. Curcio, L. Piccardi, F. Ferlazzo, A.M. Giannini, C. Burattini and F. Bisgena. LED lighting effect on sleep, sleepiness, mood and vigor. 2016 IEEE 16th International Conference on Environment and Electrical Engineering (EEEIC). 2016, 1–5.

[42] F. Hölker, T. Moss, B. Griefahn, W. Kloas, C. Voigt, D. Henckel, A. Hänel, P. Kappeler, S. Voelker, A. Schwope, S. Franke, D. Uhrlandt, J. Fischer, R. Klenke, C. Wolter. The dark side of light: A transdisciplinary research agenda for light pollution policy. *Ecology and Society.* 15, 2010, 13.

[43] L.G. Carvalheiro, W.E. Kunin, P. Keil, J. Aguirre-Gutiérrez, W.N. Ellis, R. Fox, Q. Groom, S. Hennekens, W.V. Landuyt, D. Maes, F.V. de Meutter, D Michez, P. Rasmont, B. Ode, S.G. Potts, M. Reemer, S.P.M. Roberts, J. Schaminée, M.F.W. De Vries and J.C. Biesmeijer. Species richness declines and biotic homogenisation have slowed down for NW-European pollinators and plants. *Ecology Letters.* 16(7), 2013, 870–878.

[44] J. Bennie, T.W. Davies, D. Cruse, R. Inger and K.J. Gaston. Cascading effects of artificial light at night: resource-mediated control of herbivores in a grassland ecosystem. *Philosophical Transactions of the Royal Society of London B: Biological Sciences.* 370(1667), 2015, 20140131–20140131.

[45] A. Da Silva, J.M. Samplonius, E. Schlicht, M. Valcu, and B. Kempenaers. Artificial night lighting rather than traffic noise affects the daily timing of dawn and dusk singing in common European songbirds. *Behavioral Ecology.* 25(5), 2014, 1037–1047.

[46] C. Steinmeyer, H. Schielzeth, J.C. Mueller and B. Kempenaers. Variation in sleep behaviour in free-living blue tits, Cyanistes caeruleus: Effects of sex, age and environment. *Animal Behaviour.* 80(5), 2010, 853–864.

[47] M.S. Faid, N.N.M. Shariff and Z.S. Hamidi. The risk of light pollution on sustainability. *ASM Science Journal.* 12(2), 2019, 134–142.

[48] Z. Song and X. Li. Hazards, causes and legal governance measures of China's urban light pollution. *Nature Environment and Pollution Technology.* 16(3), 2017, 975–980.

CHAPTER 7

Environmental Aspects of Light Pollution

Reyhaneh Barzegar and M. Barzegar Gerdroodbary

Babol Noshirvani University of Technology

CONTENTS

7.1 THE EFFECT OF LIGHT ON PLANTS

Plants are affected by the biotic and abiotic signals they receive from the environment. Biotic signals include insect attacks and pathogens and scratches caused by larger organisms. On the other hand, temperature changes, changes in water availability, required initial constraints, osmotic pressures, and changes in ambient light are controlled by the abiotic signal. What is further explored in this chapter is the photoreceptors of plants that detect light and thus affect the physiological responses of plants to the amount of light. The effect of light spectrum quality, intensity, and duration on plant growth and thus estimating suitable artificial light for use in places such as greenhouses are also investigated. Four types of light receivers, including phytochromes, cryptochromes, phototropins, and FKFI light receptors are present in plants. These receptors are responsible for modulating the physiological responses and physical growth of plants. In this section, a brief review of the characteristics of these receptors will be given, and a detailed review will be omitted.

DOI: 10.1201/9781003185109-7

- **Phytochromes:** The state of this receptor changes between active and inactive states based on the wavelength of light received from the environment, which is used to regulate plant growth and expansion based on ambient light. Plants use phytochromes as estimating sensors and optimize the extent and direction of their spread based on the shadow avoidance syndrome. If the ratio of red light intensity to ultra-red light intensity increases, the plant will reduce its growth in that direction to escape from the shade, and if it succeeds based on new signals, it will continue its new path toward daylight. In addition to the above, phytochromes also act as thermal signals and regulate the biological clock and many factors affecting plant growth period (such as the change of seasons that affect the flowering rate of plants in certain months) are also involved. Of course, phytochromes are not alone in doing this, but they do a significant part of the work, and other light receivers have responsibilities in this regard.

- **Cryptochromes:** In recent years, a lot of research has been done to identify and find the hand characteristics of plant light receptors that are active in the blue and ultraviolet light spectrum. This research shows that there are at least four types of behavior in plants when irradiated with these lights, which indicates the existence of at least four different types of light receptors. So far, Type 1 and Type 2 cryptochromes have been discovered, which seem to be effective in preventing stem growth when exposed to blue light. Both types of cryptochrome affect the rate of growth or suppression of plant growth in a specific direction, with the difference that at a low light intensity, Type 2 cryptochrome is more sensitive than Type 1.

- **Phototropins:** Irradiation of blue light (390–500 nm) and type A ultraviolet light (320–390 nm) lead to different responses in plants. Among these, four types of behaviors allow the plant to avoid damage under low artificial light. These behaviors include phototropism (growth or expansion and movement of the plant toward or away from the light source), change of small pores in the stems and branches of the plant in order to regulate gas exchanges by opening or closing the pores under disturbing conditions due to disturbing conditions. Transfer of chloroplasts in response to changes in light intensity (accumulation in bright spots in low light and escape of intense light to prevent damage) and ultimately the pursuit of the sun by the leaves in certain plants. They are also effective in altering the pore status and the transfer of chloroplasts, as well as possibly causing faster stem growth under blue light, which plays a key role in increasing plant calcium and the role of the plant underwater light. They play in the non-polarization of the plant skin under this light.

- **FKF1 Light Receptors:** After the complete discovery of plant DNA, it was possible to research proteins associated with each type of light receptor. It was observed that there are three specific proteins with the characteristics of "light, oxygen, voltage" similar to phototropin, which, despite the difference with phototropin, have similar chemical responses to light.

Research shows that the amount of carbon monoxide related to the one-day period of the plant is controlled by the control system of these receptors and its light-dependent function. Researchers have also been able to observe biochemical and genetic evidence of the function of these proteins as light receptors at various time intervals.

In general, optical receptors allow plants to measure and respond to four light parameters of the environment; Light spectrum quality, light intensity, light direction, and duration of light radiation. In most cases, two or more plant light receptors combine their functions and will be effective in one direction or opposite directions. These characteristics, depending on the importance of different functions, allow the plant to grow and expand from an unsuitable path to a path in the direction of light. Therefore, the absence of one of the light receptors in the plant may not have an adverse critical effect on plant performance because other receptors are usually capable of performing the functions of the absent receptor.

The response of different light receptors of a plant to the same environmental conditions may be independent of each other, intensifying each other or in different directions. Some responses are very sensitive to changes in light and occur in low light and changes in brightness invisible to the human eye, while other responses require a great deal of light. Activities affected by light include seed production, stem length increase, branch growth, change from green to flower opening stage, flower growth, fruit production, cessation of branch production, and the extent of branch expansion and division. Among the effects of light on plant physiology, the most important for artificial light is the effect of light on the plant's vital periods. Meanwhile, changes in the amount of red/infrared light, including the wavelengths that plants use for periodic functions, can pose the greatest risk to the plant. In general, four different behaviors can be observed in plants with respect to the duration of light irradiation:

- Plants that flower in short days (do not flower in long days)

- Plants whose flowers open only on long days.

- Intermediate plants that flower on days that are neither long nor short.

- Plants independent of day length whose flowers open only when the plant is mature enough (most plants).

It has also been observed that a short pulse of light (even for about 1 minute) in the middle of the night may prevent some plants from flowering. There are plants in front of you that a minute of red light on long nights causes them to flower more during the day. These different responses to the amount of red and ultraviolet light can help different plants to sell them in different seasons, especially during flowering or non-flowering time. The intensity of the effect of different light sources on plants in research was in descending order; Incandescent lamp, high-pressure sodium vapor, metal halide, cold white fluorescent, and mercury vapor.

According to these results, it can be acknowledged that without a doubt, artificial light affects different plants. The world is visible.

7.2 THE EFFECT OF LIGHT ON ANIMALS

In this section, the effects of artificial lighting at night on different animals will be studied. It should be noted that how animals absorb light (including visible and invisible activities) may be different from how humans absorb light. Measurements based on the unit of luxury are usually used only for light recognizable by the human eye and are not suitable for invisible light. Since different organs of the animal body absorb different forms of light (including different wavelengths of light), future research on light environmental pollution should respond to the light-sightedness of different animals' visual organs. Consider appropriate contact with each animal only for that species.

7.2.1 Birds

Many species of birds migrate during the night, and their attraction to firelight and artificial light during migration, especially in cloudy and seasonal air that flies shorter than the ground, is an undeniable fact. The tendency of migratory birds to light is used in some parts of the world for their hunting (such as India) and some areas for tourism (such as Africa). However, the mechanism by which birds are attracted to artificial light is not well understood. It seems that when the bird flies toward the artificial light, it loses its view of the horizon. Therefore, artificial light is used for navigation and deviates from the main flight path. This phenomenon is more common in young birds than in adult birds.

The structure of birds' visual systems are different from that of humans. Birds have seven different types of light receptors, while humans have only three types of light receptors. Bird light receptors include cone receptors, dual cylindrical receptors (unequal twins), and four types of cylindrical receptors. Additional cylindrical receptors in the bird's visual system are responsible for observing the wavelengths of ultraviolet light. In addition, birds' eyes contain oil droplets of various colors that regulate the sensitivity of light receptors. Therefore, birds see their surroundings as different from humans, which adds to the complexity of accurately recognizing the effects of artificial light on migratory birds during the night.

In addition to the effect of artificial light on night-vision migratory birds, in recent decades, there has been the possibility of disturbances in the orientation and compass performance of birds that act on magnetic receptors during migration. It has been raised due to the presence of specific wavelengths of light. In one study, three different species of blue-green light-reflecting birds were able to navigate correctly while migrating, while the yellow and red light was diverted from the correct path. However, it is not yet clear whether light radiation during migration affects the birds' natural compass or whether they fly in their original path without being affected by existing artificial light. However, it is not yet clear whether light radiation during migration affects the birds' natural compass or whether they fly in their original path without being affected by existing artificial light. Other studies in 1989 show that by reducing or limiting the white light path of lighthouses in parts of Canada, the number of migratory dead birds decreased from 200 in spring and 393 in autumn to about 19 in spring and 10 in autumn. However, the results of this research are different from other results and reports. For example, other research has shown that bright white lights that turn off or rotate have the most destructive effect, and constant red light has the least effect on bird migration.

It seems that the conflicting results of different research studies are due to the differences in the characteristics of the lamp used in lighthouses, such as its wavelength and light intensity, which have not been considered in the research. Bird mortality increased in the early 20th century when light bulbs replaced oil and gas lamps. Research published by relevant organizations has estimated that migratory bird crashes into man-made buildings and windows kill millions of birds in North America each year. Flames from offshore oil and gas rigs are also considered a serious threat to birds, and there are numerous reports of mass bird deaths near these rigs. Increasing the height of radio and television stations from the prototypes that were built (the 1940s) to the mid-1960s has killed more than a million birds because taller towers need more warning lights to be seen by aircraft. However, the results of more recent studies show a reduction in the death of migratory birds due to attraction to TV stations and towers despite their increase in height. Reasons for this include evolutionary adaptation, behavioral changes, declining migratory bird populations, changes in weather conditions, and changes in the lighting system of telecommunication towers. Bird deaths due to the effects of catastrophic towers and lighthouses have similar values over a period of time, with most deaths occurring when the moonlight is low, while deaths are approximate when the full moon is in the sky, it is not possible.

Although the effects of artificial light on the behavior of birds are unknown, general rules for artificial light suitable for these animals can be stated without considering the specific light spectrum information of the lights. If it is necessary to use lighting systems, this system should be turned off during the migratory seasons of birds, especially in special weather conditions (such as foggy nights) to prevent the absorption and death of birds. New evidence also shows that changing the red warning lights of telecommunication towers to white lights can be effective in reducing the death of migratory birds, although white-eyed white light can harass residents around the area. In addition to the above, special attention should be paid to the type of birds from a marine and terrestrial perspective. In general, the ocean environment usually has less artificial light than coastal areas, and most of the artificial light in the oceans is a point source of light that attracts seabirds from distant areas. The main sources of artificial light in the seas and oceans are ships, lighthouses, fishing boats that use light traps, as well as gas and oil platforms in the sea.

The absorption of artificial light is more common in birds that use light-based vision for feeding (such as emperor penguins). Some species of seabirds also have larger eyes than other birds to have a good view of the underwater space, which increases the sensitivity of their visual system to artificial light. At least 21 marine species (tall birds with elongated wings and hooked beaks flying over open seas) have been identified that absorb artificial light.

7.2.2 Reptiles

Reptiles, including some animals (like lizards), Lakdaran (sub-field of cattle, creeping that the body has a hardcover) and the family are snakes. Different detailed research studies [1] have been done since McFarlane first introduced the subject in 1963 to the effect of artificial light on these animals, especially sea turtles. These turtles usually nest and lay eggs on remote dark shores. Artificial light at night alters the normal behavior of female turtles in search of a suitable location, leading to problems navigating them to the ocean (a type of intrinsic orientation known as navigation).

Adverse effects of artificial light on sea turtles include measures such as the use of appropriate filters on existing or new street lights and the use of unusual street lighting systems (instead of the usual mounting system) mounted on pedestals. Turtles are more susceptible to low-wavelength light, such as purple to green, and some of these creatures are even safer than higher-wavelength light. However, further research shows that even filtered light can attract turtles and even adversely affect them by creating a phenomenon called large shadow, which causes them to move back to the sea. The response to this phenomenon has been different in the two species of tortoises under test, which is due to different spectra of light emitted on them. Other research shows that filtering high-pressure sodium vapor lamps, depending on the intensity of the light, results in fewer turtles being absorbed.

In a pilot project in Florida [2], lighting a path adjacent to the beach with a length of 1 km with the installation of LED lights in the elevation of the level crossing was amended and project results showed that the dysfunction Dryayaby turtles in the operation of the old system were improved. In addition, pedestrians, cyclists, and vehicle drivers also expressed satisfaction with the new light. For some reptiles, care must be taken not to confuse the term "night light", which is a common way of searching for food at night, with the concept of night light pollution. In reptiles with this feature – which is due to the presence of a special reflective layer in the retina of their eyes – the light emitted from the predator's eyes is reflected by the target animals and the place of hunting is determined. Therefore, light pollution can severely affect it and cause problems for animals.

Optical traps are one of the most common human tools for detecting nocturnal invertebrates because most invertebrates are highly absorbed by light at different wavelengths. Some species of reptiles, especially lizards, have taken advantage of these conditions and live in residential areas to feed on insects attracted to the light. It has also been observed that some species of snakes use this position to hunt lizards. Research [3] on reptiles' response to human presence in the environment has shown that 9 species of lizards (out of 69 species) and 1 species of snakes (out of 18 species) have changed their range of life to near artificial lights during the night. These results and other evidence suggest that snakes are lurking near artificial nocturnal lights for lizard hunting, suggesting that artificial nocturnal light can cause drastic changes in hunting and predatory activities and relationships at different times of the day and night.

7.2.3 Amphibians

Like frogs and toads, populations have plummeted around the globe. Many frogs' activities are based on their visual system, so artificial light at night can have many adverse effects on their behavior. Most well-known species of frogs use their low-intensity vision to reduce their visual acuity. Increasing the surface area of the retina and light receptors improves the specificity of multiple receptors to a neuronal neuron or the collection of multiple photons by optical receptors before transmission to neurons. To the extent that a species of tree frog has excellent visibility under very low light – even less than 0.1 lux.

Amphibians appear to have a color vision based on three colors (such as humans) that in some species, with increasing sensitivity to ultraviolet wavelengths, may also have a vision based on four colors. Most frogs have a "blue color preference" feature, and research shows that the ability to absorb blue light (less than 500 nm) causes them to dive into the water instead of moving toward land while hunting. According to the evidence, frogs are attracted to artificial light at night, which is probably due to the attraction of insects to these lights. If this phenomenon is repeated and these animals move away from their natural features, some kind of artificial mutation may occur in them.

The presence of artificial light can also affect the behavior of female frogs during mating. For example, in some species, females prefer mating in the dark to mating in the light to avoid dangers such as hunting. The fact that even a minute of light in the dark day and night can affect the production of the hormone melatonin, which in frogs has the effect of controlling activities such as color change, growth of body organs, and reproduction. Amphibians, like other species of animals, have to adapt to changes in light intensity during the day (dark and light adaptation) so that they do not diminish. It can cause temporary blindness due to the colorlessness of the visual cells, and the animal will not be able to respond to the newly received light until it returns to its original chemical state. For frogs that adapt to night darkness, the return of the eye to its original state is due to factors such as; Intensity of light, ambient brightness, and eye-adjusting properties depend and may last for several hours. Because green and red cone receptors and capillary receptors adapt to new ambient light at different times, the time for colored and non-colored frogs to recover can also be different. As far as busy passages or waterways are concerned, frogs may never be able to adapt to the environment and achieve their ideal vision because the passage of strong temporary light through these passages is so great that the receiver can return to the original state. There are no sightings of these frogs.

The best way to reduce the negative effects of artificial light on frogs is to use light bulbs only when needed (such as using motion sensors or turning off light bulbs when not needed), use LED lights, and dim the lights. The light is using a dimmer. Another solution is to prevent the penetration of light into the living areas of amphibians by using plant retaining walls around the passage. Changing the light or changing the light spectrum of light bulbs cannot be a good way to deal with these problems because most frogs and toads have color vision and different wavelengths of light have different effects on them.

Another major group of amphibians, the salamanders, are also very sensitive to changes in ambient light, especially artificial light, and their populations around the globe have declined sharply. Compared to frogs, salamanders are more sensitive to ambient light, and therefore their behavior can be considered as an indicator of the optical health of an ecosystem. According to a few studies, most species of this animal are highly light-emitting. At least a few species of salamanders have color vision, and the wavelengths of ambient light may cause the animal to shine. However, the cause of salamanders' light-repellency is not yet known and may be due to severe visual disturbance under the light or the avoidance of dry skin under the destructive effects of high light temperatures.

Sailors also use special light receptors outside the animal's eye to receive light, which can be irradiated by artificial light. For example, with long-wavelength light radiation, these creatures have difficulty navigating to the nest. Studies have shown that salamanders eat less food due to the lack of feeling safe in high light, while their activity increases significantly, which may be due to the behavioral habit of searching for food by these animals under a light. The effect of salamanders on the spectrum of artificial light varies during the various stages of animal development from infancy to adulthood. Also, like frogs, light pollution can affect these animals by altering the secretion of melatonin in salamanders.

7.2.4 Fish

More than half of the world's population lives within 100 km of the shores of the seas and oceans, and many other human activities take place near rivers and lakes. In addition, there are other temporary lights such as fish breeding ponds near the shores, tender boats, and oil rigs, etc. in these areas. Fishing by boats is usually done at night and intense light is used to attract fish. The effects of light on the behavior of aquatic animals are almost well known, but little research has been done on the effects of humans on the "day-night" periods of the sea and its effects on aquatic animals. Because daylight and darkness along with moonlight, stars, clouds, and other natural resources have a profound effect on life under the seas and oceans, light pollution can vary with the intensity, frequency, spectrum, and duration of light exposure. Waters have a profound effect on the wildlife of these environments. Some life activities in different species of fish such as tuna, salmon, catfish, Flanders and most important species (about 96% of the total population of these species) are directly dependent on the intensity of light that vital activities such as; Eating includes training in speed and movement in group movements and migration. The eyes of these fish, like humans, are composed of cylindrical cells to respond to high light intensity and cone cells to respond to low light intensity. Activities such as training in fish take place under the light that is less intense than the stimulation of cone cells. The capacity of light receptors to absorb light varies from species to species, depending on the genetics and behavioral habits of the fish. These behavioral habits may even change in the transition from freshwater to saline.

The response of fish to light can be divided into two parts: the response to luminosity (candela per square meter (and the intensity response to light) and luxury (which may be present even in one species of fish depending on such things as the general characteristics of the living environment). The period, intensity, and spectrum of ambient light (and fish characteristics change). Dams, power plants, canals, passing vehicles, and the like can cause fish sightings and increase the risk of predators. Also, these seemingly ineffective lights may prevent fish from passing through channels where they feel threatened. Research shows that mercury vapor lamps attract these animals more than other light sources due to their higher energy in blue and ultraviolet light. At first, the fish escape the light of this lamp, but after adapting to the new light of the environment, they are attracted to it and do more activity and less escape than natural light.

7.2.5 Invertebrates

Invertebrates play a key role in wildlife ecosystems as cleaners of contaminants and important members of the food chain. Therefore, the adverse effect of light pollution on them can affect all living organisms on Earth. Numerous studies [4] have been conducted on the adverse effects of light pollution on insects (in general) and in particular on butterflies and propellers (specifically). Optical traps with different light intensities have been used to study the attraction of insects to light. In addition, the effect of these traps in different conditions such as moonlight or lack of natural light has also been studied. Research has also been done on the effects of moonlight intensity at different times and other environmental conditions that can be effective in attracting insects to artificial light.

The behavior and performance of insects in relation to artificial light sources can be studied in two sections: "close to the light source" and "far from the light source". Most research [5] to date has focused on the reactions of insects near the light source, while the effects of light sources far from insects are usually assumed to be moonlight as the far source. Insect behavior in the vicinity of the lamp is seen as one of the three forms of "fixation", "restriction of movement", or "dryness". In the fixed position, the insect is not able to escape from the nearest light source without being harmed. In the case of limited movement, the insect is unable to move due to a complete loss of sense of orientation and vision and remains stationary. Drought means the insect dries up and dies. The effect of each light source on insects depends on the amount of background light. For example, under the full moon, the effect of light pollution on insects is much less. However, some species of insects behave differently in the periods before and after the full moon or during the full moon.

Research on the effect [6,7] of different types of light sources including high-pressure mercury vapor lamp, high-pressure sodium-xenon, high-pressure sodium vapor, and high-pressure mercury vapor with an ultraviolet filter on invertebrates indicates that the incandescent lamp is more effective on incandescent lamps. It is an insect, while the rate of attraction of these animals to the high-pressure sodium vapor lamp is only 45% of the rate of absorption of the high-pressure mercury vapor lamp. On the other hand, high-pressure mercury vapor lamps with ultraviolet filters show only 11% of the absorption of high-pressure mercury vapor lamps without filters, which is consistent with the results of other research on the effects of different types of lamps on insects.

Although light traps can kill insects, the problem that persists is distinguishing dead insects from insects that have lost their ability to function and react to light sources. This is especially true of the rate of insect deaths in dealing with moving vehicles. Research shows the daily deaths of 116 insects per kilometer of highway in Austria, resulting in the death of millions of insects per year due to collisions with moving vehicles. It is estimated that about one-third of these deaths are due to the attraction of insects to the light sources of vehicles.

7.3 THE EFFECT OF LIGHT ON HUMANS

Despite the use of electricity for more than a century, research into the adverse effects of light pollution on humans has been limited to recent years. The natural rhythm of the human body includes waking, fatigue, and sleep, which affect various characteristics of the body. The control center of this rhythm is the human brain, which is greatly affected by the amount of light in the environment. In fact, the sleep-wake cycle is an inherent rhythm that is defined day and night during the first years of life. This rhythm is called "circadian rhythm" which, despite its dependence on human genetics, the body's internal clock needs daily adjustment of this cycle based on light, and its departure from the regulation may cause depression, physical disorders, and even mental problems. This rhythm is controlled as shown by the two hormones melatonin and cortisol.

Melatonin is a hormone secreted by the pineal gland in response to dark and light periods of the day and night and is produced ten times as much at night. It is also a protective hormone as well as a powerful antioxidant that, in addition to various functions, also regulates the body's activities during sleep. Since melatonin helps regulate sleep and wakefulness through its light-dependent secretion and function, the conditions for its secretion must be met at midnight. It also affects the secretion of other hormones that regulate the body's 24-hour rate and pattern of performance and responses.

Most research [8] on the effects of light pollution on human health has been done on the effect of this pollution on the incidence of cancer in women, especially women who work night shifts and under artificial light. The reason for this is the effects of artificial light on the secretion of the hormone melatonin, which can greatly affect the activities of the human body at night. It also regulates the release of female sex hormones and affects menstruation, puberty, and menopause. Children have the highest levels of melatonin overnight, and the amount of this hormone decreases with age. On the other hand, the secretion of melatonin in the body in the early hours of the morning (when the darkness of the air has reached its peak) is at its maximum and stops when exposed to light, even in very small amounts.

Based on studies in laboratory mice, the researchers concluded that the minimum amount of light needed to stop melatonin secretion in mice was only about twice that of moonlight. Even in nocturnal animals (such as owls, bats, etc.), this vital hormone is stopped by abnormal light. Decreased nocturnal melatonin secretion increases the growth of tumors and cancer cells by exposing mice to even weak light at night. In humans, melatonin secretion also depends on the color of the light, its intensity, and the duration of its exposure. Interestingly, this effect is transmitted even with closed eyes. In a study performed on eight blindfolded subjects exposed to white light at an intensity of 2,000 lux for 60 minutes between 12 p.m. and 2 a.m., a decrease in melatonin secretion was observed in two patients. Sleep at night in an environment without pure darkness and with light such as; Television, monitor screens, reading and sleeping lights, or light coming from the street and outdoors may stop melatonin secretion.

Another function of melatonin is to reduce estrogen production at night. Estrogen increases the risk of malignant cancers. Also, when periods of melatonin secretion are disrupted by artificial light, the secretion of the female hormone estrogen increases from the ovaries, one of the most sensitive tumors of which is breast cancer. According to researchers, continuous night shift work increases the risk of breast cancer by 50%. In addition, employees who work night shifts for at least three nights a month and continue this process for 15 years or more are 35% more likely to develop colorectal cancer.

Low levels of melatonin cause a person to fall asleep earlier and wake up earlier, to the point that this disrupted pattern can lead to resentment. For this reason, taking pills containing this substance under the supervision of a doctor can cause a sleep pattern in people who suffer from insomnia due to low melatonin levels – such as the elderly and children whose sleep patterns are disturbed by their sleep patterns. However, excessive or incorrect use of these pills may cause the body's physiological rhythm and rhythm to be disturbed and increase anxiety and irritability, and using it during the day can lead to drowsiness. This hormone is effective in the prevention and treatment of some cancers (especially hormone-dependent cases such as breast cancer and prostate cancer) and can drastically reduce the growth rate of cancer cells. It also increases the positive effect of anti-cancer drugs and reduces their side effects.

In general, the prevalence and incidence of breast cancer in the blind without any perception of light is lower than in those blinds who are even slightly more sensitive to light. In fact, there is more melatonin in blind women, which reduces the risk of developing the disease by 36%. It can be added to this research that the incidence of breast cancer in industrialized countries, which use artificial light to make nighttime light the same as daylight, is far higher than in developing and non-industrialized countries. Constant sleeping in bright lights and working and waking up at night can increase the risk of heart disease, depression, and breast cancer in women.

Most research [9] has been done on the effects of artificial light during the night and the effect of not having enough darkness at night is less discussed. Few studies have been conducted on the effects of unwanted night light entering from outside the building (light rape) on humans. Studies in the Czech Republic show that 5% of the population receives artificial night light from outside. The building has been cited as one of the two main causes of sleep problems. 7% of people also complained about the amount of light entering the building that is not controllable, and 20% also reported that the amount of light pollution in their bedrooms with the use of special equipment, cast, and the desired dark conditions. Also, 5% of people are dissatisfied with the loss of morning light at the beginning of the sunrise, which is due to the presence of artificial lights at night.

The importance of night darkness in human life has also been researched by hunters. However, it is not clear whether the effects are due to the glare of light at night or sleeping in daylight. Various studies [10–12] on the effects of light pollution on humans show symptoms such as; fatigue, stress, increased headache, decreased sexual activity, and increased anxiety. A number of these studies [13–16] have attributed the increase in anxiety and hypertension in employees of some departments to fluorescent light, which has consequences such as; There has also been an increase in work errors due to increased stress. Also, exposure to the flashing light of the streets and environmental advertisements increases the rate of nervous attacks in people with migraines.

The color of light also has a significant effect on the health of living things and the earth's ecosystem. According to studies, exposure to monochromatic red light for 3 and 4 hours with a brightness of 100 lux – which is an acceptable value for a living room – stops the secretion of melatonin by 50%, which time for candlelight is 66, 60, and 60 minutes. And the 58-W fluorescent lamp is 15 minutes.

Light technology has made great strides since the 1960s, and new light sources have long exposed humans to blue light wavelengths, which further reduce melatonin secretion. The light spectrum of today's lighting systems in offices is not perfect compared to the spectrum of sunlight. Because the human eye has evolved over millions of years in the sun with full-spectrum and different color temperatures and uses it to regulate its body's circadian clock, people who spend many hours in artificial light always face a phenomenon called "biological darkness". Melatonin-free blood – seen in people exposed to direct sunlight – can stimulate and grow tumors, while melatonin-rich blood, which is found in people exposed to total darkness, can slow growth. Artificial light should be designed to have the least inconsistency with the 24-hour rhythm of the human body.

7.4 CONCLUSION

There are practical suggestions for the proper design of indoor and outdoor lighting that can be used for human health and the environment. Among these suggestions are the following:

- The best light for lighting an environment is natural daylight. If there are no windows in the room, non-glowing white wavelength light should be used with the maximum possible resemblance to the spectrum of the sun to evoke a natural daylight state for the body.

- The environment should be completely dark during sleep at night. Also, if there is no possibility of absolute darkness, it is better to remove the blue wavelength from the light and shift the color of the light to yellow or orange. For example, studying under a neon lamp instead of a fluorescent lamp will prevent the eye from being exposed to blue light.

- Absolute darkness during sleep is very important. The TV and bedside lamp should not be turned on while sleeping.

- In the passages, a lighting system should be installed according to the standard to minimize the entry of light into the houses (light intrusion). Due to the limited spectrum of blue light in high-pressure sodium vapor lamps, the use of this lamp in passages is recommended.

- The combination of halogen with iron and mercury vapor emits a wider range of blue light. Low-pressure sodium vapor light is less disturbed in melatonin secretion due to having the highest light efficiency compared to other light sources and its monochromatic spectrum (only yellow light with a wavelength of about 590 nm). Astronomers also prefer this lamp due to the convenience of the single-color light filter.

REFERENCE

1. Mizon, Bob; *Light pollution: responses and remedies*. New York: Springer Science & Business Media, 2012.
2. International Dark Sky Association; *Fighting light pollution: smart lighting solutions for individuals and communities*. Mechanicsburg, PA: Stackpole Books, 2012.
3. Corbelli, Edvige; Palla, Francesco; Zinnecker, Hans; "The Initial Mass Function 50 Years Later"; In *Astrophysics and space science library*, edited by Edvige Corbelli, Francesco Palla, Hans Zinnecker, New York: Springer Science & Business Media, Vol. 327; 2007.
4. Rich, Catherine; Longcore, Travis; *Ecological consequences of artificial night lighting*. Washington, DC: Island Press, 2013.

5. Narisada, Kohei; Schreuder, Duco; "Light Pollution Handbook"; In *Astrophysics and space science library*. The Netherlands: Springer Science & Business Media Vol. 322; 2013.

6. Cayrel, Roger; Fisher, A. J.; de Boer, J. B.; *Guidelines for minimizing urban sky glow near astronomical observatories*. Vol. 1. International Astronomical Union; 1980. https://cie.co.at/publications/guidelines-minimizing-urban-sky-glow-near-astronomical-observatories-joint-publication

7. Pollard, Nigel. "Guide on the Limitation of the Effects of Obtrusive Light from Outdoor Lighting Installations"; In *Symposium-international astronomical union*, Vol. 196, pp. 77–80. Cambridge University Press; 2001. https://cie.co.at/publications/guide-limitation-effects-obtrusive-light-outdoor-lighting-installations-2nd-edition

8. Louv, Richard. *Last child in the woods: saving our children from nature-deficit disorder*. New York: Algonquin Books; 2008.

9. Bedrosian, TA; Vaughn, CA; Galan, A; Daye, G; Weil, ZM; Nelson, RJ; "Nocturnal light exposure impairs affective responses in a wavelength-dependent manner"; *The Journal of Neuroscience* 33: 13081–13087; 2013.

10. Perkin, EK; Hölker, F; Richardson, JS; Sadler, JP; Wolter, C; Tockner, K; "The influence of artificial light on stream and riparian ecosystems: questions, challenges, and perspectives"; *Ecosphere* 2: art122; 2011.

11. Kyba, CCM; Lolkema, DE; "A community standard for the recording of sky glow data"; *Astronomy & Geophysics* 53: 6.17–6.18; 2012.

12. Navara, KJ; Nelson, RJ; "The dark side of light at night: physiological, epidemiological, and ecological consequences"; *Journal of Pineal Research* 43: 215–224; 2007.

13. Kavčič, P; Rojc, B; Dolenc-Grošelj, L; Claustrat, B; Fujs, K; Poljak, M; "The impact of sleep deprivation and nighttime light exposure on clock gene expression in humans"; *Croatian Medical Journal* 52(5): 594–603; 2011.

14. Kantermann, T; Roenneberg, T; "Is light at night a health risk factor or a health risk predictor?"; *Chronobiol International* 26: 1069–1074; 2009.

15. Falchi, F; Cinzano, P; Elvidge, CD; Keith, DM; Haim, A; "Limiting the impact of light pollution on human health, environment and stellar visibility"; *Journal of Environmental Management* 92: 2714–2722; 2011.

16. Brainard, GC; Hanifin, JP; Greeson, JM; Byrne, B; Glickman, G; Gerner, E; Rollag, MD; "Action spectrum for melatonin regulation in humans: evidence for a novel circadian photoreceptor"; *Journal of Neuroscience*, 21: 6405–6412; 2001.

CHAPTER 8

Social, Economic, and Ecological Impacts of Light Pollution

Zobaidul Kabir

The University of Newcastle

Mahfuz Kabir

Bangladesh Institute of International and Strategic Studies (BIISS)

Halima Khatun

Bangladesh Livestock Research Institute (BLRI)

CONTENTS

DOI: 10.1201/9781003185109-8

8.1 INTRODUCTION

With increasing populations, urban expansion, economic development, and more efficient lighting technologies, light pollution continues to affect urban cities. As a result, the patterns of change in nighttime lighting are becoming increasingly complex, and most of the research on light pollution has been concentrated in developed countries. However, light pollution is occurring in many developing countries too due to rapid urbanization, transportation, industrialization, and economic growth [1]. Light pollution and its subsequent negative impacts on the environment and society is a new challenge of our time [2]. Light pollution at night are social, economic, and environmental concerns as recognized by scientific communities. The excessive use of artificial light causes light pollution. The diffusion of artificial light from sources affects the night sky and biodiversity in both rural and urban areas although the intensity of diffusion of artificial light is relatively more in urban and industrial areas than in rural areas. By scattering in the atmosphere, diffusion of artificial light creates skyglow thus affecting the natural view of the sky. The level and variability of light pollution however vary from one place to another depending on the intensity of lighting, design, and technology.

Artificial light is an important invention of human civilization. The illumination by artificial sources is undeniably necessary for the augmentation of real prospects for socio-economic development [3]. The lighting of public and private spaces is necessary to carry out development activities at night to contribute to economic growth and this also has made dwelling places more livable due to safety [2]. Artificial light is a requirement for security during dark nights and an attraction. Beyond the cost of artificial lighting include financial and energy resources, the overflow of lighting in urban areas causes pollution on natural nights and the disappearance of stars. Apart from the invisibility of stars, meteors, and milkways in the sky, the scientific community is progressively concerned about the direct and indirect of artificial lighting on biodiversity, society, and human health [3].

However, artificial light can be used excessively without any reason and the design of outdoor night lighting systems may be found to be faulty as this does not always consider the effective need for light (where, when, how). This may result in an excessive diffusion of light and waste of energy and money, and negative impacts on the night environment and terrestrial and marine ecosystems. In general, the intensity, composition, location, and spatial and temporal characteristics of natural light affect biological mechanisms regulate and determine the survival of living species. There is a linkage among light and the circadian clock, photosynthesis, orientation, visual perception, and others that use natural night light as a resource and as an information source [4]. Artificial light pollution may alter these biological mechanisms and affect both flora and fauna negatively in addition to human being.

While excessive use of artificial light at night (ALAN) pollutes the environment, the demand for the use of artificial light has been continuously increasing particularly in emerging economies for more economic activities and safe residence and transportation at night. Furthermore, the operational costs of lighting are decreasing due to the improvement of technological innovation and therefore there is an overall increase of light even if it is not needed [5]. It has been observed that in the 50 years, the growth of lighting in urban areas has been growing exponentially [3].

Artificial light pollution serves human society by extending working time and leisure. Artificial light, however, has significantly altered the natural nighttime environment [6]. The pollution is not only caused by the excessive use of artificial lights but also by the rays scattered by particles in the atmosphere taking shape of artificial skyglow [7]. The skyglow is the most visible consequence of light pollution. Researchers show that artificial light pollution covers more than 80% of the whole world and that 99% of places in the United States are affected by some degree of light pollution [8]. Scientists have started to examine the effects of even moderate levels of light pollution on humans and other animals due to the prevalence of artificial lighting.

Overall, the necessity of artificial lighting at night in any of the countries in the world cannot be ignored as it helps to the enhancement of commerce, promotion of social and cultural activities, and ensuring public safety. Nevertheless, the potential impacts of artificial light or light pollution are acknowledged, and the effects are widespread. The potential impacts of light pollution may include but are not limited to (i) Impact on human health that might be caused mainly through the disruption of circadian biological rhythms or sleep, (ii) Impact on the aesthetic value of dark nights in the absence of starry night sky, (iii) Behavioral change of wildlife and plants, (iv) Connection of visual physiology, vehicle headlamps, nighttime lighting schemes, and harmful glare, (v) Economic impacts such as more cost of energy due to unnecessary electric light, and (vi) Impact of lighting with a varying degree at night on wildlife and vegetation.

This chapter aims to understand the social, economic, and environmental impacts of artificial light pollution. This chapter is divided into four sections. Section 8.1 is an introduction to light pollution and its impacts based on available literature. Section 8.2 describes the social, economic, and environmental impacts. A discussion on the overall impacts is proved by Section 8.3. This is followed by a conclusion.

8.2 LIGHT POLLUTION AND ITS IMPACTS

8.2.1 What is Light Pollution?

Human civilization began with the discovery of fire and the use of fire for lighting helped human beings to make their life more comfortable and extend their daytime activities during the night [9]. The discovery of artificial lights was recognized as a revolution that advanced human civilization unprecedently. However, it has been well recognized that the introduction of such anthropogenic light sources is polluting the ambient quality of natural dark nights. Light pollution can be defined as the changes in the natural level of the nightscape produced by excessive man-made light sources [8]. From an astronomic perspective, light pollution occurs when the astronomical bodies in the sky are not properly observable due to the scattering of artificial light in the atmosphere caused by artificial lighting. From an ecological point of view, light pollution denotes the alteration of natural light patterns at night that affects flora, fauna, and human being [10]. Light pollution is the result of excessive use of light and the intensity of light occurs through the accumulation of lights from the built environment such as from buildings, industries, and streets.

8.2.2 What are The Sources of Light Pollution?

The sources of natural light at night are the moon, stars, aurorae, cloud covers, and reflective surrounding surfaces such as ice and snow. On the other hand, light pollution is caused by the illumination of excessive lighting usually in urban and industrial areas. In addition to over-lighting, the trespass of light and glare can result from commercial, industrial, and residential areas [10]. One of the key sources of light pollution is "skyglow", flood-lit buildings and skyscrapers, streetlights, motorways, sports facilities, security lighting within manufacturing areas, fishing boats, cruise ships, vehicles lamps, airports, commercial centers, offshore oil platforms flares, and undersea research vessels, industrial lighting, and advertisement lighting [11]. Artificial light emits from these sources and sprinkles off atmosphere dust or aerosol particles and is mirrored to Earth as skyglow. The advancement of technology for artificial lighting produces cheaper and more efficient bulbs or tube lights and this may encourage the immense use of artificial lights.

8.2.3 Social Impact of Light Pollution

The adverse impact of ambient light pollution on human health is a significant concern because severe illness is linked with increased individual exposure to artificial lights. The most notable harm to human health is chrono-disruption or the disruption of the circadian rhythms of the organisms. Humans are adapted to the natural lights of day and night physiologically, which led to developing a neuron network in the brain to regulate physiological responses to the exposure of light [12]. Such responses include neuronal activity patterns of brain waves, cell regulation, and the generation and discharge of hormones [13]. These functions occur naturally at daylight while exposure to ALAN can interrupt this process.

The diminishing level of melatonin can create the most acute chronodisruption-related illness since the hormone regulates the pattern of human sleep-wake. The production of melatonin and its circulation throughout the human body necessitates long periods of darkness which naturally comes at night in the absence of artificial light [14]. The human body's regular sleep cycles are disrupted by light pollution. Higher levels of anxiety and depression, as well as obesity, are linked to decreased melatonin emanating from sleep disorders. These increase risks of cardiovascular diseases, diabetes, and gastrointestinal disorders [15] in the human body.

Previous literature on the impacts of artificial light on human health paid attention to nighttime exposures of individuals, which are confined to indoor workspace and residential contexts [16]. The studies demonstrate a linkage between increased exposure to artificial light and higher risks of cancer. An instance of ALAN effects can be the interruption of the circadian clock and suppression of melatonin which is produced in the presence of natural light or darkness at night. Disturbance of the circadian clock exerts negative impacts on human health, such as by rising risk of cancer and obesity among other diseases directly and indirectly [17]. Specifically, women's increased risks of breast, colorectal, and lung cancers and men's high risks of prostate, lung, colon, bladder, and pancreatic cancers [18] are found to be related to exposures to ALAN because of shift work or home-based personal behavior. The highest level of melatonin is produced hormone by the pineal gland in the evening. The hormone works as an antioxidant, which prevents the development of cancer in the human body. The concentration of artificial light promotes cancerous diseases in the human body [19]. Studies demonstrated that the prevalence of breast cancer has increased among women working shifts, which can be attributed to exposure to ALAN-time.

Impatience and anxiety of the population living in cities are caused by bright lights. Lightened-up roadside advertisements are considered to be hazardous to human health and highly disturbing to mental health. Bright lights of roadside advertisements and even streetlights jeopardize road safety by the glittering and diverting attention of the drivers [20]. Bright areas create glare, which is an unpleasant sensation in vision. It is likely to create an adverse impact on human visual perception. Viewing an intense source of artificial light can lead to discomfort glare, which can cause exhaustion, irritation, and even pain to residents and visitors. Conversely, disability glare blights the perception of objects, but it does not necessarily create discomfort. However, it can indirectly cause irritation, distraction, or a reduced ability to watch necessary information because of a reduction in visibility during cycling, walking, driving, which may result in injury and accidents [21].

8.2.3.1 Impacts on Livability

Traditionally, increasing levels of ambient artificial light are not perceived as a problem. Rather, the absence of light at nighttime is considered as an indication of poverty, underdevelopment, and backwardness. In other words, regions and locations lit with bright electric lights indicate urban, developed, safe, and healthy locales.

Light trespass[1] into private gardens, balconies, terraces, kitchens, and living rooms from external sources can exert negative effects on the use of evening, recreation, and privacy of residential buildings and spaces. Light intrusion into bedrooms via windows can reduce the quality and length of sleep at nighttime, which can affect human health and wellbeing negatively. Light trespass occurs if the lighting is placed adjacent to a building and on street poles that are very high for residential areas. It also takes place from the lighting of the decorative building, security apparatus or posts, lighting of sports facilities, advertising, etc. Moreover, the colorful, and vibrant dynamic lighting of a building can damage its quality, ambiance, and adjacent environment [21].

8.2.3.2 The Socio-Cultural Impacts

Many scientific and cultural services and facilities are diminished because of the loss of natural darkness through artificial light [22]. ALAN "closes the window" on the starry sky and one-third of humanity can no longer make out the Milky Way [8]. ALAN erodes darkness and deteriorates the natural relationship between humans and nature (with other humans and nonhumans) that is developed through historical, literary, philosophical, religious landscape, or artistic resources [23]. Thus, ALAN gradually leads to the disappearance of the experience of nature from the human mind and stimulates generational loss of

---

[1]It is also recognized as light spill or light intrusion.

memory about the natural environment. Moreover, ALAN erodes the natural conviviality and intimacy of humans with the occupation of public spaces, and the perception of the world through insight [24].

8.2.3.3 Aesthetic Impacts

Light pollution particularly skyglow in urban areas has drawn attention to the aesthetic value of darkness or the value of the night sky [25,26]. Light pollution particularly at night affects aesthetic view such as the visibility of the star-lit night sky. The beautiful sky view at night full of stars or the beauty of a moonlit night has been lost in many parts of the world, particularly in urban areas and industrial zones. It is well recognized that looking at the starry sky and galaxies may have a positive impact on the mind. This may inspire a sense of wonder and make the mind ecstatic. "In an important sense, aesthetic experience of this kind can bring home some of the ways we cannot place ourselves over and above nature" [27]. The feelings of immensity and vastness of sky full of stars from an incredible distance may wonder us.

The loss of this aesthetic view has occurred increasingly over the last century, but nobody cared about this issue except the astronomers [3]. It has been reported that the urban environment is more responsible for the impact on natural darkness or night view than any other source and the extent of natural night view is found to be more visible in the countryside than that in the urban environment. This is because light pollution is not high in the countryside. It is observed that people living in the urban environment tend to enjoy an aesthetic view of natural darkness where light pollution is relatively low. In developed countries majority of the people cannot see the stars that are visible from dark locations [28]. Because of the skyglow caused by the light pollution, more than one-third of humanity including 60% of Europeans and nearly 80% of North Americans experience light-polluted nights and cannot enjoy Milky Way and other natural phenomena that occur in the sky at night [8].

In response to protecting the aesthetic value relating to the dark sky, it is imperative to create a high share of urban green space that may reduce the skyglow that the skyward flux of non-natural light from urban areas [29]. The urban planners and designers are thinking about the darkness of cities which is aesthetically suitable [30]. They are following the aesthetic principles that may guide them to decide for the future urban environment with suitable darkness at night although dark skies for different cities may be different given the different contextual elements. For example, the modestly lit and unlit green areas, such as urban parks, outdoor recreation areas, seashores, and the outskirts of housing areas may provide pleasant experiences of natural darkness for city dwellers. For city people, the Finnish Government created adequate green space, and some have large areas of forests nearby city limits [31].

8.2.3.4 Lighting, Crime, and Safety

One of the key aspect of social impacts is crime and safety. Although the common perception is the reduction of crime due to artificial light pollution, there is no clear evidence that outdoor lighting may deter crime. Outdoor lighting may psychologically develop our perception that outdoor lighting may make us safer, but this perception may vary given the context of a particular country or city.

According to [32], outdoor lighting may not prevent crime. Rather an outdoor lighting may cost a lot of money. These findings are based on empirical evidence from 62 municipalities in England and Wales. They collected data on crime in 62 local authorities and analyzed and found there is no strong relationship between the reduction of crime and increased use of outdoor lights. The local communities of these municipalities may have a similar experience on crime and safety in the absence of outdoor lighting. They suggest that by considering the risks of crime carefully, local authorities may reduce costs associated with the installation of streetlights. This initiative may save energy too. Interestingly, outdoor lighting may reduce the feelings of safety and privacy by making victims and targeted property easier to locate. Most property crimes occur in the daytime rather than nighttime [33].

8.2.4 Economic Impacts

Considerable costs are incurred due to light pollution, which includes adverse impacts on public health, ecosystem, and wildlife. The waste of energy because of poor and improper lighting creates an additional burden of emission of CO_2 and increases global warming [28]. For example, it has been estimated in the USA that 30% of all outdoor lighting alone is wasted. This adds around US 3.3 billion in addition to the release of 21 million tons of carbon dioxide (CO_2) per year. It is necessary to plant 875 million trees annually to offset this amount of CO_2.

Improper lighting of physical infrastructures, such as roads and bridges, can give rise to numerous economic problems. For instance, light into adjacent water from a road or bridge can intrude into the habitat and ecosystem of fish and other living organisms [34]. It damages biodiversity and destructs the food chain by breaking off navigation, foraging, mating, rest, and predation of fishes within the water body. Light trespass is also likely to impact plants negatively by various means, such as delaying the loss of foliage, generating a second bloom in autumn, and increasing the growth of branches [28]. Often the fish sanctuaries in the river can be affected by the excessive lighting from streets or other sources. This may directly affect the livelihood of fishermen since there is a possibility of migration of fishes to other places or reduction of fish stock due to disruption of hatching, breeding, and feeding.

A significant area of wastage of resources is the massive emission of artificial lights on redundant places, which has considerable negative economic and environmental consequences. For instance, 3.6 trillion kWh of electricity was produced in the USA in 2018 [35]. Of this electricity, 19% was used for lighting while 16.85% of the lighting was dedicated for outdoor illumination, mostly in streets and parking areas. If 30% of outdoor lighting is considered to be poor or creates light pollution,[2] it implies that about 35 billion kWh of electricity was wasted for this purpose that cost US$3.5 billion in 2018 [36]. It was equivalent to the wastage of about 968 million gallons of 3.66 billion liters of gasoline [37]. Thus,

---

[2]Around 30% of electricity used for outdoor lighting was wasted as light pollution in the USA in 2005.

light pollution causes a considerable waste of energy directly and creates public health and ecological concerns, which challenges economic sustainability.

8.2.5 Ecological Impact

8.2.5.1 Impact on Wildlife

Artificial light is a key driving force and beneficial for living things in the biosphere in addition to an important source of energy. Importantly, there is always varying length of day and night all over the globe and therefore adaptation for wildlife to this variation caused by moon and seasons are important [26]. Artificial light may affect entire ecosystems globally and light attracts many insects, and the insects are killed when they are attracted by light, and thereby the population of insects is reduced. Some species that depend on insects for food may suffer from the shortage of food due to the decline of the insect population. In many cases, food chains can affect unusually especially when insects are trapped by lights and predators kill them or eat them [2].

Light pollution can negatively affect natural organisms by causing orientation or disorientation, and attraction or repulsion from the bright night sky. These may mostly affect the natural behavior and population size of flora and fauna [38]. There are direct and indirect impacts of light pollution on biodiversity. Studies show that light pollution is a threat to biodiversity where artificial lighting may affect the behavioral and psychological processes of several species. It is well recognized that all kinds of living things including flora and fauna on land and underwater have experienced and adapted to natural lights as well as their sensitivity to varying degrees of lights. The biological adaptation and behavioral pattern of all living things to natural light conditions including the sun, moon, and stars have grown over billions of years. Therefore, the changes in the lighting environment may significantly affect the harmonization of the diurnal clock in addition to the endogenous (e.g., biological, and psychological) instrument that tracks and forecasts variation in the external light/dark cycles [39]. It has been reported that light pollution may affect all living species by stimulating their hormone production, regulating circadian rhythms, and affecting phototropism in plants [3].

For animals and insects, the disruption of daily and seasonal activities such as mating, breeding, incubation, scavenging, hatching, offspring production, migration, movement, and nesting are common impacts on animals and insects [38]. For example, it is evident that when moonlight increases, most of the nocturnal or nighttime mammals show their reaction to the increasing moonlight. They limit their movements and hunting activities by decreasing the boundary of open areas. They even reduce the total time of activities. On the other hand, they increase their time and movement for hunting activities when the darkest periods of the night come [40].

Artificial lighting can be sensitive to the living species and may affect their pattern of living including food, habitat, and adaptation to nature in many ways. Numerous investigations show that artificial light may affect ecosystems at nighttime. The effects may include the decline of species significantly and could play a significant role in the decline of species while the role of these species can be very important for the conservation of the ecosystem. One of the major impacts of light pollution is the disorientation of nocturnal animals particularly those who are attracted to artificial lights. Some of the examples are as follows:

- Artificial lights may significantly affect the sea turtle population. When new turtle hatchlings come out from their nests at dark night, they scan their surroundings for visual signs to follow and crawl ways from the dark. These visual signs direct themselves toward the brightest and lowermost skyline and ultimately lead them to go the seas. The artificial lights at night may confuse the fragile hatchlings to find the sea and thereby reduce the chances of the survival of hatchlings [41].

- Light pollution may highly affect seabirds such as albatrosses, petrels if artificial light is available close to breeding colonies of these birds. One of the key impacts of artificial light, in this case, is that the birds may leave their breeding colonies, and this may result in the loss of birds due to a lack of suitable breeding places [42]. Another key impact is that artificial light may confuse the hatchlings those separate from each other for the first time from their colonies. The proper dispersion of the hatchlings is important for their survival [43]. Furthermore, birds that migrate or hunt at nighttime move following the moonlight and starlight. In this case, illumination from artificial lighting systems may confuse them to go to the targeted places or affect hunting activities. The artificial light may lead them to different places that can be dangerous for them. It has been noticed that every year millions of birds die hitting with unnecessarily illuminated constructions and towers. The migration of birds usually depends on signals timely coming from seasonal calendars. Artificial light causes them to migrate untimely either a bit earlier or later. Therefore, the birds may miss the ideal climatic condition for nesting, foraging, and other behaviors.

- Artificial light harms the reproductive fitness of animals that are sensitive to light. It has been evidenced that the male *Lithobates clamitans* or commonly known as green frogs living in an artificially illuminated water body spend less time calling for mates than the ones dwelling in naturally dark water bodies. Similarly, *Operophtera brumata* known as male winter months, that live in a well-lit environment, have a low attraction toward their female pheromones [44]. There is a strong relationship between the aquatic biotic world and the artificial lighting that can be called the cascade effect. Cascade effect indicates that artificial light pollution may attract the aquatic insects to the light and the hunting time of some species living in the aquatic environment, particularly predators and scavengers are altered where artificial lights exist, and some night-active riparian spiders extend their activity into the day.

It is well recognized that invertebrates make up most of the biodiversity on earth and are therefore vital to ecosystems. Artificial lighting may attract a wide range of invertebrates including moths. It is to be noted that moths are easily attracted by artificial lights. The blue, green, and ultraviolet lights have short wavelengths with high frequencies and these lights are highly attractive to most insects [45]. These insects are then eaten by other species such as birds and many of the insects die when they come to contact with bulbs. The artificial light thereby may affect the richness and abundance of moths and other insects [46].

8.2.5.2 Impact on Plant Species

The phenology of plants such as the timing of leaf-out and flowering is sensitive to the variation of light and the shifting of phenological characteristics may produce a "cascading effect" on many species [47,48]. Indicate that light is a source of information and has considerable influence on the biological and ecological functions of plants. Artificial light pollution affects plant species since plant species seemed to be very sensitive to the varying degree of light. It is noticeable that artificial light is sufficiently bright and may alter phenological shifts by inducing physiological responses in plants [47]. The development of phenological characteristics of plants can be disturbed and delayed significantly due to light pollution. While temperature plays a leading role in shifts of plant phenological phases at the spatial scale, artificial light pollution may have a repressive effect [49]. The effects such as leaves of some plants may come out quicker than usual time, loss of leaves may occur later than usual time in addition to a longer period for growth and degradation in structure or composition particularly for the plants that are extensively exposed to artificial lights.

Light pollution affects the biological rhythms of some bird species including daily as well as seasonal rhythms [50]. One of the key rhythmic aspects of a bird or any animal is sleep where "sleep plays a vital role in 'cleaning up' the brain from toxic metabolites accumulated during the active period" [51]. There is an interconnection between sleep, light, and circadian rhythm. Therefore, artificial light pollution may strongly affect the sleep patterns of wildlife such as bird species [52]. Consequently, light pollution affects the temporal behavior of birds, such as the duration of sleep of a bird and hunting activities. Light pollution may affect the orienteering of birds as well as other species of similar nature during their journey. During the journey of birds, the artificial light may lead them to a landing in a location that is different from the intended one. Furthermore, in the areas where birds live (e.g., forests or sea beaches) and not possible for them to avoid nearby artificial lighting, the daytime bird species may be active at night due to changes in biorhythms [53].

Plant species appeared to be more sensitive to artificial light pollution, and phenology advancement was hindered more prominently and even delay phenomenon exhibited when the color level showed stronger sky brightness. Linear mixed models indicate that although temperature plays a dominant role in shifts of plant phenological phases at the spatial scale, the inhibitory effect of artificial light pollution is evident considering the interactions. Plant phenology, the timing of periodic plant developmental stages [47], such as leaf-out and flowering, is sensitive to variation of environmental controls and its shifts could produce a cascading effect on most species [54]. Light as a resource and source of information contributes momentous physiological and ecological functions to plants [47]. Nevertheless, bright artificial light sufficiently alters phenological shifts by inducing physiological responses in plants.

8.3 DISCUSSION

The impact of light pollution includes social, cultural, health ecological, economic, and live-ability. The excessive use of light disrupts the aesthetic value of the sky at night where the sky looks beautiful due to stars, the moon, and other natural phenomena. One of the major social impacts is impacts on the health of both human beings and wildlife. Disturbance of sleep of human being due to light may change the biorhythmic cycle. It is imperative to know that the excessive use of artificial light not only affects the social and cultural life but also flora and fauna in many ways. Artificial light is important for modern society, but the excessive impacts of light pollution are largely ignored. Table 8.1 displays the summary of the impacts of artificial light found in this study.

To address the impacts of artificial light pollution, it is imperative to take initiatives by governments. The government needs to take policy measures with a view to the improvement of technology and reduce the excessive use of light. For example, to reduce light pollution, the government may determine the number of illuminated areas where light pollution is high. Also, the government can map the areas with artificial lighting and take necessary actions where light pollution affects humans, wildlife, and plants. To reduce or maintain the recommended level of outdoor lights such as street or highway lights, national or regional guidelines on the level of lighting should be in place. Furthermore, to reduce the light pollution and trespassing light from luminaries, shielding of luminaries can be considered. Similarly, for the reduction of the duration of the illumination from various sources, innovative design of lighting is important. Innovative technologies, for example, those are controllable by the public or use and activation of sensors may reduce the duration of illumination. It is also necessary to monitor the skyglow or the intensity of lighting in an urban and industrial area. This will provide information and understanding about the extent of light pollution. Finally, in addition to policies and guidelines, regulations need to be developed in countries where applicable to manage light pollution.

There are some challenges from governments for the management of light pollution. This may include a lack of awareness among policymakers, professionals, and general

TABLE 8.1 Impacts of Light Pollution

Aspects	Impacts
Social	Disruption of sleep, cancer may be caused by artificial light. Impact on agricultural production may create the risk of food security. Aesthetic view of sky at night is affected. Recreational impacts are affected by ARP where the recreation is more enjoyable at dark night with natural light rather than AL.
Economic	Excessive use of light more than it is required is a wastage, and it is associated with monetary cost, and misuse of natural resources. The agricultural production and plant lives and other resources such as fisheries can be affected by ALP and thereby affect the income of many people. Global warming and investment of money to address the impacts of climate change due GW.
Ecological	Disruption of biological rhythm of wildlife. Loss of population of some wildlife such as tortoise. Disruption of food chain due to the loss of population insects those are easily attracted by light. Disorientation of birds due to ALP when they move at night. Impacts on plant species can be significant due to variation of daytime and dark nighttime due artificial light pollution. Phenological advancement can be hindered.

people who use light about the negative impacts of artificial light. It is to be noted that the invention of blue-rich lighting may intensify the light pollution and cause more impacts. The increase of blue-rich lighting, therefore, needs to be controlled by law or policies to avoid the use. Also, it is difficult to change the habits of people relating to the use of lights, and redesigning and replace of existing lighting systems is costly. The governments need to realize the negative impacts of light pollution, invest money for research for new technologies and consider the advice of professionals relating to good practices for light pollution management. Education and awareness about the impact of artificial light pollution are also important. Therefore, global community needs a greater understanding the long-run impacts of excessive use of artificial lighting and the negative impacts of lighting.

8.4 CONCLUSIONS

This chapter aimed to identify the social, economic, and environmental (ecological) impacts of light pollution. The social impacts of artificial light pollution include health hazards including sleep disorders and disturbance of sleep, disturbance of aesthetic view of the sky at night due to skyglow, road accidents at night due to confusion caused by streetlights. The economic impacts of lighting are the wastage of money due to excessive use of light and climate change impacts due to the emission of greenhouse gases (GHGs). To address the climate change impacts money is necessary to mitigate. The human health affected by artificial light also has an economic cost. The environmental or ecological impacts include impacts on wildlife such as reduction of insect population, tortoise, and birds which travel at night. The phenology of plant species is affected by light pollution and this may affect food shortage and another natural phenomenon.

The chapter describes some challenges including the lack of awareness among policymakers and the users of light. In addition, the lack of technology and replacement of existing design is also a challenge. To manage light pollution, however, it is necessary to develop policies, standards, guidelines, and regulations by the governments. Also, awareness and education may play an important role to reduce the impacts of light pollution.

REFERENCES

[1] Han, P., Huang, J., Li, R., Wang, L., Hu, Y., Wang, J., and Wei Huang, W., 2014. Monitoring trends in light pollution in China based on nighttime satellite imager, *Remote Sensing*, Vol. 6: 5541–5558.

[2] Apro, A., 2020. Light pollution as environmental problem appearance in national core curriculum and in other educational documents, *Journal of Applied Technical and Educational Sciences*, Vol. 10 (3): 147–156.

[3] Zissis, G., 2020. Sustainable lighting and light pollution: a critical issue for the present generation, a challenge to the future, *Sustainability*, Vol. 12: 4552.

[4] Dodd, A.N., Belbin, F.E., Frank, A., and Webb, A.A.R., 2015. Interactions between circadian clocks and photosynthesis for the temporal and spatial coordination of metabolism, *Frontiers in Plant Science*, Vol. 6: 245–253.

[5] Massette, L., 2020. Drivers of artificial light at night variability in urban, rural, and remote areas, *Journal of Quantitative Spectroscopy & Radiative Transfer*, Vol. 255: 107250.

[6] Sanders, D., Frago, E., Kehoe, R., Patterson, C., and Gaston, K.J., 2020. A meta-analysis of biological impacts of artificial light at night, *Natural Ecology & Evolution*, Vol. 5: 74–81.

[7] Aube, M., 2015. Physical behaviour of anthropogenic light propagation into the nocturnal environment. *Philosophical Transactions of the Royal Society B: Biological Sciences*, Vol. 370: 20140117.

[8] Falchi, F., Cinzano, P., Duriscoe, D., Kyba, C.C.M., Elvidge, C.D., Baugh, K., Portnov, B.A., Rybnikova, N.A., and Riccardo Furgoni, R., 2016. The new world atlas of artificial night sky brightness. *Sciences Advances*, Vol. 2: 1600377.

[9] Patterson, W., 2015. *Electricity vs Fire: The Fight for Our Future*. Walt Patterson: Amersham, pp. 172.

[10] Katz, Y. and Levin, N., 2016. Quantifying urban light pollution- a comparison between field measurement and EROS-B imagery, *Remote Sensing of Environment*, Vol. 177: 65–77.

[11] Ngarambe, J., Lim. H.S., and Kim, G., 2018. Light pollution: is there an environmental Kuznets curve? *Sustainable Cities and Society*, Vol. 42: 337–343.

[12] Bedrosian, T.A. and Nelson, R.J., 2017. Timing of light exposure affects mood and brain circuits, *Citation: Translational Psychiatry*, Vol. 7: 1017.

[13] Haim A. and Zubidat, A.E., 2015. Artificial light at night: melatonin as a media- tor between the environment and epigenome, *Philosophical Transactions of the Royal Society B: Biological Sciences*, Vol. 370(1667): 20140121.

[14] Reiter, R.J., Tan, D.X., Sanchez-Barcelo, E., Mediavilla, M.D., Gitto, E., and Ahmet Korkmaz, A., 2011. Circadian mechanisms in the regulation of melatonin synthesis: disruption with light at night and the pathophysiological consequences, *Journal of Experimental and Integrative Medicine*, Vol. 1(1): 13–22.

[15] Munzel, T., Hahad, O., and Andreas Daibe, A., 2021. The dark side of nocturnal light pollution. Outdoor light at night increases risk of coronary heart disease, *European Heart Journal*, Vol. 42(8): 831–834.

[16] Nadybal, S.M., Collins, T.W., and Grineski, S.E., 2020. Light pollution inequities in the continental United States: a distributive environmental justice analysis, *Environmental Research*, Vol. 189: 109959.

[17] Svechkina, A., Portnov, B.A., and Trop, T., 2020. The impact of artificial light at night on human and ecosystem health: a systematic literature review. *Landscape Ecololgy*, Vol. 35: 1725–1742.

[18] Touitou, Y., Reinberg, A., and Touitou, D., 2017. Association between light at night, melatonin secretion, sleep deprivation, and the internal clock: health impacts and mechanisms of circadian disruption. *Life Science*, Vol. 173: 94–106.

[19] Muhamad, F.S., Nur Nafhatun, S.M., and Zety, H.S., 2019. The risk of light pollution on sustainability. *ASM Science Journal*, Vol. 1(12):134–142.

[20] Lyytimäki, J. and Rinne, J., 2013. Voices for the darkness: online survey on public perceptions on light pollution as an environmental problem, *Journal of Integrative Environmental Sciences*, Vol. 10(2): 127–139.

[21] Zielinska-Dabkowska, K.M., Xavia, K., and Bobkowska, K., 2020. Assessment of citizens' actions against light pollution with guidelines for future initiatives, *Sustainability*, Vol. 12: 4997.

[22] Challéat, S., Barré, K., Laforge, A., Lapostolle, D., Franchomme, M., Sirami, C., Viol, I.L., Milian, J., and Kerbiriou, C., 2020. Grasping darkness: the dark ecological network as a social-ecological framework to limit the impacts of light pollution on biodiversity. *Ecology and Society, Resilience Alliance*, Vol. 26(1): 15.

[23] Gallic, S.L. and Pritchard, S.B., 2019. Light(s) and darkness(es): looking back, looking forward, *Journal of Energy History*, Vol. 2: 1–16.

[24] Shaw, R., 2018. Nocturnal ecologies and infrastructures. In *Nocturnal City*, Chapter-3; Shaw, R. and Shaw, R., Eds.; Taylor and Francis; London, pp. 18.

[25] Gallaway T., 2014. The value of the night sky. In *Urban Lighting, Light Pollution, and Society*; Meier, J., Hasenöhrl, U., Krause, K., and Pottharst, M., Eds.; Taylor & Francis: New York, NY.

[26] Stone, T., 2021. Re-envisioning the nocturnal sublime: on the ethics and aesthetics of night-time lighting, *Topoi*, Vol. 40: 481–491.

[27] Brady, E., 2013. *The Sublime in Modern Philosophy: Aesthetics, Ethics, and Nature*. University of Cambridge Press: Cambridge, UK.

[28] Gallaway, T., Olsen, R.N., and Mitchell, D.M., 2010. The economics of global light pollution, *Ecological Economics*, Vol. 69(3): 658–665.

[29] Biggs, J.D., Fouch, T., Bilki, F., and Zadnik, M.G., 2012. Measuring and mapping the night sky brightness of Perth, Western Australia, *Monthly Notices of the Royal Astronomical Society*, Vol. 421: 1450–1464.

[30] Edensor, T., 2015. The gloomy city: rethinking the relationship between light and dark, *Urban Study*, Vol. 52(3): 422–438.

[31] Soderman, T. and Saarela, S.R., 2011. *Sustainable Urban Regions, Criteria, and Indicators for Land Use Planning, Suomenympa¨risto¨25/2011*. Suomen ympa¨risto¨keskus: Helsink, Finland.

[32] Steinbach, R., Perkins, C., Thompson, L., Shane Johnson, S., Armstrong, B., Green, J., Grundy, C., Wilkinson, P., and Edwards, P., 2015. The effect of reduced street lighting on road casualties and crime in England and Wales: controlled interrupted time series analysis, *Journal of Epidemiology and Community Health*, Vol. 69: 1118–1124.

[33] American Medical Association, 2012. Report of the council on science and public health, House of Delegates Handbook (A-16) - Combined (darksky.org), USA.

[34] Khorram, A., Yusefi, M., and Keykha, S., 2014. Light pollution, a world problem. *Health Scope*, Vol. 3(4): 24065.

[35] EIA, 2018. Electric sales, revenue, and average price - energy information administration, available at https://www.eia.gov/electricity/sales_revenue price, viewed on 16/11/2021.

[36] Faid, M.S., Nafhatun, N., Shariff, M., and Hamidi, Z.S., 2019. The risk of light pollution on sustainability, *ASM Science Journal*, Vol. 12(2): 134–142.

[37] Department of Physics, & Florida Atlantic University, 2014. Light pollution hurts the environment, hides the night sky, viewed on 16/11/2021.

[38] Karnowski, R.L., Limpus, C., Pendoley, K., and Hamann, M., 2014. Influence of industrial light pollution on the sea-finding behaviour of flatback turtle hatchlings, *Widlife Research*, Vol. 41: 421–434.

[39] Dominoni, D.M., 2015. The effects of light pollution on biological rhythms of birds: an integrated, mechanistic perspective, *Journal of Ornithology*, Vol. 156: 409–418.

[40] Beirer, P., 2006. Effects of artificial night lighting on terrestrial mammals. In *Ecological Consequences of Artificial Night Lighting*; Rich, C. and Longcore, T., Eds.; Island Press: Washington, DC.

[41] Dimitriadis, C., Fournari–Konstantinidou, I., Sourbès, L., Koutsoubas, D., and Mazaris, AD., 2018. Reduction of sea turtle population recruitment caused by nightlight: evidence from the Mediterranean region, *Ocean & Coastal Management*, Vol. 153: 108–115.

[42] Sultana, J., Borg, J.J., Gauci, C., and Falzon, V., 2011. *The Breeding Birds of Malta*; Bird Life Malta: Ta' Xbiex, Malta.

[43] Le Corre, M., Ollivier, A., Ribes, S., and Louventin, P., 2002. Light-induced mortality of petrels: a 4-year study from Reunion Island (Indian Ocean). *Biological Conservation*, Vol. 105: 93–102.

[44] Manfrin, A., Singer, G., Larsen, S., Weiß, N., van Grunsven, R.H.A., Weiß, N.-S., Wohlfahrt, S., Monaghan, M.T., and Hölker, F., 2017. Artificial light at night affects organism flux across ecosystem boundaries and drives community structure in the recipient ecosystem. *Frontiers in Environmental Science*, Vol. 5: 61–72.

[45] Bruce-White, C. and Shardlow, M.A., 2011. *Review of the Impact of Artificial Light on Invertebrates*; Buglife–The Invertebrate Conservation Trust: Peterborough, UK.

[46] Owens, A.C.S., Cochard, P., Durrant, J., Farnworth, B., Perkin, E.K., and Seymoure, B., 2020. Light pollution is a driver of insect declines, *Biological Conservation*, Vol. 241: 108259.

[47] Meng, L., Mao, J., Zhou, Y., Richardson, A.D., Lee, X., Thornton, P.E., Ricciuto, D.M., Li, X., Dai, Y., Shi, X., and Jia, G., 2020. Urban warming advances spring phenology but reduces the response of phenology to temperature in the conterminous United States. *Proceedings of the National Academy of Sciences USA*, Vol. 117: 4228–4233.

[48] Bennie, J., Davies, T.W., Cruse, D., Gaston, K.J., and Swenson, N., 2016. Ecological effects of artificial light at night on wild plants. *Journal of Ecology*, Vol. 104: 611–620.

[49] Lian, X., Jiao, L., Zhong, J., Jia, Q., Liu, J., and Liu, Z., 2021. Artificial light pollution inhibits plant phenology advance induced by climate warming, *Environmental Pollution*, Vol. 291: 118110.

[50] Dominoni, D.M., Smit, J.A.H., Visser, M.E., and Halfwerk, W., 2020. Multisensory pollution: artificial light at night and anthropogenic noise has interactive effects on activity patterns of great tits (Parus major). *Environmental Pollution*, Vol. 256: 113314.

[51] Xie, L., Kang, H., Xu, Q., Chen, M.J., Lao, Y., Thiyagarajan, M., O'Donnell, J., Christensen, D.J., Nicholson, C., IIiff, J.J., Takano, T., Deane, R., and Nedergaard, M., 2013. Sleep drives metabolite clearance from the adult brain. *Science*, Vol. 342: 373–377.

[52] Archer, S.N., Laing, E.E., Moller-Levet, C.S., van der Veen, D.R., Bucca, G., Lazar Alpar S., A.S., Santhi, N., Slak, A., Kabiljo, R., von Schantz, M., Smith, C.P., and Dijk, D., 2014. Mistimed sleep disrupts circadian regulation of the human transcriptome. *Proceedings of the National Academy of Sciences USA*, Vol. 111: E682–E691.

[53] Kyle, G.H., Cecilia, N., Benjamin, M., Frank, A., Adriana, M., and Andrew, F., 2019. Bright lights in the big cities: migratory birds' exposure to artificial light. *Frontiers in Ecology and the Environment*, Vol. 17(4): 209–214.

[54] Li, D., Stucky, B.J., Deck, J., Baiser, B., and Guralnick, R.P., 2019. The effect of urbanization on plant phenology depends on regional temperature, *Nature Ecology & Evolution*, Vol. 3: 1661–1667.

CHAPTER 9

Smart Nanomaterials Usage for Artificial Skydomes

Ajit Behera

National Institute of Technology

Mantu Kumar Mahalik

Indian Institute of Technology

CONTENTS

9.1 INTRODUCTION

Skydome is a hemisphere or sometimes sphere with a defined pattern on its inner face for simulating the sky or having a similar backdrop around a three-dimensional play-field. Skydomes are popularly used in coliseum, stadiums, amphitheaters, rinks, an arena, ballpark, grandstand, colosseum, civic center, fieldhouse, recreation ground, Metrodome, terrasse, Wembley, clubroom, bandshell, gymnasium, etc. Traditional skydomes as well as other light systems are not equipped to control the intensity of light so the circulation of excess light in the premises fuels the light pollution [1–3]. These categories of pollution are

DOI: 10.1201/9781003185109-9

Remote area sky view **Developed area sky view**

FIGURE 9.1 Differences can be visualized from a rural-area sky and an urban-area sky due to light pollution.

present all the time, however, its effects are highlighted in the nighttime in contrast to the darkness. Less than a decade, everybody was able to observe an eye-catching starry sky at night. Currently, most of the people on earth are unable to observe the Milky Way where they are living. A simple effect of light pollution has been shown in Figure 9.1. Skyglow is a responsible factor for which the urban area population could not see thousands of stars as a comparison with unpolluted sky in rural areas. It is reported that approximately 99% of the people in the United States and Europe cannot sight the beauty of the sky in a night environment. This shows that people in urban areas cannot see the Milky Way because of light pollution, in which one part is skydome illumination. Moreover, manmade light in the night hour is not only interfering with our sight of the natural sky, but it also negatively impacts our environment, our health, and our ecosystem stability [4]. The "World Atlas of Artificial Night Sky Brightness," presented that, 80% of the earth's citizen lives under skyglow [5]. This further shows that it is essential to study the effect of skydomes on the economy as well as on the natural environment as many people in the 21st century suffer from light pollution due to skydomes.

However, the region affected by human-originated lighting continues to expand around the globe. As per the main aftereffect of urbanization, light pollution is condemned for compromising night crepuscule, human and plant health, and destroying aesthetic environments [6]. Dark in the appropriate place is always required in night time for peaceful sleeping and relaxing as well as it requires for the plant ecosystems. "Dark sky" does not mean "dark ground;" it is the dark sky without emitting the excess light from the ground. Light pollution refers to any unplanned utilization of lighting systems in the nighttime as well as the daytime and is detected as an issue in all levels of our community [7]. Light pollution adversely impacts our surroundings, wildlife activities, and the value of individuals. However, a growing body related to the illuminated night sky is proportional to various negative effects including [8,9]:

- Increase in energy consumption,
- Mistreat human health,

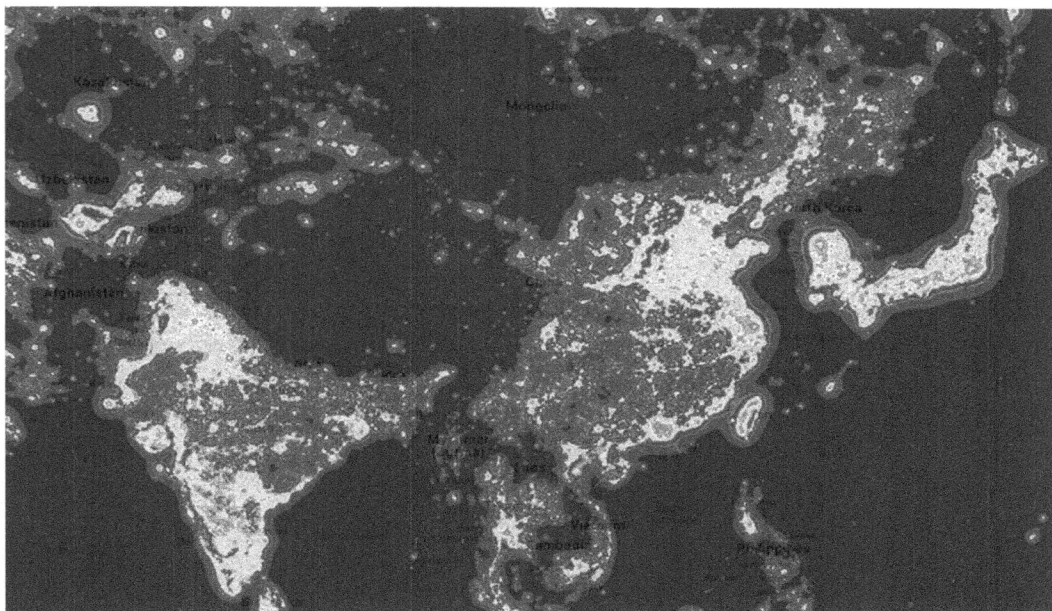

FIGURE 9.2   Lack of clear view from the satellite due to light pollution with air pollution.

- Affecting the animal normal life,

- Disrupting the ecosystem,

- Increasing misdemeanor and safety threat,

- No clear view can be collected by satellite (Figure 9.2).

Every day needs a night. Light pollution influences every community. Unfortunately, light pollution is growing exponentially. The lives of humans have originated according to the rhythms of the natural light/dark rotation of day and night. The expansion of artificial light means that no one encounters real dark nights. It affects our value of life by blocking our approach to the amazement of a pretty night sky [10]. Scientists indicated that artificial light at night can adversely impact an individual's health by enlarging the risk of stoutness, sleep disorders, stress, interrupted hormone and metabolism, diseases that develop in the thyroid, testicles, ovaries, and adrenal glands, etc. [11,12]. A growing number of researchers, environmental specialists, building planners, and smart city concerns are adopting steps to protect the actual night. Each of them tries to implement concrete solutions to oppose light pollution locally, nationally, and internationally [13,14]. All around us, this unintended pollutant wreaks silent havoc. In addition to the effects on animals and plants, excess light is also very vulnerable to mass destruction. In most cases, artificial light in the nocturnal surrounding is illuminated enough to elicit physiological responses that affect their natural phenomena, reproduction habit, and food consumption chain. The characteristic activity and ecological activities of herbivores and pollinators are also greatly affected by manmade light [15,16]. Constant exposure to light affects a plant's photosynthesis, as the plant requires a balance of sunlight and darkness to survive. Skyglow effects, in turn, can affect agricultural production rates, particularly in agricultural areas near major urban centers [17].

Therefore, realizing the ecological reactions of manmade light at night is important to detect the full consequence of individual activities on ecosystems. The skyglow is usually light pollution and is the unpolarized light that adds to the moonlight, decreasing the polarization wave. Human being incapable to recognize this motif, but some arthropods can. It is believed that navigation of many nocturnal organisms is possible with the help of polarized signals of scattered moonlight [18]. As the skyglow is a largely unpolarized wave, it can overwhelm the moon's weaker signal, generating an uncomfortable situation that results in an interrupted and impossible navigation. Fainter views such as zodiacal light and the Andromeda galaxy are almost impracticable to see using telescopes [19]. Fireflies, for example, gradually decreased in number and most of the region completely vanish due to intense artificial illumination at night. Night migratory birds along with other nocturnal creatures have also lost their course after being misguided by the illuminated sky [20]. Each year, thousands of migratory and shorebirds are finished off by interrupting human-made illumination at night.

Light pollution is alarming a severe effect on the aquatic ecosystems due to growing the risk of dangerous algal blooms. Illuminated lighthouse, cruise ships light, fishing fleets, cruise ships, and offshore platforms light disrupt the world's oceans with artificial night-time lighting. Skyglow made a major problem for astronomers and environmental researchers due to a decrease in contrast in the night sky for the exact point of vision. Darkness cycles are crucial for repair or/and recovery from daytime surrounding stress. Ground-level ozone is an important worldwide pollutant generated by photochemical reactions of precursors (NO_x and hydrocarbons) associated with road traffic. Ground-level ozone affects severely vegetation, including foliar damage. More leaf damage can be observed in northern latitude plants, a phenomenon attributed to long day lengths interfering with repair and defense processes. Research on these points proved that the continuous artificial light at night, estimated the levels of 1 $\mu mol/m^2s^1$ PPFD (Photosynthetic Photon Flux Density) by the use of fluorescent lamp (approximately 74 lux/lamp) and roadside vegetation under various high-intensity lights), produced a significant increase in foliar damage

FIGURE 9.3 Both the skydomes are causing light pollution that is responsible for the severe negative impact on the ecosystems.

because of the ozone in clover species. This cycle is likely repeated by the phytochrome pathway, as it is introduced in overnight exposure to red light and returns to far-red light by the subsequent exposure. The collective outcome of ozone damage and nighttime illumination on vegetation are severe [21]. Figure 9.3 is showing the amount of light that losses constantly in a particular area and that affects the surroundings negatively. In addition, the mutual expansion of both light pollution and air pollution gives a warning to the night sky.

9.2 DOES LIGHT POLLUTION AFFECT THE ECONOMY?

A light source that releases excess light or shines when it is not required is wasteful. Generally, the light projected in the skydome is directly upward released. It was found that more than 50% of used light is emitted out of having no use and is considered as light waste. Wasting the light is equivalent to wasting electric energy and has enormous economic and environmental consequences (Figure 9.4) [22,23]. To generate this extra-emitted energy, coal-fueled power plants releases around 15 million tons of CO_2 to surroundings per year causing air pollution, in addition, coal residue causes soil and water contamination [24]. As per the International Dark-Sky Association (IDA) estimation, 1/3 of all lighting is wasted costs around $2.2 billion per year. In an average year in the U.S. alone, outdoor lighting uses about 120 TW-hours of energy [25,26]. IDA estimated about 30% of outdoor lighting in the U.S. are wasted due to unshielded light system. That adds up to $3.3 billion and releases 21 million tons of CO_2 annually. To compensate for this CO_2 emission from light pollution, it should be planted 875 million trees annually. Planting more trees not only absorbs CO_2 emissions available in the atmosphere due to lightning and human activities but also provides us clean air for breathing in our day-to-day life. Moreover, many people who depend on nature get benefits from planting trees in terms of food and wood. Hence, we should largely conserve the forest via planting more trees as it is a part of biodiversity and also provides biomass energy for the people that they often use for their cooking purpose. As long as the larger benefits that we receive from the forest, we should not downplay the biodiversity. This is because protecting biodiversity is central for an economy to prosper and bring environmental sustainability in the long run. Otherwise, economies can have no chance of sustainable growth and development owing to the lack

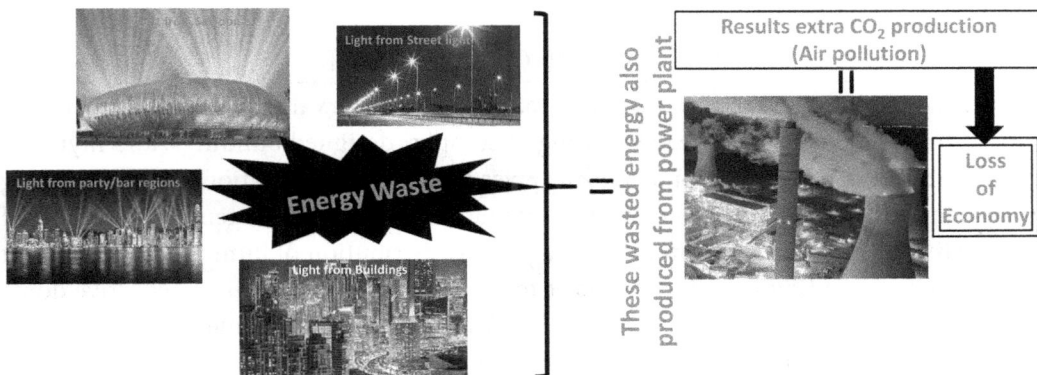

FIGURE 9.4 Representation of light pollution directly related to the economy.

of sustainable biodiversity and eco-friendly ecosystems. Light pollution is a phenomenon related not only to a specific pollution source but also to the broad accumulative effect of multi-source pollution levels [27]. Much of the outdoor illumination used at night is inefficient, too bright, poorly focused, poorly shielded, and most cases useless [28]. These light systems used to create it is wasted spreading it across the sky rather than focusing on the actual objects or/and areas people want to illuminate. If wasted energy is not managed well and light pollution level keeps on increasing in the atmosphere, then it can create climate change and global warming. The threat of climate change becomes catastrophic to the survival of the people and other species living on the surface and oceans. Therefore, environmental quality protection is essential for people and other species to survive. To have sustainable ecosystems, we should have enough energy efficiency and conservation from responsible consumers, producers, and governments of an economy. Perhaps, installing quality skydome light and outside illumination could cut energy utilization by 60%–70%, save billions of dollars and cut CO_2 emissions in the surface and atmosphere [29]. This shows that the policymakers in developed and developing economies need to consider quality skydome light while designing climate change mitigating policy.

9.3 EFFECT OF SKYDOME

Skydome produces light clouds including *Glare* (excessive brightness causes visual discomfort), *Skyglow* (brightening of the night sky), *Light trespass* (light falling where it is not intended or needed), *Clutter* (bright, confusing, and excessive groupings of light sources). There are two types of light scattering observed near skydome causing skyglow that is, Rayleigh scattering and aerosols scattering. Rayleigh scattering is more in short-wavelength light whereas aerosol scattering is less [30]. The sky looks blue in the daytime by the Rayleigh scattering and the presence of more aerosols turn the sky appearance less blue or whiter. Most of the urban region's sky is dominated by aerosol scattering due to industrial activities, thermal power generation, fly ash generation, automobiles, and farming [31]. Despite the strong wavelength dependence of Rayleigh scattering, the effect on skyglow for actual light sources is less. However, the short wavelengths scattered heavily result in greater extinction: the effects approximately become balanced when the observation point is near the light source [32].

9.4 ARTIFICIAL SKY AND CONTROL OF THE LIGHT

The astronomer has used artificial skies for many years for the validation of daylight calculations, in the assessment of the daylighting of complex designed buildings, and mythical study. Assessment of daylight for a building gives an energy conservation plan by reducing the demand for electric light. A hemispherical matt reflective dome is generally considered as the artificial sky where we can find out a proper distribution of luminance throughout the enclosed region (Figure 9.5). It is easier to get a uniform sky within a reflective dome than an overcast sky. This is possible due to reflection within the dome having the effect of reducing spatial luminance gradient by superimposing a uniform wash of indirect light onto the pattern of direct incident illumination. In artificial skies of the matt reflective dome type, a greater reflectance of the dome is generally observed as per the requirement.

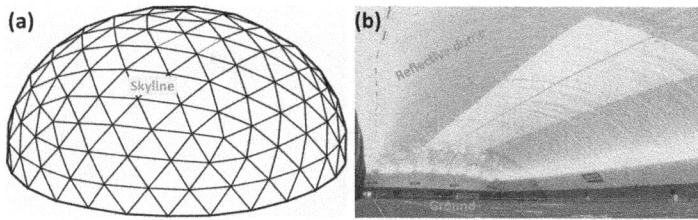

FIGURE 9.5   Artificial skydomes: (a) Hemisphere shape. (b) Cylindrical shape.

Less lamp power is needed to achieve the required illuminance which gives rise to an energy-saving mode and reduces the cooling issue. The effect of inner-reflection at any position in the hemisphere can be calculated as per Equation 9.1.

$$L = \rho E \qquad (9.1)$$

where L is the luminance, ρ is reflectance, and E is illuminance flux $= (E_D + E_I)$. E_D is the direct illuminance component and E_I is the indirect illuminance component. By Sumpner's principle, the indirect illuminance component is evenly distributed over the hemisphere surface.

Again, Matt's artificial skydomes are associated with parallax errors due to the luminance of each element of the sky, which is independent of the view direction. Same as the real sky luminance that depends on the view-direction, the luminance distribution of a non-uniform matt sky cannot be correct for more than one viewpoint. The parallax errors can be rectified, when the luminance of each element of the skydome is allowed to alter properly with respect to the view direction. The new pattern of sky luminance was no longer uniform but has a distribution that follows Equation 9.2.

$$L_\alpha = L_0 \left(1 + 2 Sin\alpha \right) \qquad (9.2)$$

where L_α is luminance at an angle α inside the dome and L_0 is the luminance at zero angles of the dome surface to the ground. The parallax error could well be intensified if the overcast sky distribution is replaced by a clear sky or an average sky characterized by sharper gradients of luminance. It could also be intensified if the matt sky surfaces were replaced by a matrix of spotlights with relatively concentrated intensity distributions. artificial skies on these lines can obtain higher freedom in luminance distribution only at the expense of a greater liability to parallax errors.

In addition to the narrow winding streets and central opening courtyard building, covering the streets or openings is a strategy that complements the traditional architectural plan. In residential areas, shading the facades of buildings is often achieved as a result of the cantilevered volumes of the projecting latticework or mashrabiyya. Shade is also brought to the commercial streets and tight alleyways by means of various types of urban roofing, including temporary shading devices. For a single building or courtyard arrangement. Illuminated shade is often obtained by architectural elements such as projecting roofs,

FIGURE 9.6   Shading effects in the daytime.

covered loggias, open galleries, and supplementary plants [33]. Skydome is also required in the daytime in various high-stair buildings for the uniform illumination of the light. Shade-prone area in the daytime has been shown in Figure 9.6.

The concept of an artificial sky was nucleated from the heliodon's constraint to provide a steady illumination in its surrounding. An artificial sky is generally applied for the architectural design to investigate daylight in buildings and rooms. Architecture engineers and researchers, lighting engineers, lighting designers, automobile and aircraft engineering use the simulated illumination equipment for several intentions. Different equipment is adopted in the architecture laboratories and natural lighting investigation laboratories. Illumination system designers and engineers use the virtual sky to calculate the light quality, to study the visibility of parts in the cockpit in automotive and aerospace engineering. Generally, indoor lighting of buildings is investigated using physical models at the design stage by observing and evaluating physical models of lighting conditions with the natural sky, but luminance varies frequently and consistent results are strenuous to get, hence virtual sky made the standard route to estimate the entrance of the light in day time [34]. The virtual sky can reproduce average and analytical skies and is not constrained by the natural sky's weather context. There are three routes to form skylight: direct illumination, reflection, or diffusion. Normally, the virtual sky has spherical shapes. In most cases, the artificial sky integrates with an artificial sun to replicate sunlight. By calculating the daylight input using a virtual sky, architecture and engineers can minimize energy use by synchronizing the light system. Daylight analysis supports the planning of passive houses, zero-energy houses, and the planning environmentally friendly buildings. To counteract legibility issues caused by glare and screen discoloration under ambient light conditions in automotive displays, the artificial sky provides a bright environment that permits designers and engineers to control the area in demand [35].

9.5 TYPE OF SKYDOME

9.5.1 Mirror Box Skydome

A mirror box is a virtual sky made up of an illuminated roof and reflector (mirror) walls, applied for reproducing a flat or overcast sky. The walls of the room are covered with vertically placed flat mirrors that generate a luminous roof image through reflection and counter-reflection. A uniform luminance distribution is generated in the mirror box from the light reflections of the mirror walls and a corresponding calculation of the CIE standard overcast sky is simulated. The illuminator is the white diffuser, which is backlit by multiple lamps to diffuse the light throughout the enclosure using sensors. At CEPT University, an artificial mirror box for daylight analysis is installed in the laboratories of the Advanced Buildings and Energy Research Center. In the University Living Lab for the Net Zero Energy Building, the analysis room consists of an artificial mirror box skydome to do light investigation [36].

9.5.2 Reflectors Skydome

The reflecting skydome consists of a large reflective surface to reproduce a uniform sky. The dome illuminating system is shaped to simulate the sky patterns that differ from a normal overcast sky. The reflective surface is to skylight distributions and assesses daylight on a large scale equipped with a revolving tabletop. The artificial sun is equipped to replicate the sunlight. Reflective skydome liners are more adaptable during their use, as compared to mirror box skydomes. In 1981 The Lawrence Berkeley Laboratory, California, USA, made a 7.32 m diameter reflective dome designed to reproduce flat skies, overcast skies, and various brightness distributions in clear skies. The 1.5 m diameter solar simulator is used. The metal dome was held up by a 7-foot-high cylindrical plywood wall, allowing large models to be brought in and out through the large doors. A highly reflective white paint was sprayed on the inside surface to achieve reflectance up to 80%. The lighting system, consisting of highly efficient fluorescent lamps and ballasts, offers an illuminance of approx. 5,000 lx with an even sky, 3,500 lx with an overcast sky, and over 6,000 lx with regularly clear skies. Large-scale models up to 6 feet in diameter can be installed with the ability to rotate the entire facility. The Central Research Institute of Industrial Buildings, Perovo, Moscow, Russia investigated the artificial sky and lighting system of the new laboratory of the Central Research Institute of Building Physics, Moscow. A 9 m diameter skydome with 16 illuminators having similar luminance is associated with a 0.9 m diameter solar simulator and an out-of-sky parabolic mirror. The University of Michigan, Ann Arbor, MI, USA made a 9.2 m diameter virtual sky to measure overcast, steady, and clear sky conditions [37].

9.5.3 Illustration of Artificial Skydome

The artificial dome replicates the sky equipped to execute the scanning process at any time, anywhere on the earth. To reduce the cost and space, the artificial dome uses robotics and precise operating systems. The simulation results are only calculated by a computer screen after combining multiple simulations. It offers natural light simulations on large scale on a revolving stage with the help of a virtual sky and sun simulator. The virtual skydome was discovered in the early 1990s, since direct perception in the artificial dome is not possible, the tool is widely used by scientists and not made for an architect. Laboratory for Solar Energy and Building Physics of the EPFL LESO-PB, Vaud has developed a scanning sky simulator as a pattern for other sky simulators that allows exact reproduction of the light distribution of all types of the sky [38].

9.6 VARIOUS TYPES OF HELIODON

A heliodon is a system for controlling the angle between a target ground and a ray of light on a horizontal plane at given latitude and the sunlight. Heliodons are primarily used by building makers and the architecture profession. By adopting an ideal building on the heliodon's flat surface and controlling the angle between light and surface, the concerned professional can observe how the building would appear in the three-dimensional sunbeam at different dates and times of the day. Heliodons are made in such a way to mimic the latitude, calculating daytime, and date. They have specified north-south direction on their surface to orient the patterns. Some heliodons are very long, in a high ceiling for metro rail. In general, setting the date presents the greatest difficulty for designing the heliodon, whereas the light source presents the most usability issues. The parallel incoming rays of the sun are not easy to replicate with artificial light at a useful scale. Various types of heliodon are present such as manual tabletop heliodon, manual sun emulator heliodon, robotic heliodon with a fixed light source, and robotic heliodon with fixed model [39].

Manual Tabletop Heliodons are applied to analyze sun shading at any variation of latitude and time. In this heliodon, the stage is attached to a traditional table that rotates based on an architectural model. To replicate the time of the year, the single light source uses a ribbon marked with months of the year and attached to the edge of the door. Generally, this setup is used in interior spaces with lamps and exterior spaces with direct sunlight to get an accurate result. Outdoors, a sundial operates the orientation of the model stand. Department of Architecture at the University of Hong Kong applied a tabletop heliodon for the solar design of a larger-scale model. *The manual heliodon* is made up of a flat table with a large-scale model on top whereas the table lies stationary with only sun lamps in motion. It helps as a teaching tool for architects, planners, and developers. The heliodon is used to teach solar geometry and solar responsive design principles in science museums. Without reliance on external sky conditions, it is simple to evaluate sun shading analysis at any latitude. This type of heliodon is very intuitive to adjust and operate. Manual sun emulator heliodon is used in Texas Christian University, Texas [40,41].

Robotic Heliodon with Fixed Light Source is the most accurate sunlight simulator. It helps to evaluate scale models in an enclosure with a fixed light source with the support of a robotic platform. This heliodon is an auto-operated physical model that is accurately placed with the help of computers around two axes. The robotic heliodon can process frequency tests and evaluations on bigger and larger models than the manual ones to produce precise results for various investigations. Robotic heliodons are used individually or integrated with artificial skydomes for designing the lighting system as well as for research purposes. In addition to its use with the artificial sky, the integrated tool can replicate both the sun and the sky for higher precision with daylight analysis. The fixed scale model can be bigger and heavier models than the other types, which permit the source to go around the model to get the experimental results and perform presentations [42].

9.7 CONTROLLING THE SKYDOME LIGHT

As the skydome produces a huge amount of light waste, controlling the illumination is highly essential to protect the natural environment. It is easier to reduce the skydome light pollution by controlling its lightbulb. The night environment is affected much more due to light pollution, hence nighttime light precaution is highly important. Below is a list of various actions that are to be adopted to restrict light pollution in the atmosphere [43–46].

1. Eliminate the upward released unwanted decorative lighting systems.

2. Using light-emitting diode (LED) and compact fluorescent (CFL) can help to decrease the energy used and protect from the surrounding exposure.

3. Avoid blue lights at night. Higher color temperatures mean bluer light. Generally, the blue-rich white light is responsible for greater glare that interrupts the human vision and worsens the skyglow due to its significantly larger geographic reach. Therefore, to reduce the blue light illumination, it is recommended to use the warm or filtered LED (CCT < 3,000 K; S/P ratio < 1.2) light. As per the American Medical Association report in 2016, it is recommended to use the lighting with 3,000 K color temperature and below. Warm-colored bulbs are always preferred based on the Kelven chat (Figure 9.7), without compromising the comfort visibility.

4. Wireless sensor embedded lighting systems used for dimmers, motion sensors, and timers, can help to fix the optimum illumination levels to save more energy.

5. Outdoor lighting is shielded in such a way to minimize glare and light. Also, use the covered bulbs that will face downwards.

6. Switched off unwanted indoor lighting in no use place in the building. All parts of the working place should be covered with a wireless sensor system to switch off the light automatically in no-work areas.

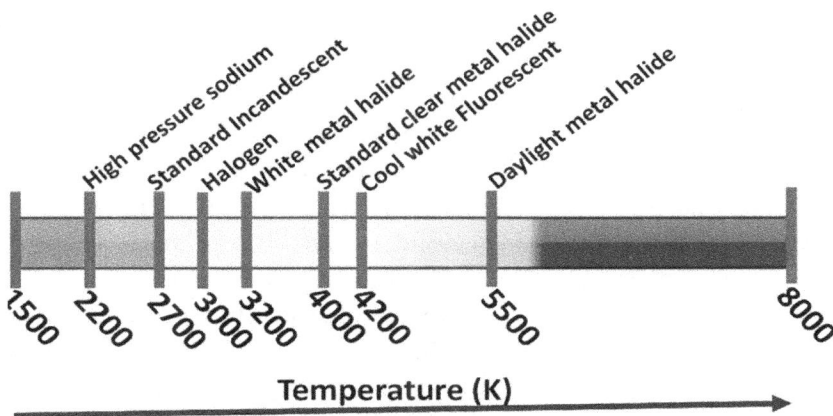

FIGURE 9.7   Kelven chat of various types of light with temperature.

7. When the moon shines bright, there is no need for outer lighting. Automated timers and systems with sensors are fixed to optimize the lights in addition to moonlight. An actuator is fixed to auto-switch off when the surrounding is naturally bright.

9.8 SKYDOME MATERIALS FOR IMPROVED LIGHTING

Several materials are used for illumination purposes or light absorption purposes to control the light from the source according to the demand, mainly focused on light pollution control. Surface plasmon resonant nanoparticles are used for the control glazing in the form of laminated self-adhesive films and bulk films [43]. Materials that can be used here are: transparent polymeric materials or spectral-selective glass plates, metal/oxide sandwiched multilayer thin films, stacking of duplex or triplet d-polymer thin layers, conductive nanoparticles dispersed polymer, iron particle doped glass, and NIR absorbing ceramic insulator. The angular selective thin films, laser grooved or cavity polymer panels, and holographic film can be used for directional control of solar and visible transmittance [44,45]. For illumination with reduced overall transmittance, a translucent polymer is used, while conventional micro-scale pigments are used in polymethyl methacrylate and polycarbonate, embedded particulates of TiO_2, ZnS, $CaCO_3$, and $BaSO_4$. For the illumination with greater transmittance, transparent refractive index microparticles (TRIMM) dopants are used in polymethyl methacrylate and polycarbonate sheets. For energy-efficient light transport systems, the materials used are hollow-mirrored light guides comprised of a cylindrical shaft with transparent entrance dome and lined with specular high reflective Ag or Al, or super high reflective bi-layer polymer stack, angular selective mirror light guide. Cylinders or flat ribbons Ultra clear solid polymer light guides are used to provide up to 40 m from the entrance to end light luminaires [46–48]. Multicolored glass or polymer fiber, single-cored ultra-clear flexible polymer, and fibers coupled to a 3-color stack of luminescent solar concentrators are also used. For controlled light distribution, a guide up to a distance of 20–30 m from the input is used with doped flexible polymer light guide TRIMM. There is uniform output from localized bright sources with low loss. Those are LEDs bonded with the light guides, TRIMM embedded polymeric films and light guides, full-color mixer special short light guides that emit white light without color spots from multiple single-color LEDs. In most cases, nanoparticles doped polymer foils are used in windows [49,50]. These windows are transparent, with little or no haze at visible wavelengths, which cannot be obtained when microparticles larger than about 100 nm in diameter are used. An example of transmission spectra is for laminated glass with a 0.7 mm thick laminate embedded with spherical nanoparticles of $In-SnO_2$ with 0.2% by weight. It is compared to the spectrum of clear laminated glazing and a general theoretical model curve. Other nanoparticles that are investigated for similar solar control glazing contain $SbSnO_2$ and LaB_6. Layers of LaB_6 doped laminate films are already used in office and shopping mall windows in hot climates [51]. LaB_6, $InSnO_2$, and $SbSnO_2$ nanoparticles differ in the peak position of their surface plasmon absorption bands, which are observed at successively longer wavelengths in the near-infrared. Thus, LaB6 blocks more solar energy but also slightly reduces transmission at visible wavelengths, which can be beneficial for glare reduction. $SbSnO_2$ and $InSnO_2$ nanoparticles have almost no visible effects

[52]. Homogenized dispersed nanoparticle with the matrix material is highly essential to maintain the transparency otherwise in the case of agglomerated nanoparticle in matrix yield haze appearance.

9.9 SMART NANOMATERIAL IN SKYDOME LIGHT TUNING

There are many ways to do filtration media and abortion media on the skydome surface. One way is the use of nanomaterial, which is just so dense that it mechanically blocks what it should. Commercially tuning the light intensity is crucial. In addition, the smart windows constructed on flexible transparent electrodes can alter the transmittance in response to a thermal or electric field to adjust the intensity of the light [53,54]. It is still a great challenge to manufacture the large-scale flexible transparent smart window. The careful selection of glass of the dome should be done based on the ordered glass parameters such as the transmission number of visible light, coefficient of reflection, coefficient of transmission, coefficient of absorption, coefficient of heat transmission, coefficient of thermal expansion, glass thickness, weight, maximum size, wind resistance, and corrosion resistance [56]. Advanced nano-film on the dome inner surface can reduce light pollution [57]. Depending on the type of film, visual light transmittance can be reduced by 45%–50% and can also decrease light pollution [58,59]. When light interacts with nanoparticles and other tiny structures, many interesting and even dramatic physical effects can occur. Silica nanorods are deposited at an angle of precisely 45° on top of a thin film of AlN to reduce light pollution [60–62]. Negative refraction of the dome inner material results in the transmitted wave emerging on the same side of the interface.

9.10 SUMMARY

Excess amounts of artificial light are considered pollution because of its severe effect on nature that expands gradually as a systemic disruption. Manufactured lighting of our planet is increasing rapidly and the growth of the intensity is about 2% per year, creating a significant impact on climate change. This chapter successfully describes the cause of light pollution from the skydome light and the effect of nanomaterials and smart materials embedded in the skydome construction to reduce the light pollution significantly. However, this chapter suggests that the governments, residential households, and business communities in developed and developing economies should use nanomaterials in the construction of quality skydome light. Eventually, this would help economies to reduce light pollution, thereby improving the quality of the environment, which is essential for providing a sustainable life-supporting system to the people and other species living on the planet. Thus, the policymakers in developed and developing economies need to consider quality skydome light as one of the important determinants in carbon dioxide emissions function while designing the climate change mitigating policy. Finally, this chapter is not without any limitation, which can be addressed in future analysis. Researchers working in the field of climate policy could consider skydome light as the key factor of understanding carbon dioxide emissions dynamics in developed and developing countries of the world.

REFERENCES

[1] Lukas Hosek and Alexander Wilkie. An analytic model for full spectral sky-dome radiance. *ACM Trans Graph* 31, 4, Article 95 (July 2012) 9. https://doi.org/10.1145/2185520.2185591.

[2] J.A. North and M.J. Duggin Stokes vector imaging of the polarized sky-dome. *Appl Opt* 36 (1997) 723–730. https://doi.org/10.1364/AO.36.000723.

[3] Joseph T. Kider, Daniel Knowlton, Jeremy Newlin, Yining Karl Li, and Donald P. Greenberg. 2014. A framework for the experimental comparison of solar and skydome illumination. *ACM Trans Graph* 33, 6, Article 180 (November 2014) 12. https://doi.org/10.1145/2661229.2661259.

[4] K. Pothukuchi. City light or star bright: a review of urban light pollution, impacts, and planning implications. *J Plann Lit* 36, 2 (2021) 155–169. https://doi.org/10.1177/0885412220986421.

[5] https://www.darksky.org/80-of-world-population-lives-under-skyglow-new-study-finds/#:~: text=A%20groundbreaking%20new%20study%20documenting, denizens%20experiencing% 20skyglow%20at%20night, 28.12.2021.

[6] Y. Gupta, A. Singh, A. Bansal, V. Bohara, and A. Srivastava. Deploying visible light communication for alleviating light pollution. 2020 IEEE International Conference on Advanced Networks and Telecommunications Systems (ANTS) (2020) 1–4. https://doi.org/10.1109/ANTS50601.2020.9342830.

[7] Karolina M. Zielińska-Dabkowska, Kyra Xavia, and Katarzyna Bobkowska. Assessment of citizens' actions against light pollution with guidelines for future initiatives. *Sustainability* 12, 12 (2020) 4997. https://doi.org/10.3390/su12124997.

[8] A.A.A. Hussein, E. Bloem, I. Fodor, et al. Slowly seeing the light: an integrative review on ecological light pollution as a potential threat for mollusks. *Environ Sci Pollut Res* 28 (2021) 5036–5048. https://doi.org/10.1007/s11356-020-11824-7.

[9] Sibylle Schroer, Benedikt J. Huggins, Clementine Azam, and Franz Hölker. Working with inadequate tools: legislative shortcomings in protection against ecological effects of artificial light at night. *Sustainability* 12, 6 (2020) 2551. https://doi.org/10.3390/su12062551.

[10] D.M. Dominoni. The effects of light pollution on biological rhythms of birds: an integrated, mechanistic perspective. *J Ornithol* 156 (2015) 409–418. https://doi.org/10.1007/s10336-015-1196-3.

[11] D. Crawford. Light pollution: changing the situation to everyone's advantage. *Symp Int Astron Union* 196 (2001) 33–38. https://doi.org/10.1017/S0074180900163806.

[12] S. Giavi, C. Fontaine, and E. Knop. Impact of artificial light at night on diurnal plant-pollinator interactions. *Nat Commun* 12 (2021) 1690. https://doi.org/10.1038/s41467-021-22011-8.

[13] Jiangtao Du, Xin Zhang, and Derek King. An investigation into the risk of night light pollution in a glazed office building: the effect of shading solutions. *Build Environ* 145 (2018) 243–259. https://doi.org/10.1016/j.buildenv.2018.09.029.

[14] M. Smith. Controlling light pollution in Chile: a status report. *Symp Int Astron Union* 196 (2001) 39–48. https://doi.org/10.1017/S0074180900163818.

[15] Falcón Jack, Torriglia Alicia, Attia Dina, Viénot Françoise, Gronfier Claude, Behar-Cohen Francine, Martinsons Christophe, and Hicks David. Exposure to artificial light at night and the consequences for flora, fauna, and ecosystems. *Front Neurosci* 14 (2020). https://doi.org/10.3389/fnins.2020.602796.

[16] Maja Grubisic, Abraham Haim, Pramod Bhusal, Davide M. Dominoni, Katharina M.A. Gabriel, Andreas Jechow, Franziska Kupprat, Amit Lerner, Paul Marchant, William Riley, Katarina Stebelova, Roy H.A. van Grunsven, Michal Zeman, Abed E. Zubidat, and Franz Hölker. Light pollution, circadian photoreception, and melatonin in vertebrates. *Sustainability* 11, 22 (2019) 6400. https://doi.org/10.3390/su11226400.

[17] C.A. Wyse, C. Selman, M.M. Page, A.N. Coogan, and D.G. Hazlerigg. Circadian desynchrony and metabolic dysfunction; did light pollution make us fat? *Med Hypotheses* 77, 6 (2011) 1139–1144. https://doi.org/10.1016/j.mehy.2011.09.023.

[18] W.H. Walker, J.C. Walton, A.C. DeVries, et al. Circadian rhythm disruption and mental health. *Transl Psychiatry* 10, 28 (2020). https://doi.org/10.1038/s41398-020-0694-0.

[19] D. Acuña-Castroviejo, G. Escames, C. Venegas, et al. Extrapineal melatonin: sources, regulation, and potential functions. *Cell Mol Life Sci* 71 (2014) 2997–3025. https://doi.org/10.1007/s00018-014-1579-2.

[20] Sergio A. Cabrera-Cruz, Ronald P. Larkin, Maren E. Gimpel, James G. Gruber, Theodore J. Zenzal, Jr, and Jeffrey J. Buler. Potential effect of low-rise, downcast artificial lights on nocturnally migrating land birds. *Integr Comp Biol* 61, 3 (September 2021) 1216–1236. https://doi.org/10.1093/icb/icab154.

[21] David F. Karnosky, John M. Skelly, Kevin E. Percy, and Art H. Chappelka. Perspectives regarding 50 years of research on effects of tropospheric ozone air pollution on US forests. *Environ Pollut* 147, 3 (2007) 489–506. https://doi.org/10.1016/j.envpol.2006.08.043.

[22] Terrel Gallaway, Reed N. Olsen, and David M. Mitchell. The economics of global light pollution. *Ecol Econ* 69, 3 (2010) 658–665. https://doi.org/10.1016/j.ecolecon.2009.10.003.

[23] T. Hunter and D. Crawford. Economics of light pollution. *Int Astron Union Colloq* 112 (1991) 89–96. https://doi.org/10.1017/S0252921100003778.

[24] Seth Dunn. Hydrogen futures: toward a sustainable energy system. *Int J Hydrogen Energy* 27, 3 (2002) 235–264. https://doi.org/10.1016/S0360-3199(01)00131-8.

[25] See: https://www.darksky.org/light-pollution/energy-waste/, 02.01.2022.

[26] See: https://swantree.org/light-pollution/, 02.01.2022.

[27] S.A. Cabrera-Cruz, J.A. Smolinsky, and J.J. Buler. Light pollution is greatest within migration passage areas for nocturnally-migrating birds around the world. *Sci Rep* 8 (2018) 3261. https://doi.org/10.1038/s41598-018-21577-6.

[28] See: https://www.darksky.org/light-pollution/, 03.01.2022.

[29] Baizhan Li and Runming Yao, Urbanisation and its impact on building energy consumption and efficiency in China. *Renewable Energy* 34, 9 (2009) 1994–1998. https://doi.org/10.1016/j.renene.2009.02.015.

[30] J. Brons, J. Bullough, and M. Rea. Outdoor site-lighting performance: a comprehensive and quantitative framework for assessing light pollution. *Light Res Technol* 40, 3 (2008) 201–224. https://doi.org/10.1177/1477153508094059.

[31] Chan Yong Sung. Examining the effects of vertical outdoor built environment characteristics on indoor light pollution. *Build Environ* 210 (2022) 108724. https://doi.org/10.1016/j.buildenv.2021.108724.

[32] Christian B. Luginbuhl, Paul A. Boley, and Donald R. Davis, The impact of light source spectral power distribution on sky glow. *J Quant Spectrosc Radiat Transfer* 139 (2014) 21–26. https://doi.org/10.1016/j.jqsrt.2013.12.004.

[33] http://www.nzdl.org/cgi-bin/library?e=d-00000-00---off-0cdl--00-0----0-10-0---0---0direct-10---4-------0-0l--11-en-50---20-about---00-0-1-00-0-4----0-0-11-10-0utfZz-8-10&cl=CL1.19&d=HASH28d1d3a01dbf303cded8b6.4&gt=2.

[34] https://en.wikipedia.org/wiki/Artificial_sky#:~:text=2.4%20Full%20dome-, Description, simulation%20device%20for%20various%20purposes.

[35] V. Costanzo, G. Evola, and L. Marletta, A review of daylighting strategies in schools: state of the art and expected future trends. *Buildings* 7 (2017) 41. https://doi.org/10.3390/buildings7020041.

[36] https://carbse.org/research/mirror-box-artificial-sky-constructed-at-carbse-for-daylight-analysis/

[37] D. Granados-López, M. Díez-Mediavilla, M. I. Dieste-Velasco, A. Suárez-García, and C. Alonso-Tristán. Evaluation of the vertical sky component without obstructions for daylighting in Burgos, Spain. *Appl Sci* 10 (2020) 3095. https://doi.org/10.3390/app10093095.

[38] M. Inanici. Evalution of high dynamic range image-based sky models in lighting simulation. *LEUKOS* 7, 2 (2010) 69–84. https://doi.org/10.1582/LEUKOS.2010.07.02001

[39] K.P. Cheung and S.L. Chung. A table top heliodon with a moving light source for use in an architect's office. *Int J Archit Sci* 3, 2 (2002) 51–60.

[40] Y. Sheng, T.C. Yapo, C. Young, and B. Cutler. Virtual heliodon: spatially augmented reality for architectural daylighting design, 2009 IEEE Virtual Reality Conference (2009) 63–70. https://doi.org/10.1109/VR.2009.4811000.

[41] T.J. Schoenemann, J.S. Haberl, and R.C. Hill. *Design of a Sustainable House for Residents of a Colonia in South Texas.* Energy Systems Laboratory, Texas A&M University (2002). https://hdl.handle.net/1969.1/4577.

[42] Y. Sheng, T. C. Yapo, C. Young, and B. Cutler, A spatially augmented reality sketching interface for architectural daylighting design. *IEEE Trans Visualization Comput Graph* 17, 1 (2011) 38–50. https://doi.org/10.1109/TVCG.2009.209.

[43] S. Schelm, G. B. Smith, P. D. Garrett, and W. K. Fisher. Tuning the surface-plasmon resonance in nanoparticles for glazing applications. *J Appl Phys* 97 (2005) 124314. https://doi.org/10.1063/1.1924873.

[44] https://sintef.brage.unit.no/sintef-xmlui/bitstream/handle/11250/2435706/Solar+Radiation+Glazing+Factors+for+Window+Panes+Glass+Structures+and+Electrochromic+Windows+in+Buildings+-+Measurement+and+Calculation+-+Revised+Version.pdf?sequence=3, 10.01.2022.

[45] https://esource.bizenergyadvisor.com/article/window-film, 10.01.2022.

[46] Geoffrey B. Smith. Materials and systems for efficient lighting and delivery of daylight. *Sol Energy Mater Sol Cells* 84, 1–4 (2004) 395–409. https://doi.org/10.1016/j.solmat.2004.02.047.

[47] https://www.energy.gov/energysaver/energy-performance-ratings-windows-doors-and-skylights, 10.01.2022.

[48] Helena Bulow-Hübe. Energy-efficient window systems: effects on energy use and daylight in buildings. Doctoral Dissertation. https://citeseerx.ist.psu.edu/viewdoc/download?doi=10.1.1.455.4670&rep=rep1&type=pdf.

[49] J. Arrue, A. Vieira, B. García-Ramiro, M.A. Illarramendi, F. Jiménez, and J. Zubia. Modelling of polymer optical fiber-based solar concentrators. *Methods Appl Fluoresc* 9, 3 (April 2021) 30. https://doi.org/10.1088/2050-6120/abfa6d. PMID: 33882464.

[50] Xiao Liu, Ying Xiong, Jiabin Shen, and Shaoyun Guo. Fast fabrication of a novel transparent PMMA light scattering materials with high haze by doping with ordinary polymer. *Opt Express* 23 (2015) 17793–17804.

[51] https://coek.info/pdf-electrochromics-for-smart-windows-oxide-based-thin-films-and-devices-.html, 12.01.2022.

[52] Masayuki Kanehara, Hayato Koike, Taizo Yoshinaga, and Toshiharu Teranishi. Indium tin oxide nanoparticles with compositionally tunable surface plasmon resonance frequencies in the near-IR region. *J Am Chem Soc* 131, 49 (2009) 17736–17737. https://doi.org/10.1021/ja9064415.

[53] https://www.iau.org/public/iyl/theme/lightpollution/, 05.01.2022.

[54] https://www.delmarfans.com/educate/basics/lighting-pollution/, 05.01.2022.

[55] https://www.darksky.org/light-pollution/light-pollution-solutions/, 05.01.2022.

[56] Franz Hölker, Janine Bolliger, Thomas W. Davies, Simone Giavi, Andreas Jechow, Gregor Kalinkat, Travis Longcore, Kamiel Spoelstra, Svenja Tidau, Marcel E. Visser, and Eva Knop. 11 pressing research questions on how light pollution affects biodiversity. *Front Ecol Evol* 9 (2021). https://doi.org/10.3389/fevo.2021.767177.

[57] Wei-Ran Huang, Zhen He, Jin-Long Wang, Jian-Wei Liu, and Shu-Hong Yu. Mass production of nanowire-nylon flexible transparent smart windows for PM2.5 capture. *iScience* 12 (2019) 333–341. https://doi.org/10.1016/j.isci.2019.01.014.

[58] H. Blala, L. Lang, S. Khan, et al. An analysis of process parameters in the hydroforming of a hemispherical dome made of fiber metal laminate. *Appl Compos Mater* 28 (2021) 685–704. https://doi.org/10.1007/s10443-021-09884-0.

[59] Véronique Georlette, Bette Sébastien, Brohez Sylvain, Pérez-Jiménez Rafael, Point Nicolas, and Moeyaert Véronique. Outdoor visible light communication channel modeling under smoke conditions and analogy with fog conditions. *Optics* 1, 3 (2020) 259–281. https://doi.org/10.3390/opt1030020.

[60] Stephen M. Pauley. Lighting for the human circadian clock: recent research indicates that lighting has become a public health issue. *Med Hypotheses* 63, 4 (2004) 588–596. https://doi.org/10.1016/j.mehy.2004.03.020.

[61] Rahul Rao and Cary L. Pint. Carbon nanotubes and related nanomaterials: critical advances and challenges for synthesis toward mainstream commercial applications. *ACS Nano* 12, 12 (2018) 11756–11784. https://doi.org/10.1021/acsnano.8b06511.

[62] Sang Wook Han, Il Hee Kim, Ju Hwan Kim, Hyun Ook Seo, and Young Dok Kim, Polydimethylsiloxane thin-film coating on silica nanoparticles and its influence on the properties of SiO$_2$–polyethylene composite materials. *Polymer* 138 (2018) 24–32. https://doi.org/10.1016/j.polymer.2018.01.036.

CHAPTER 10

Smart Materials and Devices for Human-Centric Lighting

Ram K. Gupta

Pittsburg State University

Tuan Anh Nguyen

Vietnam Academy of Science and Technology

CONTENTS

10.1 INTRODUCTION

Light pollution refers to the presence of unwanted, inappropriate, or excessive artificial lighting. It also refers to the presence of anthropogenic artificial light in otherwise dark conditions [1,2]. Thus, light pollution is used to refer not only to the outdoor environment and surroundings but also to artificial light indoors. For urban residents, light pollution competes with starlight in the night sky and interferes with astronomical observatories. As estimated, 83% of the world's people live under light-polluted skies and 23% of the world's land area is affected by skyglow [3].

As the main source of energy for all living organisms, we need light for life and well-being. Visible light provides/affects our vision and circadian rhythm, whereas invisible light could have an impact on human biological and mental health. Thus, inappropriate/poor lighting can be a health hazard (skin/eye disorders/diseases, headaches, sleep disorders, etc.). This book deals with smart lighting, which refers to the fully automated and

DOI: 10.1201/9781003185109-10

autonomous (monitored/controlled) lighting under various conditions as "human-centric lighting". Smart lighting is a revolutionary new field in photonics based on efficient light sources that are fully tunable in terms of such factors as spectral content, emission pattern, polarization, color temperature, and intensity (5D lighting). Smart lighting also serves in many aspects such as reduction of extra energy, reducing the source raw materials consumption leading to pollution reduction, significantly reducing the cost, and for convenience and security. The smart lighting market is estimated to grow from USD 13.4 billion in 2020 and is projected to reach USD 30.6 billion by 2025, at a CAGR of 18.0%. The need for energy-efficient products and the increasing adoption of smart photonics products in various applications are the key drivers for the growth of the smart lighting market globally. In addition, by using smart materials and various new technologies, such as the internet of things (IoT), big data, artificial intelligence (AI), lighting nowadays becomes smarter and more natural.

Human-centric lighting (HCL) refers to a human-centered lighting concept for artificial light indoors, which takes into account both the visual and the emotional and non-visual effects of lighting sources. Traditionally, HCL focuses on the adjustment of the lighting according to the natural course of daylight. Besides the main purpose of illuminance, the changing color temperature of the light is very important. With HCL, as the characteristics of natural light, the color spectrum of the light should be changed indoors throughout the day, such as the high blue components in the morning hours, warmer light at sunrise, increasing blue components at midday, decrease in blue components toward evening and high blue components with low illuminance after sunset (blue hour). Thus changing artificial light has a direct effect on the synchronization of the biological clock and the release of the sleep hormone melatonin from the pineal gland via photosensitive ganglion cells containing melanopsin. The below sections explore how advanced materials and devices can be used in the smart HCL.

10.2 INTERNET OF THINGS FOR HUMAN-CENTRIC LIGHTING

For the smart HCL, the color temperature and illuminance of indoor lighting can be autonomously controlled [4]. For this purpose, the color temperature and illuminance under the various indoor environment should be measured during specific periods to understand and evaluate the lighting environment around humans. Kim et al. [4] proposed the state of art measurement based on IoT including cloud computing. In their model, the color temperature and ambient light or illuminance are sensed, transferred to the cloud server, gathered as big data, and analyzed in cloud computing with Python. The obtained data can provide the correlation of color temperature and illuminance between the lighting of the office workplace around humans and the daylight around the window with analyzing big data. In this direction, for the space illuminance detection, Moon et al. [5] proposed a method for detecting an average illuminance value through a plurality of illuminance sensors. Since the conventional illuminance intensity detection method uses a single sensor, thus the uniformity of illuminance detection is deteriorated depending on the measurement position due to the narrow field of view (FOV) characteristic. The authors found the difference of average illuminance value was 12% using an illuminance sensor, 10.7% using five illuminance sensors, and 6.2% using an image sensor, as compared with the reference value using the color difference illuminometer.

Cupkova et al. [6] designed an intelligent lighting system that could detect humans from an IP camera, find faces, and detect emotion. This intelligent lighting system aimed to adjust the lights accordingly to the emotional result to improve the mood of people while taking into consideration the principles of color psychology and daytime. Their demonstration model was executed and assessed in the real-world environment such as in an office space (Figure 10.1). The wearable devices are used for the participants. The authors also collected the feedback from participants (via a form of a questionnaire) and evaluated the participants' well-being (via their subjective statements). They found that such systems eliminate the interference triggered by manual configuration, the dynamic modifications of lighting to create an enjoyable atmosphere, and the optimistic influence on output, fitness, and happiness of humans, among others.

The smart lighting system design usually focuses on the HCL control to provide energy-efficient and people mood rhythmic motivation lighting [7]. Yoon et al. [7] proposed the

FIGURE 10.1 Demonstration model installed in the real environment under two different light conditions. (Reproduced with permission [6]. Copyright (2019) the authors. Licensee SAGE. This article is an open access article distributed under the terms and conditions of the Creative Commons Attribution (CC BY) license (http://creativecommons.org/licenses/by/4.0/).)

HCL control using an indoor surveillance camera to improve human motivation and well-being in indoor environments. The authors used indoor surveillance camera video streams to predict the daylights and occupancy, occupancy-specific emotional features predictions using advanced computer vision techniques, and these human-centric features are transmitted to the smart building light management system. The smart indoor light management system connected with the IoT featured lighting devices and controls the light illumination of the objective human-specific lighting devices. In their experimental model, the RGB LED lighting devices are connected with IoT features open-source controllers in the network along networked video surveillance solution. These experimental results are verified with the automatic lighting control demon application, which is integrated with the Open CV framework-based computer vision methods, to predict the human-centric features. Then, from these estimated features, the lighting illumination level and colors are controlled automatically.

10.3 ADVANCED ORGANIC LIGHT-EMITTING DIODE LIGHTING FOR HUMAN-CENTRIC LIGHTING

An OLED is a light-emitting diode (LED), in which the emissive electroluminescent layer is a film of organic compound that emits light in response to an electric current. These organic films are situated between two electrodes and could be fabricated by using small molecules (SM) or polymers. OLED is a multilayered device, which typically consists of a transparent metal oxide anode, hole-transport layer (HTL), emissive layer (EML), electron-transporting layer (ETL), and metal cathode [8].

The OLED is fundamentally different from the LED. A p-n diode structure is used in LED and the conductivity of the host semiconductor is changed by doping. However, OLEDs do not use the p-n structures. Doping of OLEDs is used to increase radiative efficiency by direct modification of the quantum-mechanical optical recombination rate. Thus, in OLEDs, doping is used to determine the wavelength of photon emission. Adding mobile ions to an OLED could create a light-emitting electrochemical cell (LEC). An OLED display can be driven with a passive-matrix (PMOLED) or active-matrix (AMOLED) control scheme. In the PMOLED scheme, each row (and line) in the display is controlled sequentially, one by one, whereas AMOLED control uses a thin-film transistor (TFT) backplane to directly access and switches each pixel on or off, allowing for higher resolution and larger display sizes. Since the OLED emits visible light, the OLED display works without a backlight. Thus, OLEDs can display deep black levels and can be thinner and lighter than a liquid crystal display (LCD). OLEDs are mostly used as digital displays in model devices. However, currently, many works focus on the applications [9–11].

The first white OLED was fabricated in 1993 by Kido et al. [12] but its performance was very low as only 1 Lm/W with an external quantum efficiency of 1%. This device showed a very short lifetime of less than 1 day. In 2012, white OLED panel efficiency has reached 90 Lm/W at 1000 cd/m^2, and a tandem white OLED panel can be realized a lifetime of over 100,000 hours [8,13,14]. Wakimoto et al. [15] developed a 300×300 mm^2 size white OLED panel with 135 Lm/W and 40,000 hours lifetime at a higher brightness. For general lighting application, the light source requires a brightness of 3,000–5,000 cd/m^2, and a standard fluorescent tube achieves 70 Lm/W and 10,000 hours of life simultaneously. In addition, for a human eye-friendly purpose, the lighting source should possess a high color rendering index (CRI, Ra > 80) to reproduce the colors of an object [8]. Advanced OLEDs have been used in new emissive materials with different colors that can produce a white emission with a high Ra and a desired correlated color temperature (CCT).

Daskalakis et al. [16] converted an OLED from blue to white with Bradd Modes (Figure 10.2). It was observed that Bragg-conversion concept showed a role in the white-light generation. Color temperature regulation was accomplished by changing the thicknesses of the alternating superlattice layers and thereby the size of the dielectric distributed Bragg reflector. The device displayed a 20% improvement in external quantum efficiency and a 30-fold increase of on-shelf lifetime, as compared with the nonconverted blue OLED.

Besides, the white OLEDs are mercury-free illumination light sources and meet the requirements of the EU Waste Electrical and Electronic Equipment Regulation (WEEE) and Restriction of Certain Hazardous Substances (RoHS) directives. OLEDs are also an ideal low-temperature light source that having much cooler at around 30°C than those of the conventional light sources. Especially, OLEDs can shed clear light, such as transparent lighting panels and luminescent wallpapers. These fascinating features enable OLEDs as the next generation of an artificial lighting system. There are three main key technologies related to material chemistry in OLED: (i) Low operating voltage technology, (ii) Phosphorescent OLED technology, and (iii) Multi-photon emission (MPE) device technology [8].

Xia et al. [17] designed and fabricated OLED for HCL. The emission color of their single-cell OLED shifts in line with the broad black-body radiation and its tunability was controlled by an applied voltage. As reported, the CCT shift from 2,500 K to 15,000 K was intrinsically aroused from carrier mobility change in organic layers. The authors optimized

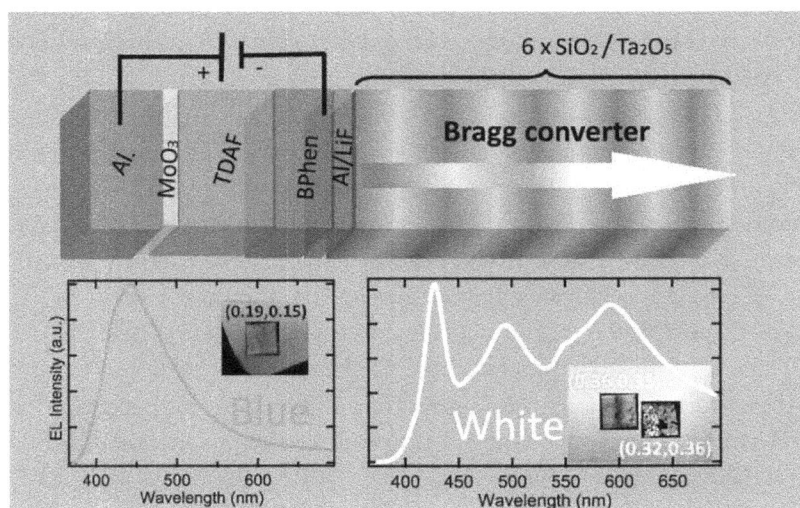

FIGURE 10.2 Schematic of the Bragg WOLED concept. (Reproduced with permission from [16]. Copyright (2019), American Chemical Society. This article is an open access article distributed under the terms and conditions of the Creative Commons Attribution (CC BY) license.)

their multi-layer OLED by calculation of carrier transport and recombination zone location. The designed OLED system could reach consistently high CRI (94–97) in a broad color range from orange to blue, providing potential in HCL applications. Park et al. [18] fabricated large-area OLED devices for lighting application. Their large-area OLED lighting panels can be integrated with a solar cell for power recycling or inorganic LEDs for emotional lightings. In the large-area (30×120 mm^2) opaque and transparent white OLED lighting panels, Park et al. [19] used grid-patterned Cr, Mo/Al/Mo, or Cu metal lines (0.15 mm in width) as auxiliary metal electrodes on an ITO anode. The authors studied the effect of the shape of the light-emitting area on luminance and heat distributions. As reported, a round-shaped OLED panel with a hexagonal metal grid exhibits highly homogeneous luminance and surface temperature distributions.

Chen et al. [20] fabricated a vertical quantum-dot light-emitting transistor (VQLET) using an organic TFT. It was observed that vertical structure provided a high current density. The vertically structured device showed a high performance with a maximum current efficiency of 37 cd/A. Choi et al. [21] investigated the slot-die coating of hybrid material structure for large-area OLED lighting panels. Their slot-die coating of an SM HTL was added with a polymer HTL to form the hybrid HTL. The authors found that OLED with the hybrid HTL showed higher luminous efficacy, compared to OLED with the SM HTL or the polymer HTL.

Light extraction from OLEDs for lighting applications through light scattering. For the large-area OLED light sources, their lifetime and overall efficiency can be increased by the optimization of light outcoupling. Bathelt et al. [22] used the scattering films on the viewer's side of the substrate to increase the outcoupling efficiency of OLEDs for lighting applications. Their experimental data showed an increase of outcoupling efficiency of about 22%. Eritt et al. [23] developed a large area manufacturing system for OLED lighting and signage applications. The devices achieved high efficiencies up to 31 cd/A on large area substrates which were comparable to devices on smaller substrates. The low light extraction efficiency of conventional OLEDs limited the lighting application of OLEDs. Therefore, to enhance the light extraction efficiencies that result in improving the luminous efficacy (LE, Lm/W) and external quantum efficiency (ηEQE, %) of OLEDs, Kim et al. [24] used the polymeric lighting extraction film (PLEF) with different geometrically profiled – the negatively nanostructured periodic semi-pyramid polydimethylsiloxane (PDMS) layers. These layers are placed directly on the backside of the green emissive bottom-emitting OLED (BE-OLED) glass substrates as an outcoupling enhancement PLEF. Through the combination of three different nanostructures on each green emissive BE-OLED, a maximum enhancement of up to 50% is achieved in the LE and ηEQE measured at the same brightness. This finding was 1.5 times higher than the reference green emissive BE-OLED without PLEF. Thus, the authors estimated the reduction of power consumption up to 30% from LE using the integrated PLEFs.

Chen et al. [25] fabricated the large-area white OLED lighting panels (72×72 mm^2) via dip-coating. Ide et al. [26] developed a high-efficiency white OLED device for lighting application. Their white OLED with high efficiency (37 Lm/W) and high-quality emission characteristics (CRI of 95) was realized by a two-unit structure with a fluorescent deep blue emissive unit and a phosphorescent green and red emissive unit. The half-decay lifetime of their white OLED at 1,000 cd/m^2 was over 40,000 h. A thin encapsulation structure for heat radiative produced a stable emission at high luminance of over 3,000 cd/m^2.

Bae et al. [27] indicated that the luminance uniformity of OLED lighting panels depends on OLED device structures of single emission layer (single-EML), 2-tandem, and 3-tandem. The experimental data showed that 3-tandem structures achieved the most uniform luminance distribution with non-uniformity of 4.1%, while single EML and 2-tandem structures accomplish 9.6%, and 6.4%, respectively, at ~1,000 cd/m^2. In addition, the simulation results ensured that a 3-tandem structure panel could allow being enlarged the panel size up to about 5,000 mm^2 for lower luminance non-uniformity than 10% without any auxiliary metal electrodes.

10.4 OPTIMIZING SPECTRA AND TUNABLE WHITE LIGHT IN LIGHT-EMITTING DIODE LIGHTING SYSTEMS FOR HUMAN-CENTRIC LIGHTING

Trivellin et al. [28] proposed an innovative white LED lighting system able to smoothly change CCT while sensibly changing the spectral power density in the blue region (460–480 nm). Their proposed tunable white LED system was high efficient (above 85 Lm/W), and was able to achieve high CRI (above 90) in a large CCT range (2,800–5,500 K), thus allowing a high comfort light in different indoor conditions. Thermal design data showed a maximum junction temperature of 85°C at 700 mA, with an equivalent thermal resistance junction to the ambient of 1.6 K/W for the series of LEDs used in the lighting system. The authors used a color mixing chamber and a parabolic reflector to design the optical emission of the light system, thus achieving a good color uniformity at different emission angles and 18° beam divergence. In addition, the specific control gear can be programmed to control the intensity and CCT of several LED lamps connected.

The proliferation of LED lighting has resulted in continued exposure to blue light, which may cause cataract formation, circadian disruption, and mood disorders [29,30]. Nikolova et al. [29] investigated the photobiological risk from the blue light of an HCL system using a tunable white LED light system. The authors presented the blue light assessment by taking into account the human eye possibilities and the subtended angle for color recognition in measurement geometry with the direction of the central axis of the observer's eye not perpendicular to the emitting surface. Blue light can be minimized in pursuit of HCL using a violet LED chip ($\lambda_{em} \approx 405$ nm) downconverted by red, green, and blue-emitting phosphors. However, there are a few phosphors efficiently can convert violet light to blue light [30]. Hariyani et al. [30] reported a new phosphor Na_2MgPO_4F: Eu^{2+}, which can be excited by a violet LED yielding an efficient, bright blue emission (Figure 10.3). Their prototype device using a 405 nm LED, Na_2MgPO_4F: Eu^{2+}, and a green and red-emitting phosphor could produce warm white light with a higher CRI than a commercial LED light bulb, while significantly reducing the blue component. Their finding demonstrated the capability of Na_2MgPO_4F: Eu^{2+} as a next-generation phosphor capable of advancing HCL.

Hu et al. [31] reported the spectral optimization for HCL by using a genetic algorithm and a modified Monte Carlo method. In their optimizing spectra for the 10-channel LED

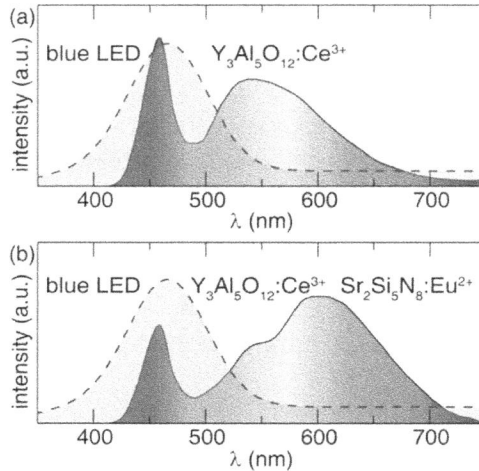

FIGURE 10.3 (a) Cold white light generated by combining a blue LED and yellow-emitting $Y_3Al_5O_{12}$:Ce^{3+}. (b) Addition of red-emitting $Sr_2Si_5N_8$:Eu^{2+}. (Reproduced with permission from [30]. Copyright (2021), American Chemical Society.)

system, the desired chromaticity, luminous flux, and circadian effect are compared and optimized. Galabov et al. [32] studied the LED luminaire for HCL for an accurate and smooth change in the radiated color temperature of the illuminator.

For smart HCL, the idealization of illuminance can be designed in various ways, such as widening the lamp's surface area, changing of luminaire's positions, and applying different methods to illuminate a surface. Duong et al. [33] studied the uplighting method and freeform optics using low-cost acrylic lenses. By changing lens shape and optimizing the factors which affect the illuminance such as the irradiation angle and the distance between the luminaire and the reflection surfaces, the authors found the most effective lens and its peaks due to each factor using the ray-tracing simulation to obtain results concerning high uniformity of illumination.

10.5 NANOSTRUCTURED LIGHT-EMITTING DIODE

A LED refers to the semiconductor light source that emits light when current flows through it. When electrons in the LED recombine with electron holes, they release energy in the form of photons. Thus, the color of the light is determined by the bandgap of the semiconductor. By the selection of different bandgaps in semiconductor materials, single-color LEDs can be obtained. In this case, emit light in a narrow band of wavelengths from near-infrared through the visible spectrum and into the ultraviolet range. White light LED can be obtained by using multiple semiconductors or a layer of light-emitting phosphor on the LED device. LEDs provide many advantages over incandescent light sources, such as lower power consumption, longer lifetime, improved physical robustness, smaller size, and faster switching.

In October 1962, TI et al. announced the first commercial LED product used a pure GaAs crystal to emit an 890 nm infrared light [34]. The first blue-violet LED using magnesium-doped gallium nitride was made in 1972. Blue LEDs consist of one or more InGaN quantum wells sandwiched between thicker layers of GaN. By varying the relative In/Ga fraction in the InGaN quantum wells, the light emission can be varied from violet to amber. Since 2001, GaN-on-silicon LEDs have been fabricated [34,35], to take advantage of existing semiconductor manufacturing infrastructure.

Regarding the white LEDs, there are two main methods to fabricate white light LED, the RGB LEDs, and phosphor-based LEDs. The first method is to use individual LEDs that emit three primary colors—red, green, and blue—and then mix all the colors to form white light. However, this method results in poor color rendering, because only three narrow bands of wavelengths of light are emitted. It also needs electronic circuits to control the blending of the colors. The other method is to use a phosphor material to convert monochromatic light (from a blue LED) to broad-spectrum white light, (similar to a fluorescent lamp). In this direction, the $Y_3Al_5O_{12}$:Ce (YAG or Ce^{3+}:YAG phosphor) cerium-doped phosphor (yellow phosphor) is suspended in the package or coated on the LED. This YAG phosphor can provide the white LEDs to appear yellow when off. A spectrum of this white LED shows blue light directly emitted by the GaN-based LED (~465 nm) and the more broadband Stokes-shifted light emitted by the Ce^{3+}:YAG phosphor (~500–700 nm). Otherwise, white LEDs may use other phosphors like manganese(IV)-doped potassium fluorosilicate (PFS) or other engineered phosphors [36].

Nano-LED refers to the nanostructured LED. Nano-LEDs are expected to have more advantages over conventional LEDs, such as lower power consumption, high efficiency, faster switching, more flexibility, and higher optical resolution. GaN nano-LED technology can be developed beyond solid-state lighting, with high efficiency, and across large areas, in combination with the possibility to merge optoelectronic-grade GaN nano-LEDs with silicon microelectronics in a hybrid approach [37]. Lee et al. [38] fabricated the flexible inorganic nanostructure LEDs using high-quality GaN/ZnO coaxial nanorod heterostructures grown directly on large graphene films (Figure 10.4). Then, these nanostructure LEDs/graphene films are readily transferred onto flexible plastic substrates, which could operate without significant deterioration of the LED performance.

Kashyout et al. [39] synthesized the nanostructured indium gallium nitride ($In_xGa_{1-x}N$) as active thin films for LED devices using both crystal growth and thermal vacuum evaporation techniques (Figure 10.5). As reported, white light generation based on $In_xGa_{1-x}N$ LED devices can be fabricated by simple and low-cost methods. Rishinaramangalam et al. [40] fabricated the electrically injected triangular-nanostripe core-shell semipolar III-nitride LEDs (TLEDs) using interferometric lithography and catalyst-free bottom-up selective-area metal-organic chemical vapor deposition (MOCVD). Their approach enables semipolar orientations on inexpensive, c-plane sapphire substrates, as compared with the planar growth on free-standing GaN substrates. As reported, the broad electroluminescence spectra, wavelength shift with increasing current density, and nonlinear light vs current characteristics are associated with the observed quantum-well nonuniformities.

Lin et al. [41] fabricated an Al oxide honeycomb nanostructure on the n-GaN surface of a thin-GaN LED. As reported, with the Al oxide honeycomb nanostructure, the total lighting output of thin-GaN LED was enhanced by 35%. The authors indicated that the

(a)

(b)

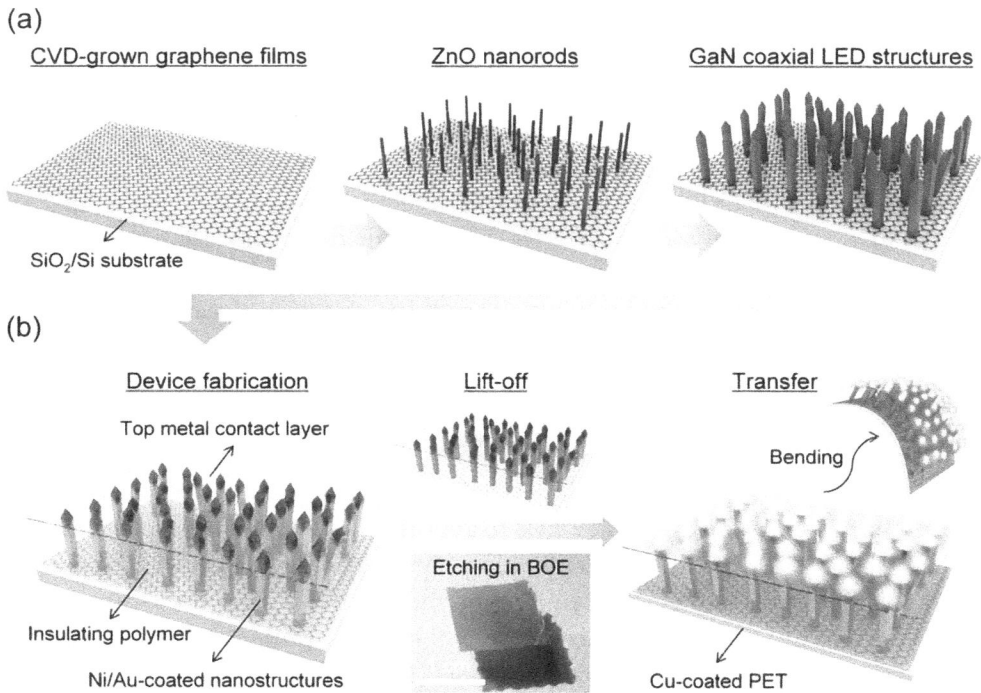

FIGURE 10.4 (a) Schematics of GaN/ZnO coaxial nanorod heterostructures growth. (b) Schematics of device fabrication. (Reproduced with permission from [38]. Copyright (2011), John Wiley and Sons.)

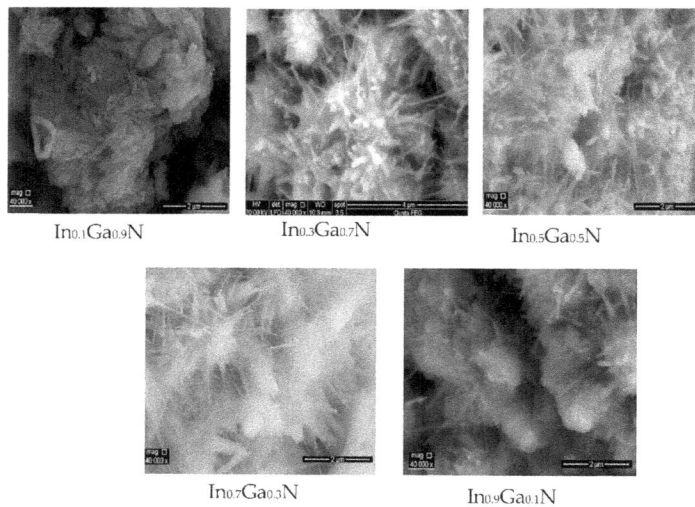

$In_{0.1}Ga_{0.9}N$ $In_{0.3}Ga_{0.7}N$ $In_{0.5}Ga_{0.5}N$

$In_{0.7}Ga_{0.3}N$ $In_{0.9}Ga_{0.1}N$

FIGURE 10.5 Field emission scanning electron microscopy (FESEM) for bulk alloys at $In_xGa_{1-x}N$ heated at 950°C ($0.1 \leq x \leq 0.9$). (Reproduced with permission from [39]. Copyright (2019), by the authors. Licensee MDPI, Basel, Switzerland. This article is an open access article distributed under the terms and conditions of the Creative Commons Attribution (CC BY) license (http://creativecommons.org/licenses/by/4.0/).)

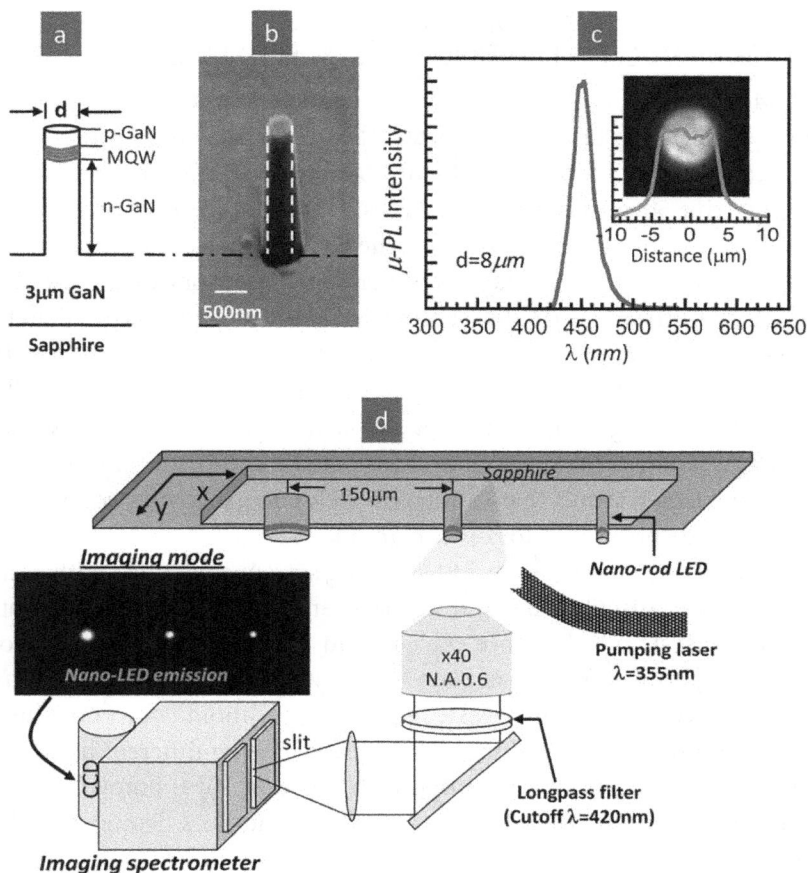

FIGURE 10.6 (a) A schematic of the nanorod LED layer structure. (b) An SEM image of a fabricated nanorod LED. (c) μ-PL spectrum of a nanorod LED. (d) A schematic of the μ-PL and imaging measurement setup. (Reproduced with permission [42]. Copyright (2011), American Chemical Society.)

networking nanowall of the Al oxide honeycomb structure acted as a waveguide to extract the light emitted to the outer medium effectively. Kuo et al. [42] reported an optically pumped nanorod LED with an ultrahigh extraction efficiency of 79% at $\lambda = 460$ nm without using either a back reflector or thin-film technology (Figure 10.6). The authors indicated that size reduction at nanoscale represented a new degree of freedom for alternating and achieving a more directed LED emission.

Fabrication of arrays of submicron light sources, that can be controlled individually, is very challenging. Bezshlyakh et al. [43] used nano-LED arrays that can directly address each array element and a self-pitch with dimensions below the wavelength of light. Their design has two geometries: a 6×6 array with 400 nm LEDs and a 2×32 line array with 200 nm LEDs. These nano-LEDs are applied as core elements of a novel on-chip super-resolution microscope. Kluczyk-Korch et al. [44] fabricated the individually switchable InGaN/GaN Nano-LED arrays as highly resolved illumination engines. Moreno et al. [45] used 200 nm GaN LED in the LED scanning transmission optical microscopy (nano-LED STOM). The authors demonstrated the possibilities for the miniaturization of on-chip-based microscopes.

10.6 NANOSENSORS, NANODEVICES, AND WIRELESS NETWORKS IN LED LIGHTING SYSTEM FOR HCL

Zhong et al. [46] used an arch-shaped flexible triboelectric nanogenerator (TENG) for instantaneously lighting up LEDs. As reported, the instantaneous output power of a single TENG device can reach as high as ~4.125 mW by finger typing, which is high enough to instantaneously drive 50 commercial blue LEDs connected in series. Tan et al. [47] used smart wireless sensors for personal control of the DC grid-powered networked LED lighting. In their work, the light-dependent resistor (LDR) sensors measured illuminance, whereas pyroelectric infrared-red (PIR) sensors detect the movements of inhabitants. Their experimental data showed that the proposed smart LED lighting system can attain about 44% energy saving as compared to the original AC fluorescent system. Similarly, Dikel et al. [48] used the LED lighting systems paired with high-resolution sensor systems for substantial energy savings. The authors demonstrated a combination of local occupancy and daylight harvesting features that resulted in 79% energy savings compared to typical prevailing energy code provisions for open-plan offices.

Huynh et al. [49] used a controllable LED lighting system embedded with ambient intelligence gathered from a distributed smart wireless sensor network (WSN) to optimize and control the lighting system to be more efficient and user-oriented. Their proposed smart WSN-based LED lighting system has been test-bedded and retrofitted into an existing workplace to conserve up to 10% of waste light energy without compromising the workplace lighting condition. For the indoor environments having different lighting levels due to daylight factors, Özçelik et al. [50] proposed a low-cost PV (photovoltaics) based and distributed sensor smart LED illuminating system. The authors demonstrated 72% more energy saving in comparison with the conventional LED illuminating system.

Chew et al. [51] designed an energy-saving controller for an intelligent LED lighting system. This energy-saving controller was capable of shaping the light output of an LED lighting system autonomously based on data received from sensors. The authors implemented an optimized smart algorithm on a controller to process the sensor feedback and employ pulse width modulation dimming to vary the brightness of the luminaire. In addition, a wireless sensor module was designed to provide accurate sensor feedback to the controller. The experimental data indicated that the controller achieved 55% energy savings in a continuous usage pattern environment and 62% energy savings in a discrete usage pattern environment under our test conditions.

Dongying and Wei [52] presented an intelligent dimming system based on LED lamps. In their work, a pyroelectric infrared detector and a photosensor are used to sample environment information. An STC-MCU (8-bit single-chip microcontroller) was used to process the information and generate pulse-width modulation (PWM) control signals to set the lamps into predefined illumination modes. Compared to traditional control modes, the system features lower power consumption and is suitable for aisles and stairwells. Cho et al. [53] developed a smart LED lighting system using a multi-sensor module and Bluetooth low energy technology. In their work, to monitor the environmental information such as ambient light intensity, temperature, and/or activity of humans or other objects, a multi-sensor module including an ambient light sensor, temperature sensor, and motion sensor was combined with a microcontroller. Thus LED lighting systems can be controlled automatically by collecting environmental data. In addition, through Bluetooth LE technology, users can monitor the environmental information on a smartphone and/or control the LED lighting system manually.

To automatically adjust the light intensity, save energy, and maintain user satisfaction, Magno et al. [54] proposed a low-cost, wireless, easy to install, adaptable, and smart LED lighting system. Their system combined motion sensors and light sensors in a low-power wireless solution using Zigbee communication. As reported the proposed smart lighting system reduced total power consumption by 55% during 6 months and up to 69% in the spring months.

Aldrich et al. [55] designed the polychromatic solid-state lighting controlled using a sensor network. The authors developed both a spectrally tunable light source and an interactive lighting testbed to study the effects of proposed systems that adjust in response to changing environmental lighting conditions and users' requirements of color and intensity. The authors indicated that it was possible to maximize luminous efficacy or the CRI for a given intensity and color temperature.

REFERENCES

1. Verheijen FJ (1985) Photopollution: artificial light optic spatial control systems fail to cope with. Incidents, causation, remedies. *Exp Biol* 44:1–44.
2. Cinzano P, Falchi F, Elvidge CD, Baugh KE (2000) The artificial night sky brightness mapped from DMSP satellite Operational Linescan System measurements. *Mon Not R Astron Soc* 318:641–657.
3. Falchi F, Cinzano P, Duriscoe D, Kyba CC, Elvidge CD, Baugh K, Portnov BA, Rybnikova NA, Furgoni R (20 21) The new world atlas of artificial night sky brightness. *Sci Adv* 2:e1600377.
4. Kim Y, Kim H, Jeon J, Kang TG (2017) IoT & Python based measurement and analysis of lighting color temperature and illuminance for humancentric lighting control. In: AIC Congress. AIC Congress, Korea, p. 161.
5. Seong-Jae M, Yeong-Seog L (2019) Image sensor module for detecting space illuminance in indoor environment. *J Korea Inst Inf Commun Eng* 23:771–778.
6. Cupkova D, Kajati E, Mocnej J, Papcun P, Koziorek J, Zolotova I (2019) Intelligent human-centric lighting for mental wellbeing improvement. *Int J Distrib Sens Networks* 15:1–9.
7. Hoon YS, Soo LK, Sang CJ, Vinayagam M, Woo LM, Gun WD, Uk KJ (2020) Indoor surveillance camera based human centric lighting control for smart building lighting management. *Int J Adv Cult Technol* 8:207–212.
8. Sasabe H, Kido J (2013) Development of high performance OLEDs for general lighting. *J Mater Chem C* 1:1699–1707.
9. Kamtekar KT, Monkman AP, Bryce MR (2010) Recent advances in white organic light-emitting materials and devices (WOLEDs). *Adv Mater* 22:572–582.
10. D'Andrade BW, Forrest SR (2004) White organic light-emitting devices for solid-state lighting. *Adv Mater* 16:1585–1595.
11. Chang Y, Lu Z (2013) White organic light-emitting diodes for solid-state lighting. *J Disp Technol* 9:459–468.
12. Junji K, Masato K, Katsutoshi N (1995) Multilayer white light-emitting organic electroluminescent device. *Science (80-)* 267:1332–1334.

13. Ide N, Yamae K, Kittichungchit V, Tsuji H, Ota M, Komoda T (2014) Development of extremely high efficacy white OLED with over 100 lm/W. *J Photopolym Sci Technol* 27:357–361.

14. Ohisa S, Takahashi N, Pu Y-J, Kido J (2014) Fabrication of light scattering structure by self-organization of a polymer: application to light out-coupling enhancement in OLEDs. *J Photopolym Sci Technol* 27:363–367.

15. Wakimoto T, Fukuda Y, Nagayama K, Yokoi A, Nakada H, Tsuchida M (1997) Organic EL cells using alkaline metal compounds as electron injection materials. *IEEE Trans Electron Devices* 44:1245–1248.

16. Daskalakis KS, Freire-Fernández F, Moilanen AJ, van Dijken S, Törmä P (2019) Converting an organic light-emitting diode from blue to white with Bragg modes. *ACS Photonics* 6:2655–2662.

17. Xia Yiren, Wan OY, Cheah Kok Wai (2016) OLED for human centric lighting. *Opt Mater Express* 6:1905–1913.

18. Park JW, Shin DC, Park SH (2011) Large-area OLED lightings and their applications. *Semicond Sci Technol* 26:34002.

19. Park J, Lee J, Noh Y-Y (2012) Optical and thermal properties of large-area OLED lightings with metallic grids. *Org Electron* 13:184–194.

20. Chen Q, Yan Y, Wu X, Lan S, Hu D, Fang Y, Lv D, Zhong J, Chen H, Guo T (2019) High-performance quantum-dot light-emitting transistors based on vertical organic thin-film transistors. *ACS Appl Mater Interfaces* 11:35888–35895.

21. Choi K-J, Lee J-Y, Shin D-K, Park J (2016) Investigation on slot-die coating of hybrid material structure for OLED lightings. *J Phys Chem Solids* 95:119–128.

22. Bathelt R, Buchhauser D, Gärditz C, Paetzold R, Wellmann P (2007) Light extraction from OLEDs for lighting applications through light scattering. *Org Electron* 8:293–299.

23. Eritt M, May C, Leo K, Toerker M, Radehaus C (2010) OLED manufacturing for large area lighting applications. *Thin Solid Films* 518:3042–3045.

24. Kim JY, Joo CW, Lee J, Woo J-C, Oh J-Y, Baek NS, Chu HY, Lee J-I (2015) Save energy on OLED lighting by a simple yet powerful technique. *RSC Adv* 5:8415–8421.

25. Chen Y, Wang J, Zhong Z, Jiang Z, Song C, Hu Z, Peng J, Wang J, Cao Y (2016) Fabricating large-area white OLED lighting panels via dip-coating. *Org Electron* 37:458–464.

26. Ide N, Tsuji H, Ito N, Matsuhisa Y, Houzumi S, Nishimori T (2010) White OLED devices and processes for lighting applications. In: Proc.SPIE. SPIE Photonics Europe, p. 772202.

27. Bae HW, Son YH, Kang BY, Lee JM, Nam H, Kwon JH (2015) Luminance uniformity study of OLED lighting panels depending on OLED device structures. *Opt Express* 23:30701–30708.

28. Trivellin N, Meneghini M, Ferretti M, Barbisan D, Lago MD, Meneghesso G, Zanoni E (2015) Effects and exploitation of tunable white light for circadian rhythm and human-centric lighting. In: 2015 IEEE 1st International Forum on Research and Technologies for Society and Industry Leveraging a better tomorrow (RTSI). pp. 154–156.

29. Nikolova K, Petrinska I, Ivanov D, Pavlov D (2019) Investigation of the photobiological risk from blue light of a human-centric lighting system—case study. In: 2019 11th Electrical Engineering Faculty Conference (BulEF). pp. 1–4.

30. Hariyani S, Brgoch J (2021) Advancing human-centric LED lighting using Na2MgPO4F:Eu2+. *ACS Appl Mater Interfaces* 13:16669–16676.

31. Hu W, Davis W (2020) Spectral optimization for human-centric lighting using a genetic algorithm and a modified Monte Carlo method. In: OSA Advanced Photonics Congress (AP). Optical Society of America, p. PvM2G.4.

32. Galabov B (2019) Experimental study of LED luminaire for Human Centric Lighting. In: 2019 11th Electrical Engineering Faculty Conference (BulEF). pp. 1–4.

33. Thi Giang D, La TL, Quoc Tien T, Hong Duong P, Tong QC (2020) A simple designed lens for human centric lighting using LEDs. *Appl Sci* 10:343.

34. Dadgar A, Alam A, Riemann T, Bläsing J, Diez A, Poschenrieder M, Strassburg M, Heuken M, Christen J, Krost A (2001) Crack-free InGaN/GaN light emitters on Si(111). *Phys Status Solidi* 188:155–158.

35. Dadgar A, Poschenrieder M, Bläsing J, Fehse K, Diez A, Krost A (2002) Thick, crack-free blue light-emitting diodes on Si(111) using low-temperature AlN interlayers and in situ SixNy masking. *Appl Phys Lett* 80:3670–3672.

36. Dutta PS, Liotta KM (2017) Full spectrum white LEDs of any color temperature with color rendering index higher than 90 using a single broad-band phosphor. *ECS J Solid State Sci Technol* 7:R3194–R3198.

37. Wasisto HS, Prades JD, Gülink J, Waag A (2019) Beyond solid-state lighting: Miniaturization, hybrid integration, and applications of GaN nano- and micro-LEDs. *Appl Phys Rev* 6:41315.

38. Lee C-H, Kim Y-J, Hong YJ, Jeon S-R, Bae S, Hong BH, Yi G-C (2011) Flexible inorganic nanostructure light-emitting diodes fabricated on graphene films. *Adv Mater* 23:4614–4619.

39. B. Kashyout AE, Fathy M, Gad S, Badr Y, A. Bishara A (2019) Synthesis of nanostructure InxGa1−xN Bulk alloys and thin films for LED devices. *Photonics* 6: 44.

40. Rishinaramangalam AK, Nami M, Fairchild MN, Shima DM, Balakrishnan G, Brueck SRJ, Feezell DF (2016) Semipolar InGaN/GaN nanostructure light-emitting diodes onc-plane sapphire. *Appl Phys Express* 9:32101.

41. Lin CL, Chen PH, Chan C-H, Lee CC, Chen C-C, Chang J-Y, Liu CY (2007) Light enhancement by the formation of an Al oxide honeycomb nanostructure on the n-GaN surface of thin-GaN light-emitting diodes. *Appl Phys Lett* 90:242106.

42. Kuo M-L, Kim Y-S, Hsieh M-L, Lin S-Y (2011) Efficient and directed nano-LED emission by a complete elimination of transverse-electric guided modes. *Nano Lett* 11:476–481.

43. Bezshlyakh DD, Spende H, Weimann T, Hinze P, Bornemann S, Gülink J, Canals J, Prades JD, Dieguez A, Waag A (2020) Directly addressable GaN-based nano-LED arrays: fabrication and electro-optical characterization. *Microsys Nanoeng* 6:88.

44. Kluczyk-Korch K, Moreno S, Canals J, Diéguez A, Gülink J, Hartmann J, Waag A, Di Carlo A, Auf der Maur M (2021) Individually switchable InGaN/GaN nano-LED arrays as highly resolved illumination engines. *Electronics* 10:1829.

45. Moreno S, Canals J, Moro V, Franch N, Vilà A, Romano-Rodriguez A, Prades JD, Bezshlyakh DD, Waag A, Kluczyk-Korch K, Auf der Maur M, Di Carlo A, Krieger S, Geleff S, Diéguez A (2021) Pursuing the diffraction limit with nano-LED scanning transmission optical microscopy. *Sensors* 21:3305.

46. Zhong J, Zhong Q, Fan F, Zhang Y, Wang S, Hu B, Wang ZL, Zhou J (2013) Finger typing driven triboelectric nanogenerator and its use for instantaneously lighting up LEDs. *Nano Energy* 2:491–497.

47. Tan YK, Huynh TP, Wang Z (2013) Smart personal sensor network control for energy saving in DC grid powered LED lighting system. *IEEE Trans Smart Grid* 4:669–676.

48. Dikel EE, Newsham GR, Xue H, Valdés JJ (2018) Potential energy savings from high-resolution sensor controls for LED lighting. *Energy Build* 158:43–53.

49. Huynh TP, Tan YK, Tseng KJ (2011) Energy-aware wireless sensor network with ambient intelligence for smart LED lighting system control. In: IECON 2011–37th Annual Conference of the IEEE Industrial Electronics Society. pp. 2923–2928.

50. Özçelik MA (2017) The design and implementation of PV-based intelligent distributed sensor LED lighting in daylight exposed room environment. *Sustain Comput Informatics Syst* 13:61–69.

51. Chew I, Kalavally V, Oo NW, Parkkinen J (2016) Design of an energy-saving controller for an intelligent LED lighting system. *Energy Build* 120:1–9.

52. Dongying J, Wei W (2010) The intelligent system for LED lighting based on STCMCU. In: 2010 International Conference on Computer and Communication Technologies in Agriculture Engineering. pp. 445–447.

53. Cho YS, Kwon J, Choi S, Park D-H (2014) Development of smart LED lighting system using multi-sensor module and bluetooth low energy technology. In: 2014 Eleventh Annual IEEE International Conference on Sensing, Communication, and Networking (SECON). pp. 191–193.
54. Magno M, Polonelli T, Benini L, Popovici E (2015) A low cost, highly scalable wireless sensor network solution to achieve smart LED light control for green buildings. *IEEE Sens J* 15:2963–2973.
55. Aldrich M, Zhao N, Paradiso J (2010) Energy efficient control of polychromatic solid state lighting using a sensor network. In: Tenth International Conference on Solid State Lighting. SPIE Optical Engineering + Applications, p. 778408.

CHAPTER 11

Wireless Nanosensors Network for Light Pollution Control

Ajit Behera, Santos Kumar Das, and Ramakrishna Biswal

National Institute of Technology

CONTENTS

11.1 INTRODUCTION

Nanosensors are one of the most advanced applications of nanotechnology. Nanosensor works in a nanonetwork to provide errorless results. Nanonetwork can be defined as a set of coupled nanomachines (machine dimension in the range of 1–100 nm), which helps to execute very simple work such as sensing, computing, storing information, and actuation. A nanosensor is a device that can be influenced by environmental stimuli at the nanoscale range by showing the peculiar behavior of nanomaterials to identify and calculate each activity. Although, the range of sensing by a single nanosensor is bounded to its enclosed surrounding with its extremely small dimension. Nanonetwork can expand its proximity by enhancing the capabilities of single nanomachines, including the complexity and the range of operation by permitting them to coordinate, share and fuse information [1]. Nanonetworks captured new applications of nanotechnology in various industries (such as biomedical, military, consumer goods, etc.) and environmental research (such as air,

DOI: 10.1201/9781003185109-11

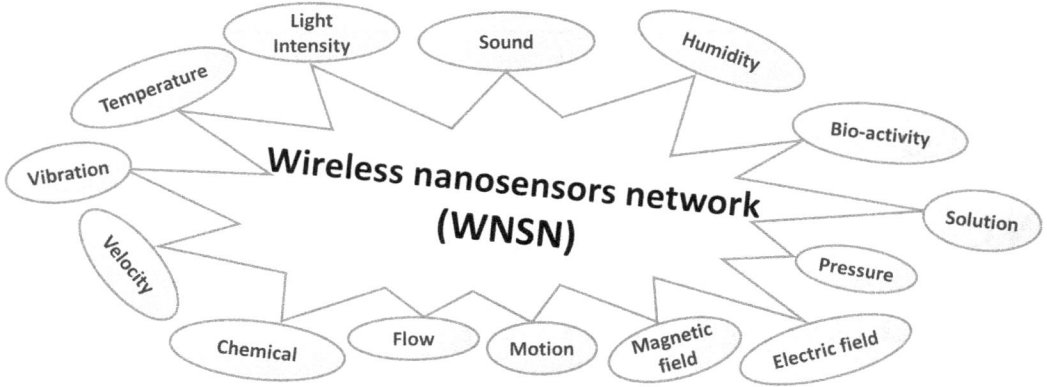

FIGURE 11.1 Various types of wireless nanosensors network operated based on various stimuli.

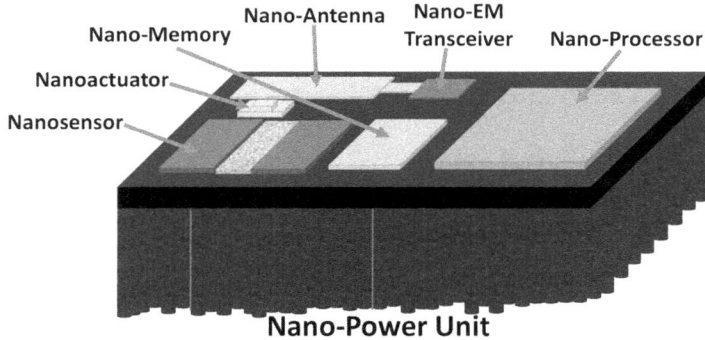

FIGURE 11.2 Nanosensor device embedded in WNSN.

water, soil quality, light pollution, etc.). Wireless nanosensor networks (WNSNs) are the assemblage of many nanosensors having information transferring capacity [2–4]. They can be used for sensing and data collecting precisely using less power consumption. Many emerging applications are there for WNSNs based on various stimuli absorption as shown in Figure 11.1. The network architecture of the WNSN system includes nano-nodes, nano-routers, nano-micro interface devices, and gateway [5,6]. Data interaction between nanosensors is necessary for widening the range in various nanosensors network applications.

A WNSN comprises a cluster of nanosensors that are utilized to cover a wide range to carry out sensing and data collection with higher resolution as well as lower energy consumption [7]. It requires a smaller infrastructure consisting of sensor nodes having a sensing range from centimeter to a few kilometers and those sensors in the network could work co-ordinately for effective monitoring [8]. The architecture of a WNSN is shown in Figure 11.2. Communication within nanosensors in a network can stretch the abilities and implementation of each nano-devices. In this case, there are two main processes are applied for data transformation: nano-electromagnetic communication and molecular communication [9,10]. To bring out the target application, each nano-node in WNSN conjoins to execute the predetermined assignment. The information generated by sensing is collected, pass to the

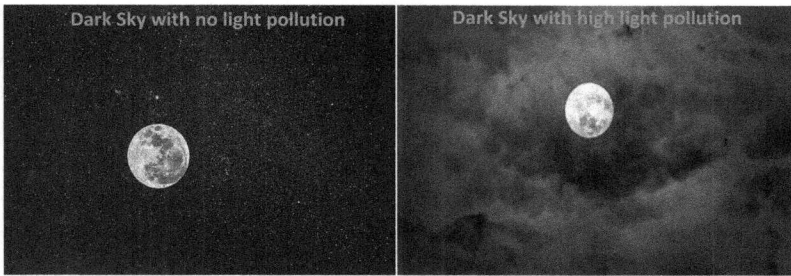

FIGURE 11.3   Observation of sky in the light-pollution-free area and light-pollution area.

nano controllers that absorb all the information and transmit to the nano-micro interface (gateway) [11]. However, this gateway transfers the accumulated information to the access point or associated sink module for transmission to the outside service provider.

WNSN plays a crucial role to detect light pollution and gives information to control it. Excess light in a particular area coming from a source having no use is considered light pollution. Excessive light exposure may also affect the alertness and mood of a human being as well as animals affecting their circadian rhythms. Medical research suggests that light pollution may cause a variety of adverse health effects. This may include an increase in anxiety, physiological stress, fatigue, and headache [12]. Light trespass is a type of light pollution that is observed when unwanted/undesirable light comes inside another property and became a responsible parameter to cause problems (such as sleep deprivation) [13]. Not having enough sleep can further add to health problems and deterioration of the quality of life. Glare is another form of light pollution that arises due to bad outdoor lighting, which can create unsafe driving conditions. It can often blind drivers or pedestrians partially and contribute to accidents. Aside from human beings, artificial light also affects other organisms and ecosystems [14]. It can cause physiological harm, alter competitive interactions, upset navigation, and change predator-prey relations. Light pollution also promotes algae bloom, which can kill the lake's plants and deteriorate water quality [15]. In addition to that, light pollution is the cause of all other pollution such as carbon emissions. There are over 317 million streetlights globally and the quantity will increase to 363 million as per the estimation for the year 2027 [16]. They all use electricity generated by burning fossil fuels. The carbon emissions (most of which were generated by burning fossil fuels) accounted for 32.5 gigatons in the year 2017. By looking at the global energy demand, it will continue to rise in the coming years unless drastic measures are taken. An increase in carbon emissions means a rise in the earth's surface temperature, which leads to harmful effects on our planet's ecosystems, biodiversity, and livelihood [17]. Figure 11.3 shows the effect of observation in the night beauty sky at the light pollution-free area as compared with light pollution area.

11.2 NANOSENSORS AND LIGHT POLLUTION

Nanosensors enable the establishment of a new era of technologies that can smoothly interpret at the nanoscale and determine unique processes, which are not possible at the macroscale. Metal and metal oxide nanoparticles, nanowires, thin films, CNT, polymers, and biomaterials are used in nanosensor fabrication [18,19]. Light pollution has a remarkable impact on humans and the ecosystem, although these impacts are strenuous to estimate. It is, therefore, crucial to act within the framework of a longer-term vision that includes global changes at social, cultural, organizational, economic, and technological levels. It is also crucial to take solid actions that can have a short-term effect on reducing light pollution to improve sustainability. For these requirements, wireless nanosensor plays a great role [20–22]. For example, currently developed constructions are empowered to incorporate the most eco-friendly and energy-efficient lighting equipment with auto-sense switches for shut-off. Due to light pollution, along with human health, the ecosystem is also affected adversely. Typical problems that arise by the light pollution are [23–26]:

- Sleep deprivation led to hormonal variation contributes to depression, mental illness, cardiovascular disease, cancer, obesity, etc.

- Adversely impact of light pollution on physical fitness as well as mental fitness and degrade the value of life.

- Adversely affect the growth and health of wild animals and ecosystems due to disturbance of their natural day/night round.

- Inability to observe the advantages of enjoying the beauty of the night sky, star systems, nature-related recreational activities such as animal/bird watching, inbound tourism prospects, etc.

- Disturbance in the balance with nature/wildlife contact.

- Road stress results in accidents due to blinding by the light glare from other vehicles. Many wild animal's death happened during road crossing is due to light glare.

- Lessen the security due to thieves and other criminals those take advantage of light exposure and mostly light glare blinds their victim.

- Decrease the fruitfulness of farmland crops and animals (including bees and moths) due to interruption of their usual day/night rhythm. This is due to the effect of unusual light on their photosynthesis, growth, etc. (for plants), and sleep patterns, hormones, etc. (for animals).

- Degrade the resilience of ecosystems that results in a loss of ecosystem services, and increases in nonindigenous/invasive species.

- Excess light tends to wastage of energy, whether or not the energy is produced from fossil fuels. Most used fossil fuels in energy harvest dispose more CO_2 and cause severe air pollution and climate change. If the energy is produced from nuclear power, it generates nuclear waste that causes questionable disposal of nuclear waste in the environment.

Energy wasted is money wasted meaning that these resources cannot be used for other priority areas thus adding to a waste economy.

11.3 THE ACTION OF NANOSENSORS TO DETECT THE LIGHT POLLUTION

Nanotechnology provides a glazing direction to the industries to develop nanoscale systems having embedded activities, such as sensing, computing, data storing, data transferring, and actuation. A most peculiar application of nanonetworks is light pollution. Nowadays, in the area of environmental research and applications, such as ecosystem sustainability, monitoring health cycle and plant growth systems, biodiversity control, air pollution control the intensity of light from various sources are measured [27]. In the case of light pollution, the sensor plays a vital role to detect the intensity of light that is exceeding the required threshold intensity limit at a particular distance. Application specified sensors are operated based on the light of sight communication [28].

Light of sight is generally a class of broadcasting that can transfer and collect data only where transmitter and receiver node are given each other without any deterrent between them. Energy absorption is a crucial task for a wireless body area sensor network. The sensing power of a sensor is based on energy capture for various types of communications [29]. Data collected from the typical passive nanosensor equipment for wireless communication that placed in a particular region using a THz frequency range preferentially through micro unmanned vehicle [30]. Here, to quantify the total lag time (t_{tot}) between the nanosensor entrance time in the service area and acquiring the sensor report at the flying gateway, it is essential to measure the time (t_{ch}) by summing the time required to recollect the energy and the lagging time for data transmission (t_{sg}) and can be expressed as Equation 11.1.

$$t_{tot} = t_{ch} + t_{sg} \tag{11.1}$$

The data transmission time, t_{sg} is the collective index based on the size of the energy transferred, processes used for data transmission, time of execution. It is generally estimated as $t_{sg} = 10$ ms. The energy accumulation period by the nanosensor can be found in Equation 11.2.

$$t_{ch} = \frac{E_{tot}}{P_{rx}r_c} \tag{11.2}$$

where, E_{tot} is the total energy needed for data transfer of a single sensor report; P_{rx} is the amount of energy to be received in a second by the passive nanosensor. It is important to include the transformation coefficient of electromagnetic energy into electric energy $r_c = 0.5$ to measure the power required to transfer the single sensor report to the flying gateway by the passive nanosensor unit [31]. To execute the acquired information, and the encapsulation of information and transfer the packet to the micro unmanned vehicle, E_{tot} can be modified as Equation 11.3.

$$E_{tot} = E_s + E_p + E_{packet\text{-}tx} \tag{11.3}$$

where E_s is the energy demand used for calculating the indicator value for nanosensor device; E_p is the energy required for processing the data; $E_{packet\text{-}tx}$ is the energy required for transferring the information set to the flying gateway. Ultra-low-power-smart-visual-sensor were used for modeling for the energy absorption to get optimal energy consumption for the nanosensors [32]. Therefore, the value of E_s is 1.06 µJ and E_p is 0.73 µJ. To measure the energy consumed to transfer each information set, based on the volume and can be known using Equation 11.4.

$$E_{\text{packet-tx}} = N_{\text{bits}} W E_{\text{pulse-tx}} \tag{11.4}$$

where, N_{bits} is the bit count retained in the transferred packet; W is the code weight (the possibility of transferring the pulse "1" in place of mute mode "0"), $E_{\text{pulse-tx}}$ is the consumed energy to transfer a single pulse. In the THz range, numerous parameters are there that influence the energy of the received signal, which is explained in Equation 11.5.

$$P_{\text{rx}} = \frac{P_{\text{tx}} G(f)}{A(f)} + N_{\text{T}}(f) + N_{\text{mol}}(f) \tag{11.5}$$

where P_{tx} is the power of the transmitted signal; $G(f)$ is the gain parameter of the antenna; $A(f)$ denotes total attenuation ratio; $N_{\text{T}}(f)$ is the noise due to thermal energy; $N_{\text{mol}}(f)$ is the noise due to molecular-based absorption. To measure the attenuation ratio, it is important to include the signal attenuation during the propagation in space, A_{fspl} and the molecular-based absorption, A_{mol}. Hence, the gross attenuation ratio is given in Equation 11.6.

$$A(f) = A_{\text{fspl}} + A_{\text{mol}}(f) \tag{11.6}$$

The signal attenuation during the transmission in space is expressed by Equation 11.7,

$$A_{\text{fspl}}(f) = \left(\frac{4\Pi f d}{c}\right)^2 \tag{11.7}$$

where d is the source-receiver distance; f is the frequency of transmission, and c is the velocity of the light.

The characteristic of THz range application in wireless communications is due to molecular absorption that results in molecular vibration and orientation. The EM energy absorbed by the molecule during the transmission of signal at frequencies nearly equal to the resonant frequency of the molecule results in the absorption of a part of the signal and generation of noise $N_{\text{mol}}(f)$ due to the inter-molecular kinetic energy. The molecular absorption coefficient measured the capacity of energy absorption by a molecule and is estimated by its physical properties (molecular communication, spatial orientation, etc.). Molecular absorption losses are calculated as a function of the transmission frequency, the receiving-transmitting distance, and the conditions applied in the environment in which the signal transmits [24]. The coefficient of molecular absorption can be explained by Equation 11.8.

$$A_{\text{mol}}(f) = e^{k(f)d} \tag{11.8}$$

where k is the average absorption coefficient.

It was found that in certain frequency ranges there is a larger absorption observed and leads to a remarkable reduction of communication gap [33]. But, the presence of transparency windows lowers the absorption capacity. It should be noted that the average absorption coefficient does not depend on the transmission distance, but depends only on the environmental conditions and the frequency of the transmitted signal. If the values of transmittance are less than 94.5% (equivalent to ambient absorption coefficient > 5.5%), the noise of the molecular absorption, N_{mol}, becomes the same as that of the maximum value of −203.89 dB/Hz (≈10–20 W/Hz) that indicates to the Johnson-Nyquist thermal noise floor. Here, higher transferring speeds are not required for sensor reporting, and hence no requirement of a broadband channel, a bandwidth of 100 kHz per nanodevice [34,35].

11.4 REDUCTION OF LIGHT POLLUTION BY WNSN

The only viable solution that exists today is to upgrade the existing lighting infrastructure to intelligent lighting infrastructure. The streetlights are the biggest source of light pollution, making them smart is a wise move. Intelligent street lighting adjusts the brightness of streetlights as per the movement of pedestrians, cyclists, and cars. Currently, smart street lighting solutions are adopted sensor-based technology [36]. This sensor-based, smart street lighting solution dims the streetlights when no one is detected but brightens the streetlights to a predefined level when a human presence is detected. This means that when the lighting requirement is low, for example, during the hours between midnight and five in the morning when there is hardly anyone around, the streetlights can be scheduled to burn at low brightness levels. Thus, such an innovative street lighting solution radically saves energy, reducing CO_2 emissions in thermal plants and curbing light pollution while keeping citizen safety and contentment at the center [37]. Another benefit of upgrading to intelligent lighting is laying a strong foundation for a smart city. Existing streetlights already form a city-wide network. Adding light sensors, controllers and gateways will upgrade the ordinary city-wide network into a smart city-wide network. A smart street illumination system saves up to 80% of energy [38]. Delft University of Technology (TU Delft) has reported that the smart street illumination system on its campus uses up to 80% less electric energy than the traditional systems and is proved the cheaper maintenance of the system [39]. Wireless communication between the streetlights and a control room is a great success nowadays [40]. In addition, to control light pollution, the following practices should be adopted [41–45]:

Application of lowest wattage lamp: in most commercial applications, the maximum wattage used is 250 W. Here better to use blue to warm illumination and the bulbs with low watt power.

Use motion sensor-controlled lighting: during the entrance of the place, the light will glow and if no one is in that place then there will be automatically switched off.

Incorporate lighting curfews: street illumination is equipped in such a way on selected highways to enhance the visibility of the surrounding in the nighttime. Street illumination is generally warranted for the daily traffic volume. However, in the early morning hours, the traffic volumes are generally so low that to diminish the requirement for roadway lighting.

Use automatic timers or motion detectors set for all outdoor lights: motion detection sensor to be placed in the outdoor and embedded with the light system. If no motion, then after a few seconds lights will switch off.

Minimize exterior building lighting: unnecessary exterior light systems should be illuminated.

Eliminate landscape lighting: Landscape light is generally up directional to highlight something. Better to use eliminate it.

Adopt a shielded lighting plan: The light system should be pointed in such a way that it will direct within the periphery of the target.

Use solar-powered lighting: Solar power light system only uses the sunlight to generate electricity. Therefore, there is no fear of CO_2 production.

Switched off late-night lighting: Reduce or turn off night hour decorative illumination and outdoor lighting if the regions are not underutilized.

Install reflectors at driveway outline: Place reflectors for street guide in place of installing many lights. Reflectors are always free to use, economic, and no need for an external power supply.

11.5 ARTIFICIAL INTELLIGENCE AND DEEP LEARNING FOR SWSN

Artificial intelligence (AI) within WNSN has been considered as one of the most important communication systems because it consumes less power, increases node density of the network, cost-effective, speeds up as the flow of data increases, and increases the data transmission rate of communication. Various types of AI techniques are artificial neural networks (ANN), genetics algorithms (GA), Fuzzy logic, particle swarm optimization, ant colony optimization, Cuckoo optimization, etc. [46–48]. AI-based techniques help in the enhancement of network lifetime and throughput. ANN is one of the AI techniques that are inspired by the human nervous system, which is interconnected through neurons and has capabilities to understand and learn by training and can be applied further to determine complex trends. All other types of AI systems are operated in such a way as to give optimum results in a network. To construct big data, many entities to be incorporated in an order are the data collection, storage, processing, analysis, and observation. The research focused on the basic feature of dealing with big data to embed in the social web for beneficial applications and new technologies. Figure 11.4 shows a fundamental design of a big data system in a WSNS. Here, a sink node accumulates entire data from sensor nodes to store in temporary storage. Then, the aggregated data are manipulated using a big data structure in the main storage. Finally, the transferred information is operated by big data platforms corresponding to various applications [49].

Generally, the data collection by the sensors can enhance rapidly. Traditional information technologies for data execution, storing, and real-time reporting are too expensive. However, most of the events observed periodically and are largely unusual data or minor variations gives raise a larger waste of data storage. These unnecessary data are of no use and redundant. Therefore, contrary to the conventional WSNS, it is necessary to collect and transfer a large amount of data with the reduction of the data latency in WSN-based big data systems [50]. To make energy-efficient, it is necessary to make a data redundancy efficient system.

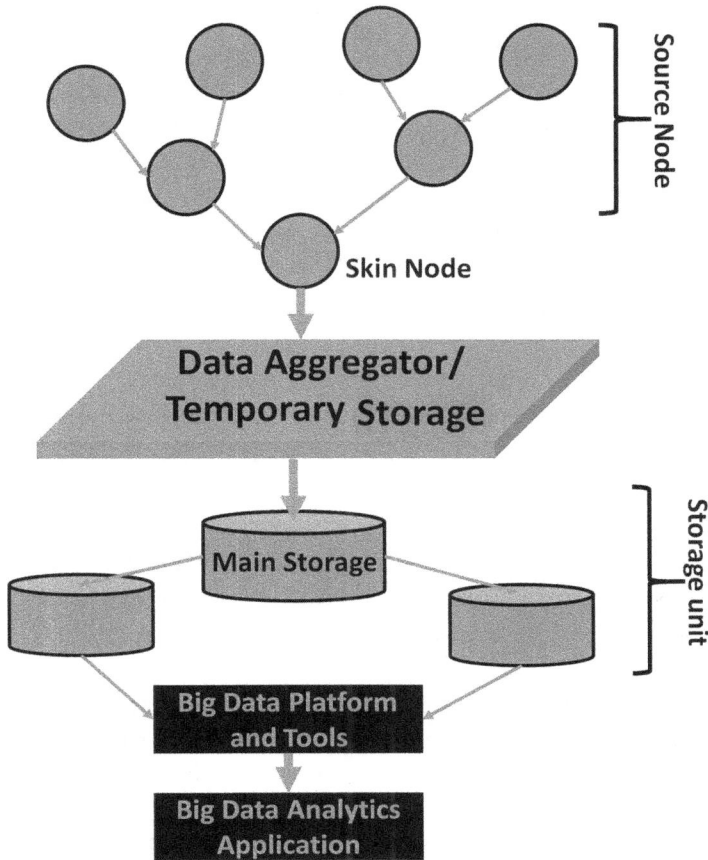

FIGURE 11.4   Typical design of a big data system.

The interlink between the big data systems and the WSNS is based on the data processing techniques in the network. For the WNSN side, it would save their limited resources, whereas, at the same time, the big data system receiving clean, non-redundant, and relevant data would reduce excessive data volume. Thus, it would reduce overload by discovering values from these data rapidly. The architecture for gathering, storage, and analysis of data generated by WNSN for monitoring light pollution in a smart city is in the progress stage. The conventional route of distributing, operating, and managing network systems cannot meet the current need, and the deep learning technology formulates to work out these problems. With the above information, it is clear that deep learning technology can enhance the network in the field of WNSN. In addition, the above deep learning technology generally has a multilayer-network structure rarely involving reinforcement learning [51].

11.5.1 AI in Routing and Traffic Management

Currently, AI technology became an enthralling subject in almost all research fields including scientists and engineers. It is a novel subject for the intelligence of machines. The main purpose to use an AI system with WNSN is to improve systems that reproduce the smart capabilities of individuals. Distributed AI is depending on the concept of multi-agent systems that works in a group, facilitate the interpretation with each other to assign the work, functionalize their network, and execute knowledge. Routing allows networks to transport packets to various nodes as per the requirement. A routing mechanism is responsible for data transfer between any endpoints through the network and the network executes in such a way to handle overload conditions, lightly loaded conditions, and fluctuating traffic patterns. Few AI-based algorithms are introduced to address quality of service (QoS) traffic routing and classification in Software-Defined Networking (SDN). Many AI-based works that address routing issues fall into the category of algorithms such as distributed AI and agent-based routing, shortest path algorithms, and algorithmic resource allocation technique [52]. The quality of experience in networks quantifies the service given by the network as per customer demand. QoS, Aware Adaptive Routing (QAR) in multi-layer hierarchical SDN with reinforcement learning approach can be used. The QoS algorithm depends on the expertise as various requirements for packet delay, loss, and throughput and is synchronized by an efficient transmission that accompanies each specified application. The QAR algorithm using reinforcement learning can be performed by considering long-term income, action systems, quality function, efficient routing, QoS deployment, SoftMax action choice policy, and start-action-reward-state-action techniques for enhancing the quality and Markov decision techniques with QoS-aware reward functionality.

A well-organized algorithm is capable to detect QoS classes easily, and directing to the optimized route even addition of any new applications, decreasing the requirement to maintain a real-time update in the number of applications on the Internet. The associate network comprises two branches, the first is the local traffic identifier branch at the SD switches at the network edge and the second is the global traffic classifier at the network controller. This branch provides three advantages: (i) SD switches are very simple and are embedded with lightweight elephant flow identification, (ii) network controller checks the accuracy and flexibility of the QoS classifier, (iii) the entire framework builds with a modular framework construction principle, which facilitates the modification at any part of that framework. In SDN, the QoS classification solutions are used to allow fine-grained QoS-aware traffic engineering. The framework real-time and adaptively classifies a traffic flow into a QoS category without having to identify the target application where the traffic flow is originated. The TC engine that is present in the centralized SDN controller shows effective network monitoring with low overhead, lesser switch alteration, detects significant QoS flows, classifies QoS-aware traffic, and activates services like application discovery by using deep packet inspection running in the network controller. This system comprises two important techniques that are the detection of elephant flows and collection of statistics and extraction of features. These two processes are operated by two routes. The first route engaged to dig out significant QoS flows in any new input flows, whereas the second route engaged with QoS-aware traffic classification and associated complex functionalization [53]. The machine learning algorithm adopted in this process helps to train the complex network and categorize the inputs (lead/lag time data or inter-arrival period, packet condition, and protocol data).

11.5.2 AI in Security and Admission Control

A compromised network can be utilized to ingress sensitive data, security attacks, or shut down the Internet to achieve a safe network link at any time. As technology becomes more sophisticated and needs a faster and more friendly network, security must be modified according to the growth of possible threats. Security measures employed in networks must be capable to face the new threats, whereas trained as per the previously detected threats and adapt to provide new ways to counter those threats. Network security's ability to learn to adapt to new threats can be reinforced by AI and machine learning algorithms. Machine learning is the best practice for this problem whether the threats are old or new that follow a specific pattern. Security services introduced at the network interface in SDN include flow control, auto-security handling that decouples network management and data transmission management, information routing, authentic identification, and privacy shield. Network monitoring is reinforced by SDN to detect anomalies when investigating new threats using the OpenFlow protocol. Equipment running the OpenFlow protocol has a privacy control mechanism for network-level security to ensure information is safe and privacy is maintained. It does this by detecting the network function and monitoring suspicious characters in the network system [54]. One of the main roles of the SDN controller is admission control (AC) during the connection approach. When the network is heavily used, the AC controller handles the service demand. If resources are accessible, AC receives the demand and otherwise discards the demand. The AC methods used are threshold-based and they use minimum, maximum, exclusive, and non-exclusive limits for the resource share that is fixed by the network operator. Setting the threshold is problematic because the optimal design is based on the state of network traffic that alters each time. Algorithms used for CA can be categories of standard care and worst situations. Unfavorable cases are characterized by maximum and minimum execution, where the malicious attack is the worst possible connection. The average fall of operation is good, but performance on a specific opponent is not guaranteed. An online meta-algorithm depends on previous resolutions obtained from various AC algorithms. To overcome this problem, a strategic expect meta-algorithm (SEA) was implemented that selects the algorithm with quantitative consecutive steps and does not recheck its preference on each new connection. SEA is depending on the actual gain achieved by each algorithm when it was chosen. The standard expected execution of SEA is always higher than the unique execution acquired by the algorithm systematically selected by SEA itself [55].

11.6 ADVANTAGES AND LIMITATIONS OF WNSN USED TO CONTROL LIGHT POLLUTION

Network of the nanosensor is capable to cover a wide range of areas to detect and send information about the light intensity to the control unit. Advantages of these nanosensors network are [56–60]:

- Nanosensors are small, portable, and can place in the preferable area

- Frequent response due to receiving energy efficiently because of the larger surface area of nanodevice

- A large number of data is collected at a time from various points at a place to perpetuate the accuracy of the information

Along with the above advantages received from the nanoscale sensor, it also has different limitations as given below:

- To accumulate the data, a large number of nanoscale machines to be installed

- Frequent maintenance is required in air pollution area

- Restricted resources of processor, energy handled, and memory storage

- Limited modeling and computational ability that able to show simple operations

- Restricted within the communication range

11.7 SUMMARY

In this current age, WNSNs are successfully adopted to control light pollution, which is a novel step in the developed area. Working principles of nanosensor system and their structural integrity has been discussed in the chapter. How light pollution can be detected and can be controlled was also discussed briefly. AI systems reinforced the WNSN efficiently and the current research proposes the successful utilization of light pollution. The advantages, as well as the limitations, are given here to provide a future scope of research for light pollution data acquisition and control.

REFERENCES

[1] A.O. Balghusoon, S. Mahfoudh, Routing protocols for wireless nanosensor networks and internet of nano things: A comprehensive survey, *IEEE Access*, Volume. 8, 2020, 200724–200748, https://doi.org/10.1109/ACCESS.2020.3035646.

[2] A. Galal, Xavier Hesselbach, Nano-networks communication architecture: Modeling and functions, *Nano Communication Networks*, Volume 17, 2018, 45–62, https://doi.org/10.1016/j.nancom.2018.07.001.

[3] P. Wang, J.M. Jornet, M.G.A. Malik, N. Akkari, I.F. Akyildiz, Energy and spectrum-aware MAC protocol for perpetual wireless nanosensor networks in the Terahertz Band, *Ad Hoc Networks*, Volume 11, Issue 8, 2013, 2541–2555, https://doi.org/10.1016/j.adhoc.2013.07.002.

[4] M. Kocaoglu, O. B. Akan, Minimum energy coding for wireless nanosensor networks, 2012 Proceedings IEEE INFOCOM, 2012, pp. 2826–2830, https://doi.org/10.1109/INFCOM.2012.6195709.

[5] A. Rahim, P. Malone, Intrusion detection system for wireless nano sensor networks, 8th International Conference for Internet Technology and Secured Transactions (ICITST-2013), 2013, pp. 327–330, https://doi.org/10.1109/ICITST.2013.6750215.

[6] I.F. Akyildiz, J. Miquel Jornet, Electromagnetic wireless nanosensor networks, *Nano Communication Networks*, Volume 1, Issue 1, 2010, 3–19, https://doi.org/10.1016/j.nancom.2010.04.001.

[7] M. Pierobon, J.M. Jornet, N. Akkari, A routing framework for energy harvesting wireless nanosensor networks in the Terahertz Band. *Wireless Network*, Volume 20, 2014, 1169–1183, https://doi.org/10.1007/s11276-013-0665-y.

[8] S.J. Lee, C. Jung, K. Choi, S. Kim, Design of wireless nanosensor networks for intrabody application, *International Journal of Distributed Sensor Networks*, July 2015, https://doi.org/10.1155/2015/176761.

[9] K. Yang, A comprehensive survey on hybrid communication in context of molecular communication and Terahertz communication for body-centric nanonetworks, *IEEE Transactions on Molecular, Biological and Multi-Scale Communications*, Volume. 6, Issue 2, November 2020, 107–133, https://doi.org/10.1109/TMBMC.2020.3017146.

[10] D. Malak, O.B. Akan, Molecular communication nanonetworks inside human body, *Nano Communication Networks*, Volume 3, Issue 1, 2012, 19–35, https://doi.org/10.1016/j.nancom.2011.10.002.

[11] F. Lemic, Survey on Terahertz nanocommunication and networking: A top-down perspective, *IEEE Journal on Selected Areas in Communications*, Volume 39, Issue 6, June 2021, 1506–1543, https://doi.org/10.1109/JSAC.2021.3071837.

[12] J.A. Casey, H.C. Wilcox, A.G. Hirsch, Associations of unconventional natural gas development with depression symptoms and disordered sleep in Pennsylvania. *Scientific Reports* Volume 8, 2018, 11375. https://doi.org/10.1038/s41598-018-29747-2.

[13] K.J. Gaston, T.W. Davies, J. Bennie, J. Hopkins, Review: Reducing the ecological consequences of night-time light pollution: Options and developments, *Journal of Applied Ecology*, Volume 49, Issue 6, December 2012, 1256–1266, https://doi.org/10.1111/j.1365-2664.2012.02212.x.

[14] D. Crawford, Light pollution-A problem for all of us, *International Astronomical Union Colloquium*, Volume 112, 1991, 7–10, https://doi.org/10.1017/S0252921100003572.

[15] R.K. Singhal, M. Kumar, B. Bose, Eco-physiological responses of artificial night light pollution in plants. *Russian Journal of Plant Physiology*, Volume 66, 2019, 190–202, https://doi.org/10.1134/S1021443719020134.

[16] G. Zissis, P. Dupuis, L. Canale, N. Pigenet, Smart Lighting Systems for Smart Cities. In: Lazaroiu G.C., Roscia M., Dancu V.S. (eds) *Holistic Approach for Decision Making Towards Designing Smart Cities*. Future City, vol. 18. Springer, Cham, 2021, https://doi.org/10.1007/978-3-030-85566-6_5.

[17] U. Thurairajah, J. R. Littlewood, G. Karani, A Proposed Method to Pre-qualify a Wireless Monitoring and Control System for Outdoor Lighting to Reduce Energy Use, Light Pollution, and Carbon Emissions. In: Littlewood J. R., Howlett R. J., Jain L. C. (eds) *Sustainability in Energy and Buildings 2021. Smart Innovation, Systems and Technologies*, vol. 263. Springer, Singapore, 2022, https://doi.org/10.1007/978-981-16-6269-0_31.

[18] N.M. Noah, Design and synthesis of nanostructured materials for sensor applications, *Journal of Nanomaterials*, Volume 2020, Article ID 8855321, 2020, 20, https://doi.org/10.1155/2020/8855321.

[19] R. Abdel-Karim, Y. Reda, A. Abdel-Fattah, Review—nanostructured materials-based nanosensors, *Journal of The Electrochemical Society*, Volume 167, Issue 3, 2020, 037554, https://doi.org/10.1149/1945-7111/ab67aa.

[20] M. Grubisic, A. Haim, P. Bhusal, D.M. Dominoni, K.M.A. Gabriel, A. Jechow, F. Kupprat, A. Lerner, P. Marchant, Light pollution, circadian photoreception, and melatonin in vertebrates, *Sustainability*, Volume 11, 2019, 6400, https://doi.org/10.3390/su11226400.

[21] G. Zissis, Sustainable lighting and light pollution: A critical issue for the present generation, a challenge to the future. *Sustainability*, Volume 12, 2020, 4552, https://doi.org/10.3390/su12114552.

[22] D.M. Broday, The Citi-Sense Project Collaborators, Wireless distributed environmental sensor networks for air pollution measurement-The promise and the current reality. *Sensors*, Volume 17, 2017, 2263, https://doi.org/10.3390/s17102263.

[23] A.A. A. Hussein, E. Bloem, I. Fodoret, Slowly seeing the light: an integrative review on ecological light pollution as a potential threat for mollusks. *Environmental Science and Pollution Research* Volume 28, 2021, 5036–5048, https://doi.org/10.1007/s11356-020-11824-7.

[24] M. Javaid, A. Haleem, R. P. Singh, S. Rab, R. Suman, Exploring the potential of nanosensors: A brief overview, *Sensors International*, Volume 2, 2021, 100130, https://doi.org/10.1016/j.sintl.2021.100130.

[25] J. Pauwels, I. L. Viol, Y. Bas, N. Valet, C. Kerbiriou, Adapting street lighting to limit light pollution's impacts on bats, *Global Ecology and Conservation*, Volume 28, 2021, e01648, https://doi.org/10.1016/j.gecco.2021.e01648.

[26] B. W. Brook, A. Alonso, D. A. Meneley, J. Misak, T. Blees, J.B. van Erp, Why nuclear energy is sustainable and has to be part of the energy mix, *Sustainable Materials and Technologies*, Volumes 1–2, 2014, 8–16, https://doi.org/10.1016/j.susmat.2014.11.001.

[27] S. Girotti, E. N. Ferri, M. G. Fumo, E. Maiolini, Monitoring of environmental pollutants by bioluminescent bacteria, *Analytica Chimica Acta*, Volume 608, Issue 1, 2008, 2–29, https://doi.org/10.1016/j.aca.2007.12.008.

[28] H. Lamphar, Spatio-temporal association of light pollution and urban sprawl using remote sensing imagery and GIS: A simple method based in Otsu's algorithm, *Journal of Quantitative Spectroscopy and Radiative Transfer*, Volume 251, 2020, 107060, https://doi.org/10.1016/j.jqsrt.2020.107060.

[29] L.E.M. Matheus, A.B. Vieira, L.F.M. Vieira, M.A M. Vieira, O. Gnawali, Visible light communication: Concepts, applications and challenges, *IEEE Communications Surveys & Tutorials*, Volume 21, Issue 4, Fourthquarter 2019, 3204–3237, https://doi.org/10.1109/COMST.2019.2913348.

[30] R. Pirmagomedov, R. Kirichek, M. Blinnikov, A. Koucheryavy, UAV-based gateways for wireless nanosensor networks deployed over large areas, *Computer Communications*, Volume 146, 2019, 55–62, https://doi.org/10.1016/j.comcom.2019.07.026.

[31] R. Pirmagomedov, M. Blinnikov, R. Kirichek, A. Koucheryavy, Wireless Nanosensor Network with Flying Gateway. In: Chowdhury K., Di Felice M., Matta I., Sheng B. (eds) *Wired/Wireless Internet Communications. WWIC 2018. Lecture Notes in Computer Science*, vol 10866. Springer, Cham, 2018, https://doi.org/10.1007/978-3-030-02931-9_21.

[32] I.F. Akyildiz, J.M. Jornet, C. Han, Terahertz band: Next frontier for wireless communications, *Physical Communication*, Volume 12, 2014, 16–32, https://doi.org/10.1016/j.phycom.2014.01.006.

[33] R.N. Khushaba, S. Kodagoda, D. Liu, G. Dissanayake, Electromyogram (EMG) based fingers movement recognition using neighborhood preserving analysis with QR-decomposition, 2011 Seventh International Conference on Intelligent Sensors, Sensor Networks and Information Processing, 2011, pp. 1–105, https://doi.org/10.1109/ISSNIP.2011.6146521.

[34] W. Guo, C. Mias, N. Farsad and J. Wu, Molecular versus electromagnetic wave propagation loss in macro-scale environments, *IEEE Transactions on Molecular, Biological and Multi-Scale Communications*, Volume 1, Issue 1, March 2015, 18–25, https://doi.org/10.1109/TMBMC.2015.2465517.

[35] P. Boronin, V. Petrov, D. Moltchanov, Y. Koucheryavy, J.M. Jornet, Capacity and throughput analysis of nanoscale machine communication through transparency windows in the Terahertz Band, *Nano Communication Networks*, Volume 5, Issue 3, 2014, 72–82, https://doi.org/10.1016/j.nancom.2014.06.001.

[36] E. Zarepour, M. Hassan, C. T. Chou and A. A. Adesina, Frequency hopping strategies for improving terahertz sensor network performance over composition varying channels, Proceeding of IEEE International Symposium on a World of Wireless, Mobile and Multimedia Networks 2014, 2014, pp. 1–9, doi: 10.1109/WoWMoM.2014.6918973.

[37] F. Marino, F. Leccese, S. Pizzuti, Adaptive street lighting predictive control, *Energy Procedia*, Volume 111, 2017, 790–799, https://doi.org/10.1016/j.egypro.2017.03.241.

[38] S. Bara, Light pollution and solid-state lighting: Reducing the carbon dioxide footprint is not enough, Proc. SPIE 8785, 8th Iberoamerican Optics Meeting and 11th Latin American Meeting on Optics, Lasers, and Applications, 2013, pp. 87852G, https://doi.org/10.1117/12.2025344.

[39] G. Pasolini, P. Toppan, F. Zabini, C.D. Castro, O. Andrisano, Design, deployment and evolution of heterogeneous smart public lighting systems. *Applied Sciences*, Volume 9, Issue 16, 2019, 3281, https://doi.org/10.3390/app9163281.

[40] G.G. Cerruela, I.L. Ruiz, M.Á. Gómez-Nieto, State of the art, trends and future of bluetooth low energy, near field communication and visible light communication in the development of smart cities. *Sensors*, Volume 16, Issue 11, 2016, 1968, https://doi.org/10.3390/s16111968.

[41] W.A. Jabbar, M.A.B. Yuzaidi, K.Q. Yan, U.S.B.M. Bustaman, Y. Hashim, H.T. AlAriqi, Smart and green street lighting system based on Arduino and RF wireless module, 2019 8th International Conference on Modeling Simulation and Applied Optimization (ICMSAO), 2019, pp. 1–6, https://doi.org/10.1109/ICMSAO.2019.8880451.

[42] A. Jechow, Observing the impact of WWF earth hour on urban light pollution: A case study in Berlin 2018 using differential photometry. *Sustainability*, Volume 11, Issue 3, 2019, 750, https://doi.org/10.3390/su11030750.

[43] W. Jiang, H. Guojin T. Long, H. Guo, R. Yin, W. Leng, H. Liu, G. Wang, Potentiality of using Luojia 1-01 nighttime light imagery to investigate artificial light pollution, *Sensors*, Volume 18, Issue 9, 2018, 2900, https://doi.org/10.3390/s18092900.

[44] D. Száz, D. Mihályi, A. Farkas, Polarized light pollution of matte solar panels: anti-reflective photovoltaics reduce polarized light pollution but benefit only some aquatic insects. *Journal of Insect Conservation*, Volume 20, 2016, 663–675, https://doi.org/10.1007/s10841-016-9897-3.

[45] C.D. Gălățanu, L. Canale, D.D. Lucache, G. Zissis, Reduction in light pollution by measurements according to EN 13201 standard, 2018 International Conference and Exposition on Electrical And Power Engineering (EPE), 2018, pp. 1074–1079, https://doi.org/10.1109/ICEPE.2018.8559722.

[46] E.B. Priyanka, S. Thangavel, K.M. Sagayam, Wireless network upgraded with artificial intelligence on the data aggregation towards the smart internet applications. *International Journal of System Assurance Engineering and Management*, 2021, https://doi.org/10.1007/s13198-021-01425-z.

[47] F. Filippetti, G. Franceschini, C. Tassoni, P. Vas, AI techniques in induction machines diagnosis including the speed ripple effect, *IEEE Transactions on Industry Applications*, Volume 34, Issue 1, January–February 1998, 98–108, https://doi.org/10.1109/28.658729.

[48] B. Chandra Mohan, R. Baskaran, A survey: Ant colony optimization based recent research and implementation on several engineering domain, *Expert Systems with Applications*, Volume 39, Issue 4, 2012, 4618–4627, https://doi.org/10.1016/j.eswa.2011.09.076.

[49] B.S. Kim, K. I. Kim, B. Shah, F. Chow, K.H. Kim, Wireless sensor networks for big data systems, *Sensors (Basel)*, Volume 19, Issue 7, 2019, 1565, https://doi.org/10.3390/s19071565.

[50] A.C. Djedouboum, A.A.A. Ari, A.M. Gueroui, A. Mohamadou, Z. Aliouat, Big data collection in large-scale wireless sensor networks, *Sensors*, Volume 18, 2018, 4474, https://doi.org/10.3390/s18124474.

[51] G. S. Mahalakshmi, G. Muthu Selvi, S. Sendhilkumar, P. Vijayakumar, Y. Zhu, V. Chang, Sustainable computing based deep learning framework for writing research manuscripts, *IEEE Transactions on Sustainable Computing*, Volume 4, Issue 1, 1 January–March 2019, 4–16, https://doi.org/10.1109/TSUSC.2018.2829196.

[52] Meenaxi M. Raikar, S.M. Meena, Mohammed Moin Mulla, Nagashree S. Shetti, Meghana Karanandi, Data traffic classification in Software Defined Networks (SDN) using supervised-learning, *Procedia Computer Science*, Volume 171, 2020, 2750–2759, https://doi.org/10.1016/j.procs.2020.04.299.

[53] P. Wang, S.-C. Lin, M. Luo, A framework for QoS-aware traffic classification using semi-supervised machine learning in SDNs, 2016 IEEE International Conference on Services Computing (SCC), 2016, pp. 760–765, https://doi.org/10.1109/SCC.2016.133.

[54] A. He, A Survey of artificial intelligence for cognitive radios, *IEEE Transactions on Vehicular Technology*, Volume 59, Issue 4, May 2010, 1578–1592, https://doi.org/10.1109/TVT.2010.2043968.

[55] J. Leguay, L. Maggi, M. Draief, S. Paris, S. Chouvardas, Admission control with online algorithms in SDN, NOMS 2016-2016 IEEE/IFIP Network Operations and Management Symposium, 2016, pp. 718–721, https://doi.org/10.1109/NOMS.2016.7502884.

[56] K. Czarnecka, K. Błażejczyk, T. Morita, Characteristics of light pollution-A case study of Warsaw (Poland) and Fukuoka (Japan), *Environmental Pollution*, Volume 291, 2021, 118113, https://doi.org/10.1016/j.envpol.2021.118113.

[57] J. Li, X. Yongming, C. Weiping J. Meng, S. Boyang, W. Yuyang, J. Wang, Investigation of nighttime light pollution in Nanjing, China by mapping illuminance from field observations and Luojia 1-01 imagery, *Sustainability* Volume 12, Issue 2, 2020, 681, https://doi.org/10.3390/su12020681.

[58] S. Parvez, C. Venkataraman, S. Mukherji, A review on advantages of implementing luminescence inhibition test (Vibrio fischeri) for acute toxicity prediction of chemicals, *Environment International*, Volume 32, Issue 2, 2006, 265–268, https://doi.org/10.1016/j.envint.2005.08.022.

[59] J. Ngarambe, G. Kim, Sustainable lighting policies: The contribution of advertisement and decorative lighting to local light pollution in Seoul, South Korea, *Sustainability* Volume 10, Issue 4, 2018, 1007. https://doi.org/10.3390/su10041007

[60] F. Falchi, P. Cinzano, C.D. Elvidge, D.M. Keith, A. Haim, Limiting the impact of light pollution on human health, environment and stellar visibility, *Journal of Environmental Management*, Volume 92, Issue 10, 2011, 2714–2722, https://doi.org/10.1016/j.jenvman.2011.06.029.

CHAPTER 12

Smart Wireless Sensor Networks for Street Lighting

Hadis Foladi, Ali Mir, and Ali Farmani
Lorestan University

Kwadwo Mensah-Darkwa
Kwame Nkrumah University of Science and Technology

Tuan Anh Nguyen
Vietnam Academy of Science and Technology

CONTENTS

12.1 INTRODUCTION

Light pollution is defined as excessive, misdirected, or obtrusive artificial (outside) light [1]. Light pollution has several adverse effects, including (i) Obliterating starlight in the night sky; (ii) Interfering with astronomical research; (iii) Disrupting ecosystems; (iv) Posing a health risk (skin/eye ailments/diseases, headaches, and sleep disturbances); and (v) Wasting energy. Furthermore, excessive light at night may contribute to air pollution.

NASA has published nighttime photos of the Earth. Nowadays, city lighting is more prevalent than ever before, and light is directly related to social security. However, artificial light disrupts the circadian cycle and negatively impacts human health because of

DOI: 10.1201/9781003185109-12

light absorption and repulsion. As a result, one-third of humanity can no longer view the Milky Way Galaxy from home, which is a significant cultural disadvantage and raises the expense of astronomy. Light pollution from artificial lights impacts plants, animals, and the ecological system [1].

Remote sensing for night lighting was pioneered by Levin Noam et al. [2]. According to reports, remotely sensed night light data could aid in understanding the environmental repercussions of light emissions (light pollution), particularly their effects on human health. In addition to light density, light color is pollution; for example, blue light irritates humans at night, upsets the biological clock, and disrupts the life of some insects such as fireflies, mosquitoes, and some flies [3]. One of the most visible initiatives is IDA, which identifies dark sky places primarily based on astronomy or environment. This program was started in 2001 and has granted licenses in multiple areas through 2019 [4]. Following that, some simple ways for detecting light pollution at night are provided.

12.2 LIGHT/OPTICAL SENSORS

Sensors are utilized in a variety of fields due to inherent qualities such as sensitivity, selectivity, ease of operation, and detection. These devices are evolving as detection tools and are used to identify particles in real-world samples. Sensors have been miniaturized for ease of usage [5]. The study of materials at the nanoscale has also pushed the boundary of sensor technology. The integration of nanotechnology in light-sensing systems is driven by the demand to collect accurate and precise data for analysis. The integration of nanotechnology and light sensors offers high-throughput results, enhanced detection levels, and rapid analysis. The microstructure of the sensing materials is another major component that influences sensor performance. For example, graphene-based materials with a variety of advantageous features have demonstrated high sensing efficiency in a wide range of applications. Graphene absorbs only 2.3% of the emitted-white light and transmits the remaining 97.7%. The frequency at which this substance emits light is unrelated to its absorption rate. Its sheets have a porous structure that protects against radiation pollution and is utilized in sensors technology [6].

Most wireless sensors measure scalar physical phenomena such as temperature, pressure, humidity, or location. Network distribution systems have been developed by combining low-power wireless networks and low-cost hardware such as cameras and microphones. WSNs provide a versatile and straightforward platform for measuring and interacting with the physical environment and support for some hop-by-hop mesh networks with automatic management and configuration, which are used to measure low-width physical properties such as light, sound, temperature, movement, and humidity [7]. As a result, air monitoring systems based on precise and low-cost sensors are incredibly beneficial for measuring light intensity [8].

12.3 LIGHT EMISSION DIODES

Street lighting is one of the essential programs in smart cities, and while it is necessary to promote security, it also wastes energy. The majority of street installations are inefficient and obsolete, consuming a lot of energy. There have been attempts to use light emission diodes

(LEDs) and automated management systems with smart grids. There are an estimated 300 million lights in the world, with LEDs accounting for 10%. As a result of the usage of LEDs in outdoor lighting, there is less light pollution, more consistent service, and less energy use. LED adoption has been linked to a reduction in light pollution. Furthermore, when utilized for outdoor illumination, LED technology consumes less energy. This is because, unlike fluorescent lights, LEDs can be controlled remotely and their output can be set, allowing lighting to be dimmed or turned off when not required while remaining compliant with local regulations [9].

LEDs are now employed as a new service all over the world, and yellow-light products are no longer manufactured. LEDs are used for a variety of reasons, including:

- **Light Quality:** white LED light makes items more recognized, has more and better security and presents a superior light design that is accessible with traditional sources, resulting in more lighting and less pollution.

- **Energy-Saving:** LED lighting is more efficient than traditional technologies, saving 40%–80% of energy through its design.

- **Less Maintenance:** LEDs have a longer lifespan than traditional technologies and result in lower costs.

- **Smart Light Control:** LED streetlight systems with smart light regulating systems are more accessible and offer additional benefits. It benefits municipalities while also reducing light pollution. Figure 12.1 shows the prediction of the LED market globally from 2009 to 2015 [10].

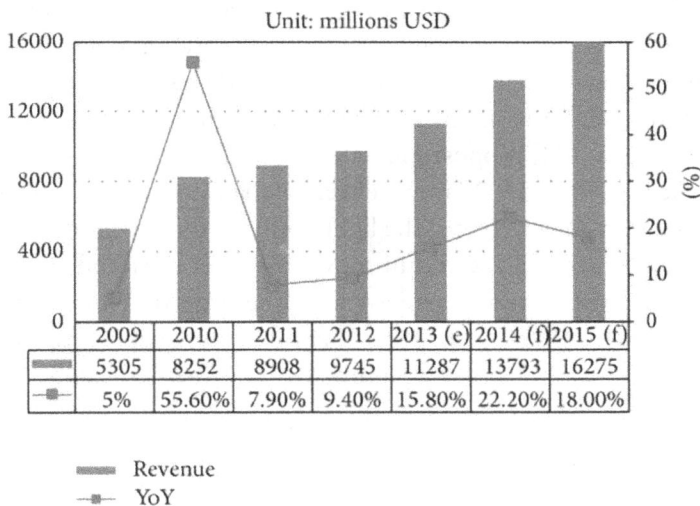

Unit: millions USD

	2009	2010	2011	2012	2013 (e)	2014 (f)	2015 (f)
Revenue	5305	8252	8908	9745	11287	13793	16275
YoY	5%	55.60%	7.90%	9.40%	15.80%	22.20%	18.00%

FIGURE 12.1 2009–2015 prediction of global high-power LED market. (Reproduced with permission [10], Copyright (2014) the authors. This article is an open access article distributed under the terms and conditions of the Creative Commons Attribution (CC BY) license.)

Some experts prefer phosphor-coated amber LEDs because low blue light LEDs, such as PC-AMBER, combine the benefits of LEDs with low blue light [11]. White-light LEDs have a significant problem in that, despite being energy efficient, they still emit blue-light spectrum. LEDs with a blue-light spectrum, according to biologists, can disrupt the natural rhythms of plants, wildlife, and the ecosystem. There is a paradox in that LED technology is environmentally hazardous but is widely employed. Astronomers understand that blue light is scattered due to Riley's scattering and has the most significant impact on the circadian rhythm. The wavelength of light affects creatures; some are sensitive to short-wavelength light, while others are sensitive to long-wavelength light. Unlike eco-friendly solutions, lighting professionals choose white light with a blue spectrum because it has the most brightness and contrast while using the least amount of energy. COOLER LEDs offer good energy efficiency, and because converting blue-LED light to white light with phosphor consumes energy, blue-LED light loses less energy. In contrast to the reasoning against LEDs, augmentations for LEDs are pretty homogenous. LEDs, according to most experts, can cut pollution to acceptable levels.

12.4 SMART LED TO SAVE ENERGY

LEDs are desirable because they offer numerous benefits, including high efficiency, extended life, compact size, low temperature, and controllability. Energy savings will be accomplished by transitioning to LED lighting. Attaining the green targets, on the other hand, requires prudent control. Traditionally, street lighting was controlled by a clock, but it is now controlled by photo-electronic cells that detect the amount of natural light. One of the concepts displayed in Figure 12.2 is the design of smart light. It is constructed from aluminum and LED-modified tubes with programmable optics and electronics. To regulate, the ambient temperature, read reflected light and LED direct light are collected and transmitted to a web page.

Juntunen et al. [13] present the smart LED luminaire and wireless technology for energy savings in pedestrian road lighting. The author used 18 LEDs and a free optic laser. LEDs are embedded in two rows and the lenses have been made by glass molding. Indeed, there is a challenge for street lighting in terms of efficiently lighting roadways with curved and twisted designs. Sun et al. [14] proposed an effective and efficient adaptive LED luminaire to meet this challenge. Their LED lamp delivers a roadway-shaped light pattern, which maximizes illumination performance. The light is efficiently and homogeneously directed only where needed, reducing glare and improving eye comfort and the visual discrimination ability of vehicle drivers and pedestrians. The proposed luminaire only requires replacing the cover plate, which is a special microlens array sheet, to produce different shape light patterns. Their LED light is collimated and then efficiently distributed on a freeform roadway by the special microlens sheet.

12.5 LIGHT POLLUTION REDUCTION DEVICES

Night lighting is an essential part of society; however, it is inefficient and contributes to light pollution. In order to save energy and reduce light pollution, the equipment should be used with care. Light pollution combines and overlaps four components: glaring, clutter, light penetration, and the sky shining. The model lighting ordinance law prevents light pollution,

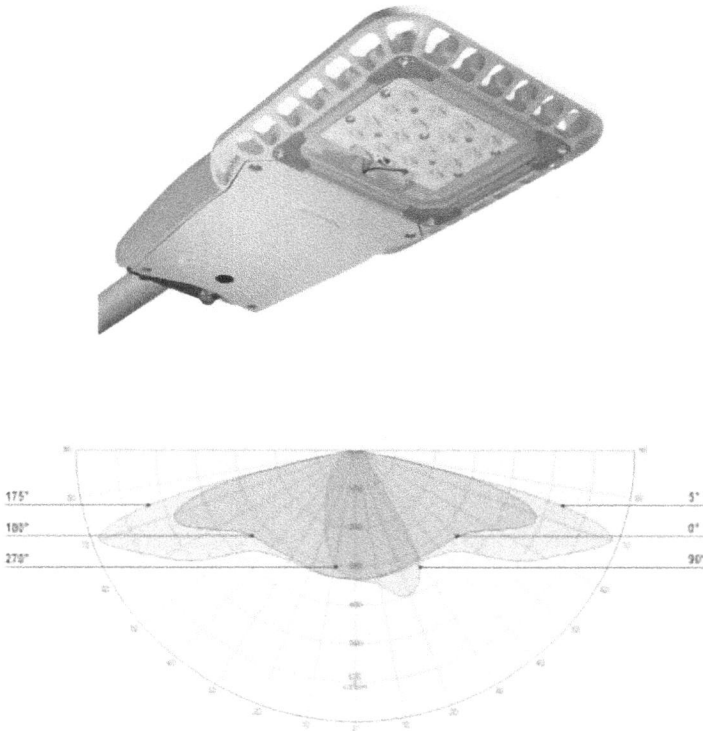

FIGURE 12.2 Smart LED light for lighting showing polar diagram. (Adapted with permission from [12] Copyright (2018). Copyright (2013) Elsevier.)

which is a helpful method in outdoor lighting and reducing light attacks and sky glare. In this method, the area is divided into five regions, and the amount of light required for each is determined. This method employs the classification of outdoor lighting fixtures to provide adequate protection. One of the strategies presented in this study is the use of light photometric data that does not allow light to be emitted at angles greater than 90°. A newly designed device, LED WARP9, which utilizes microtechnology, was used, and an unmatched control was performed on its light to ensure uniformity, hence reducing the causes of glare lighting on the ground, i.e. half of the LEDs are observed each time, resulting in 50% less glare.

IDA is a device that detects timers, remote controls, and radars, which can be used in place of continuous signals in planes. The Obstacle Collision Avoidance System is a sensor-based system that uses ground-radar technology to detect plane collisions. When it detects lights, it becomes white during the day and red during the night and darkness, and it activates a radio system as an alarm. Using towers, wind turbines, and other structures prevents flashing lights and reduces energy consumption [15].

12.6 A DRONE TO ASSESS LIGHT POLLUTION

Most designers believe that deploying curved lights is an excellent way to reduce light pollution; however, it should be noted that shining light on the ground is reflected in the sky, causing pollution. To estimate this light, simulations have been performed. One of these

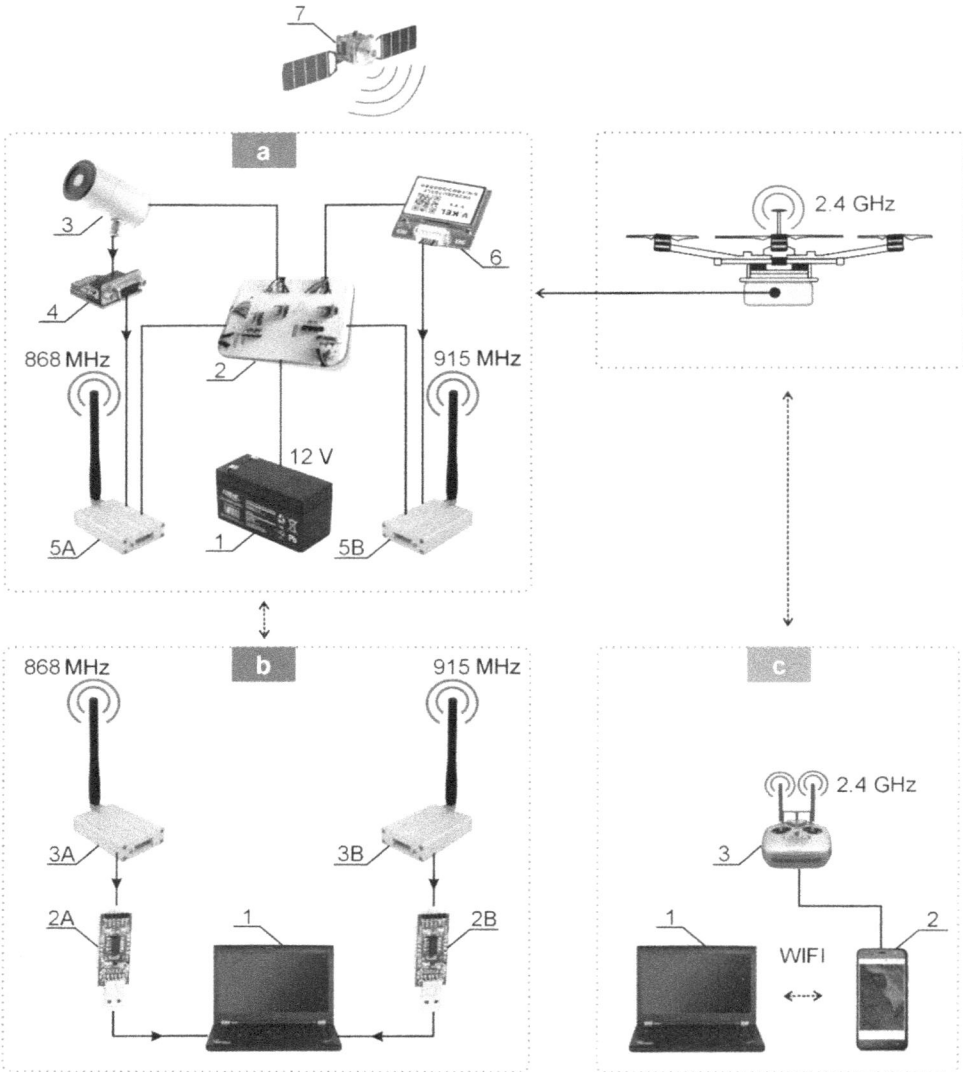

FIGURE 12.3 Explanatory wiring diagram of individual components: block (a) 1, battery; 2, printed circuit board with voltage converters installed; 3, illuminance meter; 4, RS232/UART adapter connecting the illuminance meter with the RS232 connector with the RF transmitter equipped with the URT connector; 5, transmitters; 6, GPS module; 7, satellite. Block (b) 1, PC computer; 2, UART/USB converter; 3, receivers. Block (c) 1, PC; 2, smartphone with software installed; 3, UAV controller. (Reproduced with permission [16]. Copyright (2020) the authors. This article is an open access article distributed under the terms and conditions of the Creative Commons Attribution (CC BY) license (http://creativecommons.org/licenses/by/4.0/).)

measurement systems is a drone which is equipped with a lux-meter light sensor, illuminance meter, and the GPS module (Figure 12.3). A camera with a stabilizer is mounted on the drone to collect and save data. The drone has a limited operating time and performs well in good weather conditions at 10–25° above zero, estimating artificial light pollution in the night sky [16].

12.7 SMART WIRELESS SENSOR NETWORKS FOR STREET LIGHTING

WSN was proposed by Mendalka et al. [17] for intelligent street lighting systems. Their system is made up of WSN nodes that communicate with light sources based on high-power LEDs. It can also be used for services other than lightings, such as telemetry, noise, humidity, and temperature monitoring, as well as services linked to road information systems, intelligent transportation systems, and intelligent roadways.

Fan and Guo [18] developed a street lighting system technology using the ZigBee-based wireless sensor network (WSN) for LED street lighting systems. In a later publication, Leccese et al. [19] presented a fully controlled street lighting isle based on the raspberry-Pi card, a ZigBee sensor network, and WiMAX. It was designed and organized in hierarchical layers that perform local operations to physically control the lamp posts and communicate information to another for remote control. Figure 12.4 shows the block schematic of the system architecture. Separately, each lamp post uses an electronic card for management, and a ZigBee network transmits data to a central control unit, which controls the whole isle. The central unit used the Raspberry-Pi control card.

In their work, the WiMAX connection was tested and used to remotely control the smart grid, thus overcoming the distance limitations of popular Wi-Fi networks. The isle was tested for some months in the field. As reported, the lamp post used a photovoltaic panel (for power supply) and ZigBee network/RaspBerry-Pi for the control. Furthermore, to overcome the absence of Asymmetric Digital Subscriber Line (ADSL) or 3rd Generation (3G) communication, which may not be available in some portions of the streets, the team used WiMAX technology, which allowed them to fully operate the isle from a remote station.

Badgaiyan and Sehgal [20] proposed an intelligent street lighting system based on several sensors, including pyroelectric infrared sensors (PIR) and a WSN (Figure 12.5). The authors also designed the intelligent streetlight system using the ZigBee device. In their study, the sensors or PIR sensors were utilized to detect the passing of a vehicle or

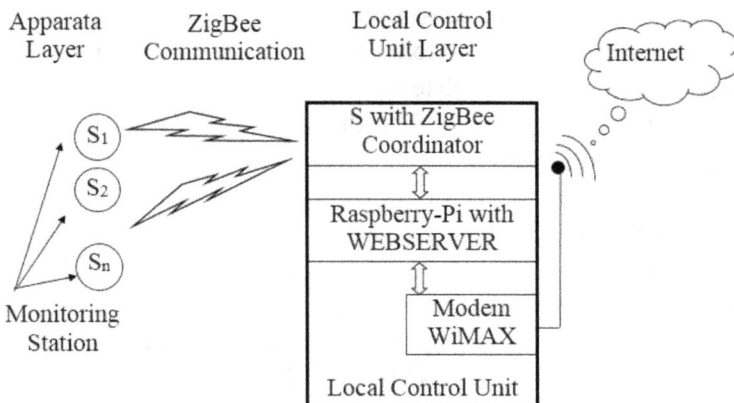

FIGURE 12.4 Block scheme of the system architecture. (Reproduced with permission [19]. Copyright (2014) the authors. This article is an open access article distributed under the terms and conditions of the Creative Commons Attribution (CC BY) license (http://creativecommons.org/licenses/by/4.0/).)

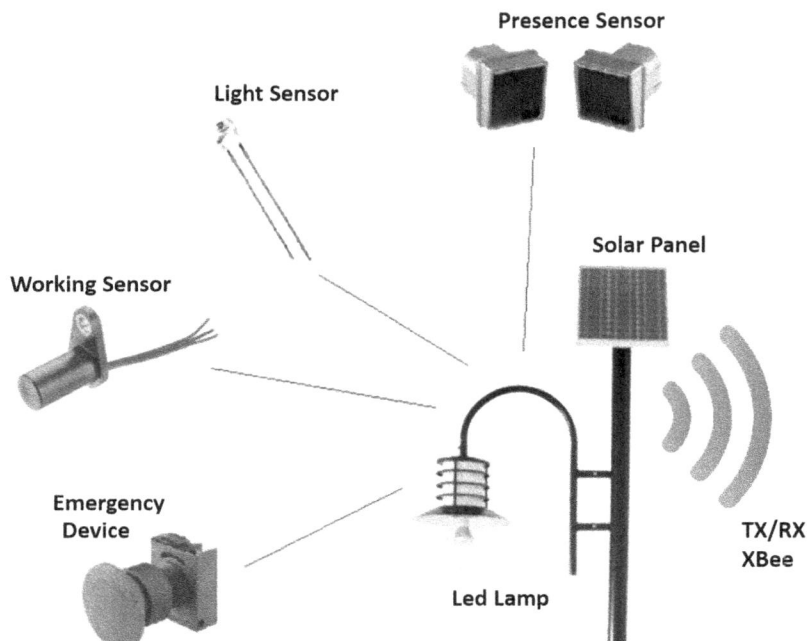

FIGURE 12.5   Scheme of lamppost with sensors. (Reproduced with permission [20]. Copyright (2015) the authors. This article is an open access article distributed under the terms and conditions of the Creative Commons Attribution (CC BY) license (http://creativecommons.org/licenses/by/4.0/).)

pedestrian, causing the lighting to turn on and off. Light sensors assess the intensity of external light and ensure a minimum level of road illumination, as required by legislation. These light sensors exhibit high sensitivity within the visible spectrum. The supervision module used sensors for improving fault management. In this regard, a Hall sensor detects when the lamp is switched on. The system then recognizes errors that are compared with the stored information. The ZigBee network reports this information to the station management unit and initiates corrective action. Figure 12.5 shows the various types of sensors.

Yusoff et al. [21] developed a complete smart street lighting system in Malaysia. Their system incorporated the WSN concept into its design, utilizing Waspmote as a sensor node, Meshlium as a gateway, and a computer as a server. For data logging purposes, the sensor node is used to monitor the running status information, amount of energy used, and dimming time of the street light. They discovered a way to save energy by regulating the light intensity using the Pulse Width Modulation technique. Lavric et al. [22] introduced a system for monitoring and controlling street lighting. To reduce the costs related to energy consumption and maintenance, dimming of the light intensity was carried out using a vehicle detection system. The system design is based on the WSN communication protocol.

Alexandru and Valentin [23] evaluated the performance of topology control algorithms integrated into a street lighting control sensor network. Their street lighting monitoring and control system was based on a WSN communication protocol. In their work, the A3, EECDS, and CDS-Rule K topology control algorithms were tested using the Atarraya simulation software.

The evaluated parameters were: i) the number of active nodes resulting from the use of topology construction algorithm; ii) the number of packets sent, and iii) the energy consumed while building the topology. Chunguo et al. [24] reported the geographical routing for WSN of street lighting monitoring and control system to monitor and control each street lighting. Their system consisted of a sensor node, a remote terminal unit (RTU), and a control center. The sensor nodes were integrated into each lighting pole and network with RTUs.

In their study, a geographical routing strategy was proposed based on the network features. The lighting nodes were divided into various clusters, according to the power substation powering scope. In each cluster, the nodes and power substation distribution is either star or triangle. The packet is forwarded by the node to the power substation, where the RTU is installed based on the distance to the RTU in star distribution. Their system uses cluster serial numbers to avoid packet dissemination to the neighboring clusters. To identify duplicate forwarded packets, the node serial number and packet timestamp were employed. The agent node in triangular distribution was used to solve the forwarding no convergence problem. As explained previously, the suggested geographical routing for street lighting systems is straightforward and does not necessitate the maintenance of network architecture or sophisticated path discovery algorithms.

Denardin et al. [25] proposed a control network for a LED street lighting system, using a WSN based on IEEE 802.15.4 standard. Their proposed control network enables disconnection of the street lighting system from the mains during the peak load time, automatically reducing its impact in the distributed power system at overload conditions. Besides, the system allows to reduce the power consumption, decrease the management cost and monitor the status information of each street lighting unit. Their network layer is implemented with a geographic routing strategy and a novel routing algorithm that offers low overhead and good scalability features.

Lavric et al. [26] presented the field testing of a street lighting monitoring and control system. Their system is built upon a large-scale WSN network that allows for remote control of street lighting lamps. Their study uses integrated Doppler sensors to detect the vehicle and help complete the power efficiency objective. When the system detects a vehicle, the light intensity of the lamps increases or decreases in the opposite situation. Fernandes et al. [27] reported the flexible WSN for smart lighting applications. Their work uses an efficient routing strategy for large areas WSN based on the IEEE802.15.4 standard. The authors designed the infrastructure for the remote control of the system and the validation in terms of the average delivery delay of the measurement/control information.

12.8 CONCLUSION

The rising urban population has resulted in the proliferation of artificial lights in cities, and this artificial light generates light pollution. This pollution is more evident at night, and one of its downsides is that it has a detrimental influence on astronomy, which has a high cost and affects plants, animals, and the entire ecosystem. Light pollution causes issues such as flicker and clutter. As a result, significant research has been provided to identify light pollution and solutions to lessen it. One of these ways is the use of sensors and smart WSN, which are frequently used to detect and reduce light pollution.

REFERENCES

[1] D.E. Lystrup, The dark side of the light: Rachel carson, light pollution, and a case for federal regulation, *Jurimetrics*. 57 (2017) 505–528. http://www.jstor.org/stable/26322762.

[2] N. Levin, C.C.M. Kyba, Q. Zhang, A. Sánchez de Miguel, M.O. Román, X. Li, B.A. Portnov, A.L. Molthan, A. Jechow, S.D. Miller, Z. Wang, R.M. Shrestha, C.D. Elvidge, Remote sensing of night lights: A review and an outlook for the future, *Remote Sens. Environ.* 237 (2020) 111443. https://doi.org/https://doi.org/10.1016/j.rse.2019.111443.

[3] D. Ziou, F. Kerouh, Estimation of light source colours for light pollution assessment, *Environ. Pollut.* 236 (2018) 844–849. https://doi.org/https://doi.org/10.1016/j.envpol.2018.02.022.

[4] A. Papalambrou, L.T. Doulos, Identifying, examining, and planning areas protected from light pollution. The case study of planning the first National Dark Sky Park in Greece, *Sustainability.* 11 (2019). https://doi.org/10.3390/su11215963.

[5] Y. Zhang, Y. Zhu, Z. Zeng, G. Zeng, R. Xiao, Y. Wang, Y. Hu, L. Tang, C. Feng, Sensors for the environmental pollutant detection: Are we already there? *Coord. Chem. Rev.* 431 (2021) 213681. https://doi.org/https://doi.org/10.1016/j.ccr.2020.213681.

[6] S. Sagadevan, M.M. Shahid, Z. Yiqiang, W.-C. Oh, T. Soga, J. Anita Lett, S.F. Alshahateet, I. Fatimah, A. Waqar, S. Paiman, M.R. Johan, Functionalized graphene-based nanocomposites for smart optoelectronic applications, *Nanotechnol. Rev.* 10 (2021) 605–635. https://doi.org/10.1515/ntrev-2021-0043.

[7] W. Wei, B. Zhou, Features detection based on a variational model in sensornets, *Int. J. Digit. Content Technol. Appl.* 4 (2010) 115–127.

[8] M. Penza, D. Suriano, M.G. Villani, L. Spinelle, M. Gerboles, Towards air quality indices in smart cities by calibrated low-cost sensors applied to networks, in: SENSORS, 2014 IEEE, 2014: pp. 2012–2017. https://doi.org/10.1109/ICSENS.2014.6985429.

[9] F. Marino, F. Leccese, S. Pizzuti, Adaptive street lighting predictive control, *Energy Procedia.* 111 (2017) 790–799. https://doi.org/https://doi.org/10.1016/j.egypro.2017.03.241.

[10] Y.S. Su, Competing in the global LED industry: The case of Taiwan, *Int. J. Photoenergy.* 2014 (2014). https://doi.org/10.1155/2014/735983.

[11] N. Schulte-Römer, J. Meier, M. Söding, E. Dannemann, The LED paradox: How light pollution challenges experts to reconsider sustainable lighting, *Sustainability.* 11 (2019) 6160.

[12] A. Djuretic, M. Kostic, Actual energy savings when replacing high-pressure sodium with LED luminaires in street lighting, *Energy.* 157 (2018) 367–378. https://doi.org/https://doi.org/10.1016/j.energy.2018.05.179.

[13] E. Juntunen, E. Tetri, O. Tapaninen, S. Yrjänä, V. Kondratyev, A. Sitomaniemi, H. Siirtola, E.-M. Sarjanoja, J. Aikio, V. Heikkinen, A smart LED luminaire for energy savings in pedestrian road lighting, *Light. Res. Technol.* 47 (2015) 103–115.

[14] C.-C. Sun, X.-H. Lee, I. Moreno, C.-H. Lee, Y.-W. Yu, T.-H. Yang, T.-Y. Chung, Design of LED street lighting adapted for free-form roads, *IEEE Photonics J.* 9 (2017) 1–13.

[15] W.S. Kardel, Rethinking how we light at night: Cutting light pollution for more sustainable nights, *J. Green Build.* 7 (2012) 3–15.

[16] P. Tabaka, Pilot measurement of illuminance in the context of light pollution performed with an unmanned aerial vehicle, *Remote Sens.* 12 (2020). https://doi.org/10.3390/rs12132124.

[17] M. Mendalka, M. Gadaj, L. Kulas, K. Nyka, WSN for intelligent street lighting system, in: 2010 2nd Int. Conf. Inf. Technol. ICIT), IEEE, 2010: pp. 99–100.

[18] C. Fan, Y. Guo, The application of a ZigBee based wireless sensor network in the LED street lamp control system, in: 2011 Int. Conf. Image Anal. Signal Process., IEEE, 2011: pp. 501–504.

[19] F. Leccese, M. Cagnetti, D. Trinca, A smart city application: A fully controlled street lighting isle based on Raspberry-Pi card, a ZigBee sensor network and WiMAX, *Sensors.* 14 (2014) 24408–24424.

[20] C. Badgaiyan, P. Sehgal, Smart street lighting system, *Int. J. Sci. Res.* 4 (2015) 271–274.

[21] Y.M. Yusoff, R. Rosli, M.U. Karnaluddin, M. Samad, Towards smart street lighting system in Malaysia, in: 2013 IEEE Symp. Wirel. Technol. Appl., IEEE, 2013: pp. 301–305.

[22] A. Lavric, V. Popa, I. Finis, The design of a street lighting monitoring and control system, in: 2012 Int. Conf. Expo. Electr. Power Eng., IEEE, 2012: pp. 314–317.

[23] L. Alexandru, P. Valentin, Performance evaluation of topology control algorithms that can be integrated into a street lighting control sensor network, in: 2013 11th RoEduNet Int. Conf., IEEE, 2013: pp. 1–4.

[24] C. Jing, L. Ren, D. Gu, Geographical routing for WSN of street lighting monitoring and control system, in: 2010 Int. Conf. Comput. Des. Appl., IEEE, 2010: pp. V3–235.

[25] G.W. Denardin, C.H. Barriquello, A. Campos, R.A. Pinto, M.A. Dalla Costa, R.N. do Prado, Control network for modern street lighting systems, in: 2011 IEEE Int. Symp. Ind. Electron., IEEE, 2011: pp. 1282–1289.

[26] A. Lavric, V. Popa, S. Sfichi, Street lighting control system based on large-scale WSN: A step towards a smart city, in: 2014 Int. Conf. Expo. Electr. Power Eng., IEEE, 2014: pp. 673–676.

[27] R.F. Fernandes, C.C. Fonseca, D. Brandão, P. Ferrari, A. Flammini, A. Vezzoli, Flexible Wireless Sensor Network for smart lighting applications, in: 2014 IEEE Int. Instrum. Meas. Technol. Conf. Proc., IEEE, 2014: pp. 434–439.

CHAPTER 13

Photonic Nanodevices and Technologies against Light Pollution

Elisangela Pacheco da Silva, Elizângela Hafemann Fragal, and Ederson Dias Pereira Duarte

UEM - State University of Maringa

Sidney A. Lourenço

Federal University of Technology (UTFPR)

Edvani C. Muniz

UEM - State University of Maringa

Federal University of Technology (UTFPR)

Federal University of Piauí

Thiago Sequinel

University of Grande Dourados

Rafael Silva

UEM - State University of Maringa

Eduardo José de Arruda

University of Grande Dourados

Luiz Fernando Gorup

Federal University of Rio Grande

Federal University of Pelotas

Federal University of Alfenas

DOI: 10.1201/9781003185109-13

Vanessa Hafemann Fragal

UEM - State University of Maringa

CONTENTS

13.1 INTRODUCTION

Artificial light has played a role in the development of human civilizations, since its first use under the form of controlled fire up to the invention of electric lighting. The purpose of light's use has changed throughout the millennia. Light has been used to help our ancestors to defend themselves from wild animals, warming, and cooking. Nowadays, artificial light has the purpose to provide nightlife in urban centers, improving the productivity in business through the implementation of night shifts, and making the city more attractive to tourists.

The acquired knowledge of lighting principles led to advancements in light source design, production, and materials, mainly focusing on brighter and more efficient light sources. These efforts brought us the latest LED (Light Emitting Diode) and laser sources. On the other hand, the civilizing progress for obtaining artificial lighting, which enabled us to develop and populate huge, high, and dense urban areas, produced now a side effect of its extensive use: the light pollution issue, in its diverse forms such as skyglow, glare, light trespass, and clutter, among others [1].

Specifically, in the visible range of the light spectrum, researchers are reporting harmful impacts of light pollution in human health, ecosystem functions, and astronomical observation. On the ecosystem side, alterations in some plants' leaf growth cycle and changing the stability of photosensitive molecules and their functionalities are observed in wild areas affected by light pollution; species of animals that use lighting differences as reference are suffering from disorientation and disorder in predator-prey relationships due to the availability of too much light in wider time ranges. Researchers have also shown that light pollution can be harmful to human health because light pollution affects the day/night biological system, the synthesis of melatonin, and/or could induce disorders in the rest of the individuals. Beyond, trends in sleep quality reduction have emerged from recent research. Science itself is also experiencing drawbacks due to the influence of light pollution in astronomical observances [1].

All the above-mentioned side effects opened windows for a new branch of research in the lighting field focused on light pollution reduction and the adequacy of lighting sources and systems for different purposes. Besides new designs and materials applied to already widely used lighting devices, the development of nanodevices against light pollution is a fertile field for further research and innovation, and its results can lead not only to scientific gains but also to economic and environmental.

In the context of global discussions about sustainable development, pushed by political pressures to increase and fasten the adoption of green technologies, the concept of smart cities has emerged. One definition of a smart city is an urban that "can use a smart solution in various sectors of society, employing data and digital technology to improve the quality of life of the citizens", in addition to adapting the use of available resources by integrating intelligent systems. The use of efficient sources of lights and photodetectors as photoresistors, photodiodes, and phototransistors in artificial light designs, boosted by its combination with other technologies like motion sensors, dimmers, and Internet of Things (IoT) networks finds a place in the projected smart cities, taking part in the goal of improving its citizens' quality of life, while also reducing the light pollution caused by the city itself [2].

The goal of this chapter is to focus on the two types of nanodevices that can be used against light pollution – the sources of lights and the photodetectors, and to make the readers familiar with them. Hence, this chapter is organized as follows. Section 13.2 brings some concepts of light, along with its terms and principles. Sections 13.3 and 13.4 demonstrate the sources of light pollution and their impacts on life and the environment. In Section 13.5 the notion of smart cities and their components are presented. Section 13.6 describes the two types of nanodevices that can be used against light pollution, specifically LEDs, photoresistors, photodiodes, and phototransistors. Still in this section, we try to answer two main questions: How do they work and how do they act against light pollution, besides bringing the recent advances in nanomaterials to enhance photodetectors' sensitivity. Section 13.7 describes the economic value of these photonic devices for the world and the impact of the COVID pandemic in the market. Finally, the future directions and challenges related to photonic nanodevice for light pollution are presented in Section 13.8.

13.2 LIGHTING PRINCIPLES AND TERMS

To talk about light pollution, a brief understanding of light itself – the core object of the pollution we are talking about – is necessary. Light is an essential part of our lives, and it can be produced naturally or by human action. Visible light is defined as "the radiation

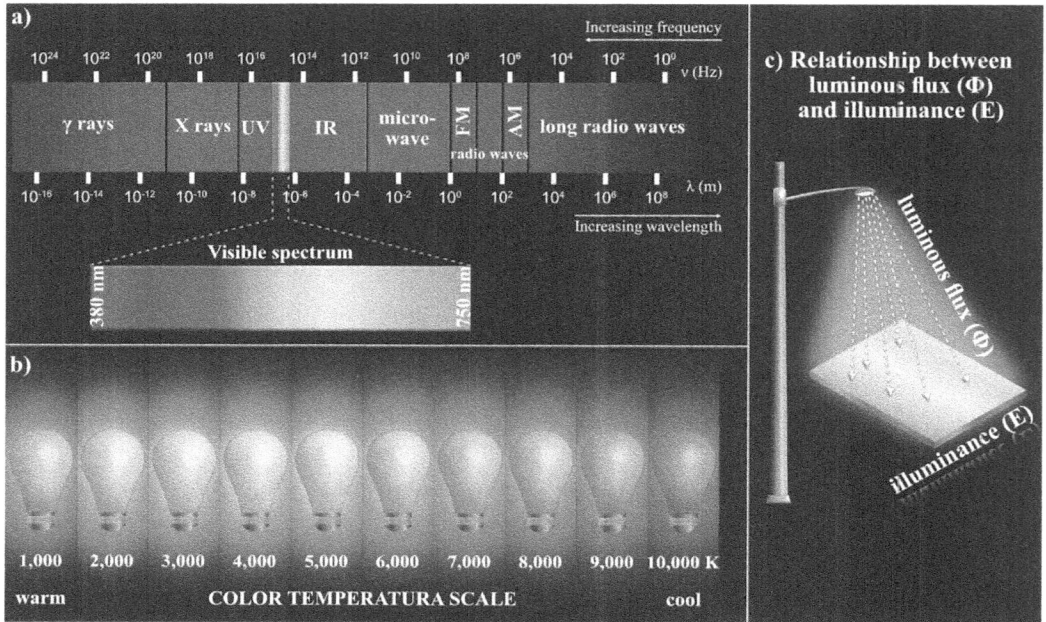

FIGURE 13.1 (a) Electromagnetic spectrum – zoom visible spectrum. (b) Range of color temperature of artificial light is from 2,000 K (warm) up to 10,000 K (cool) and (c) Relationship between luminous flux (Φ) and illuminance (E).

inside the spectrum of electromagnetic radiation which its wavelengths (λ) can be perceived by the human eye". More specifically, the wavelengths between 380 and 780 nm [3], as shown in Figure 13.1a.

The wavelength is responsible to determine the color in visible light, which can be monochromatic – a single wavelength – or polychromatic – multiples wavelengths. Color is a fundamental perception of the human eye (our light sensor) and this perception is changed by the environment and conditions to which humans are exposed. The chromatic characteristic of light is divided into correlated color temperature (CCT – given in degrees of Kelvin) and color reproduction index (CRI). The CCT range of artificial light is from 2,000 K (warm) up to 10,000 K (cool), as shown in Figure 13.1b. Meanwhile, the CRI (%) represents the ability of a light source to accurately reproduce the colors of the object it illuminates. As higher the index as accurate is the reproduction [4].

The perception of light is also a personal experience, which varies according to biological differences between the observers. Therefore, to achieve quantifiable measurements of the visual sensation caused by light, the photometry branch was developed.

The actual SI's photometry base unit is the candela (cd), the unit for luminous intensity in a given direction. In the lighting subject, other photometric measurements are also relevant, being some of them: (i) luminous flux (Φ), measured in lumens, for which the SI unit is lm; (ii) the illuminance (E), measured in lux (lx=lm/m^2); and (iii) the luminance (L), measured in candela per square meter (cd/m^2) (Figure 13.1c). Luminous flux (Φ), in general terms, is the total light flux emitted in every direction from a lamp, and its derived illuminance through the measurement of the flux emitted in all directions around the lamp that is measured in lumens (lm). The illuminance (E), or luminous flux density, is the luminous flux incident over a given area. The luminance (L) is the measurement of the luminous intensity per area which flows in a given direction. Simplifying, it is the measurement of the amount of light emitted, reflected, or simply passes by a given area, being the standard candela per square meter (cd/m^2) [3].

13.3 SOURCES OF LIGHT POLLUTION

Light pollution, like other types of pollution (such as water, air, sound, land, and others) is not a natural phenomenon. In fact, it comes from source(s) which usually is a side effect of another human activity. As appointed by the International Dark-Sky Association (IDA), "light pollution is a side effect of industrial civilization" [1]. The rapid population growth and the migration for urban areas resulted in the great consumption of artificial lights to supply the unending demands. With the advance of technology, from whale oil to LED, the cities have been becoming ever brighter. The use of new technologies not only reduced the waste energy but made the city more beautiful with an irresistible view, attracting tourists from worldwide.

In this way, light pollution is an effect caused by excessive and inappropriate use of artificial lighting, which is poorly directed, and reflected on surfaces in the space where it is inserted. Residential lighting, advertising, offices, industries, street lighting, and sports venues, among others, are the main sources of excessive lighting [1]. In urban areas, light pollution is determined by the portion of the luminous flux that does not reach the useful area. As light has some principles and terms, there are different types of light pollution, such as (i) Glare ("excessive brightness that causes visual discomfort"), (ii) Clutter ("bright, confusing and excessive groupings of light sources"), (iii) Light trespass ("light falling where it is not intended or needed) and iv) skyglow ("brightening of the night sky over inhabited areas"), Figure 13.2a. A brief description of these categories is given below [1]:

i) Glare is defined as the impairment of visual function caused by the presence of a light source located in the visual field, it can be a direct glare (direct visualization of the lamp) or an indirect glare (reflected through surfaces reflective or shiny).

ii) Clutter refers to excessive clustering of lights that can create confusion from obstacles (including those that can be configured to light), and potentially cause accidents. Clutter is particularly noticeable on roads where street lights are not well designed or where well-lit advertisements surround the roads. The motives of the person or organization that installed the lights, their placement, and design may even distract drivers and could contribute to accidents.

iii) Intrusive light or light trespass refers to that portion of the luminous flux of a luminaire that directly affects the properties and facades of buildings, where it is not intended or needed.

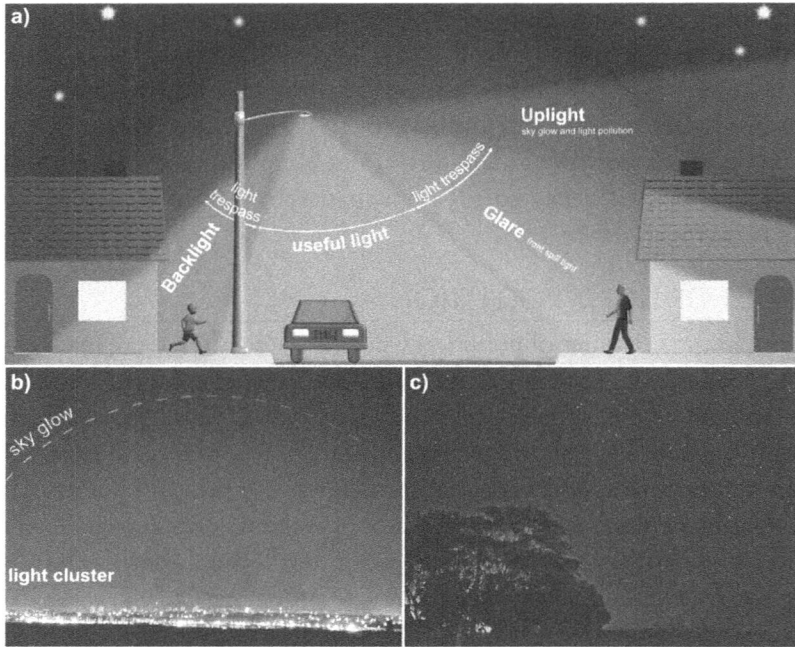

FIGURE 13.2   (a) Types of light pollution, and (b and c) Effects caused by light pollution (photos taken close to a city in Mato Grosso do Sul state, Brazil).

iv) Skyglow is an effect of all the luminous flux that flows in the sky direction, as by direct light trespass or indirectly by light reflected in the ground or water. This light is also reflected by the sky, widespread the pollution coming from cities and taking it to rural or unhabituated zones.

Some of the side effects caused by light pollution are: astronomical light pollution – reducing the number of visible stars; unshielded light – pollutes ecological and astronomers' environments, tall lighted structures – can cause animals collisions; reflected light shield-reduces astronomers' vision and disrupts natural ecosystems, Figure 13.2b.

13.4 IMPACTS OF LIGHT POLLUTION ON LIFE AND THE ENVIRONMENT

Light pollution is one of the fastest-growing and least known types of environmental degradation. People are delighted with a bright city, but hardly anyone knows the negative impact of excessive artificial light on wildlife and human health. Natural light is essential for the living organism; however, light pollution masks the natural light cycle and modified the environment's illumination. Due to the fast urbanization and exponential growth of population, more people are experiencing light pollution without noticing.

A few generations ago, the natural night sky was still visible, more extended, and darker as compared to nowadays. Presently, countless light that illuminates parking lots, streets, and stadiums lose light to the night sky, scattered by molecules of air and particles of smog. The IDA considered light pollution an unfavorable effect of artificial light that decreased visibility at night and wastes energy [1]. Examples are skyglow, glare, light trespass, and light clutter. The main factor to light pollution and skyglow is advertising panels and scenic light projectors directed upward no shielded streetlights, and excessive artificial light sources [5].

There are uncountable consequences of light pollution. The excess of light is endangering the ecosystems the human health, disturbing the work of professional astronomers, and still wasting energy. From astronomers' perspective, light pollution is a massive problem since they usually spend a lot of money building their observatories under the darkest skies. Amateur astronomers have the same problem, especially in the USA [6]. Since light pollution has increased in towns and cities, it is tough to see visible stars and other astronomical objects. Collecting data is a huge problem, time exposure photos to record a distance are limited by the sky glow to fog film and digital images. In this way, astronomy has suffered due to people no longer being interested in science.

The increase in artificial light usage nowadays is also considered a significant driver behind the alteration in the circadian rhythm of humans, plants, and animals. From a behavioral point of view, light pollution negatively affects the fauna and flora's natural and population size. Plants that are extensively exposed to artificial lights leaf out faster and consequently lose leaves more quickly with a more extended growth period and mortification in the composition [7]. Artificial night lighting may disorient animals that navigate in a dark environment. The range of anatomical adaptations to allow night vision is broad and can blind animals with an increase in light pollution.

Furthermore, predation increases with the altered night landscape, which benefits predators and diurnal reptiles that forage under artificial lights. This may disrupt the predator-prey relationship and ecosystem functions and can also affect the reproductive process [8].

Another example is associated with polarized light pollution from glass-encapsulated photovoltaic modules. Unfortunately, such polarized light pollution attracts different species of aquatic water insects by the horizontally polarized light they reflect. Polarized pollution can be harmful to aquatic insects trapped from this light signal and die before reproduction or lay their eggs in inappropriate places [9].

Most of the species in terrestrial and aquatic ecosystems use ambient light to grow and regulate their metabolism. This ecosystem behavior is strongest influenced by the ambient light level, evidenced by a slip of species into diurnal or nocturnal activity. The artificial lighting causes a sky glow over urban areas, causing several impacts on organisms' behavior, disrupting these natural cycles and physiological changes for many nocturnal species. Most organisms included humans, have been involved [10].

The mean impact of light pollution on the physiological functions of wildlife is still unknown. Most of the studies from days' birds show that they start their activity and sing earlier when exposed to light pollution. A study in 2013 shows that melatonin, a photosensitive hormone, is partially inhibited in birds exposed to light pollution. Considering the sphere of lights, birds may collide with each other, become exhausted, or be caught by predators and attract birds to boats, broadcast towers, smokestacks, and so on, resulting in direct mortality, interfering with migration routes. During the breeding period, the breeding colonies of seabirds are highly affected by light pollution. The main impact is associated with the abandonment of colonies and the disorienting effect on immature individuals who disperse from their colonies for the first time. These pieces of evidence highlight the negative impacts of light pollution concerning the biological processes of animals [7,11].

Most organisms, including humans, have developed, over time, molecular circadian systems that are adjusted by night and day cycles. Before the invention of artificial light, physiological and behavioral traits were determined by the climate and the movement of the stars, sun, and moon. For humans, epidemiological and clinical studies in urban areas show that light pollution represents a risk to human health by altering the day/night cycle [12]. Below is explained how light pollution can affect human health.

- **Circadian Disruption Hypothesis:** Light pollution can desynchronize concerning the day/night cycle.

- **Melatonin Hypothesis:** Light pollution promotes decreases pineal melatonin production and secretion. Melatonin deficiency alters biological rhythms and loss of antioxidants, anti-inflammatory, and so on.

- **Sleep Disruption Hypothesis:** Light pollution could reduce the quality and quantity of sleep. These factors can also be associated with the maintenance of internal homeostasis.

13.5 SMART CITY AGAINST LIGHT POLLUTION

Currently, light is the main focus of urban architecture thanks to technological innovations in LED lighting. Lights stimulate people's emotions and can create a happy sensation. The brighter and illuminated more beautiful is the city, the effect is awe-inspiring, and it is a strategy used to attract tourists in the hours after dark. More and more the streets are lit up, to make the city attractive, besides providing security and comfort to the pedestrian, cyclists, and vehicle drivers. However, as mentioned before, the excess of brightness is harmful to the environment and human health, in addition to contributing to the waste of energy. In this way, different technologies were developed and are being used to overcome the issues caused by excessive lighting systems, mainly to control and decrease energy consumption in smart cities.

The word *smart* gained strength in the last decades. The concept of *smart city* has been consolidated as a fundamental subject in the global discussion about sustainable development, whose global market was valued at USD 739.78 billion in 2020 and is expected to reach USD 2,036.10 billion by 2026. This intense growth is driven by the increase in the adoption of green technologies [13].

What is a smart city anyway? And what is the goal of smart cities? According to Georgescu and Popescul, a city is considered smart when it can use a smart solution in various sectors of society, employing data and digital technology to improve the life's quality of the citizens [14]. Traffic management, parking system, energy management, street lighting optimization, environmental and territorial monitoring are the goals of intelligent cities.

However, to reach these goals, information and communication technology are increasingly necessary to overcome many of the impediments of the traditional city operation.

These technologies are incorporated in the urban infrastructure to provide solutions such as centralized control rooms, intelligent transport systems, controllable lighting, sensor networks, and so forth. In this context, the IoT plays a major role in various sectors and domains. IoT has been defined as "*A global infrastructure for the information society, enabling advanced services by interconnecting (physical and virtual) things based on existing and evolving interoperable information and communication technologies*" [15]. Therefore, IoT devices are designed to make a city smarter all over the world. Some good books and reviews that bring a depth understanding of IoT and smart cities and how they are interconnected are available on the internet [16–18]. Here we focus on how green devices are interconnected to the IoT to make a city smarter, improve the street light performance, and their potential for reducing light pollution.

Streetlight is one the major source of light pollution. So, to lessen the light pollution, smart cities have employed intelligent technologies to improve their public illumination infrastructure. Among these technologies, we can highlight motion sensors, dimmers, and photonic materials (photodetectors, and LEDs) to compose IoT devices. IoT is not a single device but a collection of various of these technologies. Each one of these technologies has a specific function to form an effective IoT system in smart street lights, as described below.

- **Motion Sensors:** Motion sensors work to detect the nearby movement of people, cars, cyclists, etc., and then send to the microcontroller the information to adjust the lighting according to the requirement. They are the core component in the system of the smart street light. There are several types of motion sensors, like infrared (IR), microwave, ultrasonic, and piezoelectric sensors, each one works differently.

- **Dimmers:** Dimmer is a device that uses a simple electrical principle to reduce the brightness or switch ON and OFF the light depending on the circuit program. The modern dimmer is controlled by a digital control system, but it can be directly controlled in domestic use.

- **Photonic Devices:** Photonic device is a technology that enables the generation, transmission, processing, and detection of light and has been crucial in public illumination. They are the heart of many IoT innovations. The major applications of photonic materials to smart cities are lightning, detectors, and communications. LEDs are the main sources of light, which can reduce about 50% of energy consumption. In addition, it is possible to automatically control and adjust the optical characteristics. Photodetectors can be classified into photoresistors, photodiodes, and phototransistors, each of them can work to control the street lights in a smart system, changing their physical properties according to the light incidence and local demand.

- **Networking Communication:** Thanks to the fast internet networking communication can be used to interconnect the devices to become the city smarter. Optical fiber has been highlighted due to its fast speed without affecting the overall performance. However, a new generation of optical fibers currently being developed could outperform conventional fibers for many applications, known as photonic crystal fiber [19]. This photonic material in the future to carry the information over long distances, since its morphology can guide the light more efficiently.

The public illumination can host these devices that can monitor the city and communicate with each other through the internet. These devices are intimately interconnected with the communications technologies to provide a smart street light. In resume, the working principle of the IoT device is based on "sensing" the motion of a person, vehicle, or cyclist, sending the information for a microcontroller composed of photodetectors and dimmers, which receive the signal and control the LED illumination according to the requirement, i.e., if there no movement is detected, the intensity of LEDs can be reduced or even turned OFF. The devices collect data in real-time, and if any failure is detected, the information is transmitted to the monitoring room in a central area of the city to restore the failure. Thus, the operation cost is highly reduced, the energy is saved and the light pollution can be minimized. Figure 13.3, shows an example of smart street light, its components, and the working principle.

The more recent technology used by smart cities is the solar cell street light. The solar cell or photovoltaic cell is an electrical device that absorbs solar energy during the day-time and converts this solar energy into electrical energy that can power the lamps during the nighttime [20]. This means that solar cell street lights can produce clean and sustainable energy, reduce their energy consumption as well as their carbon footprint and consequently limit their environmental impact. In addition, solar cell street lights can help to

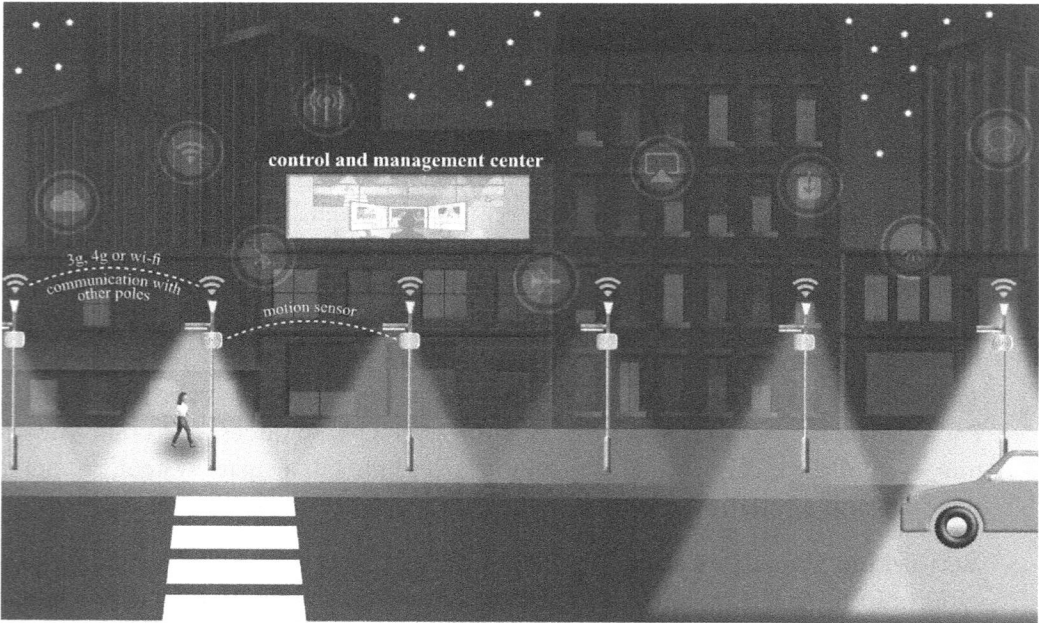

FIGURE 13.3 Schematic demonstration of how smart city acts against light pollution.

minimize light pollution, by the adjust the intensity of the light over the night, and thus, preserving the biodiversity sensitivity to excessive light.

Based on the aforementioned, in the next sections, we will focus on the main smart photonic nanodevices used against light pollution, their working principle, and the recent developments in their nanomaterials.

13.6 TYPES OF PHOTONIC DEVICES TO MITIGATE LIGHT POLLUTION

Photonic devices are components for creating, manipulating, or detecting light. Photonic encompass various types of devices, which are used in many areas from everyday life to the most advanced science. The potential application of photonic nanodevices against light pollution makes them the heart of smart cities. They can contribute to controlling light, reducing waste energy and still have an important role in the communications systems. Therefore, a brief overview of two main photonic devices, LEDs and photodetectors, is highlighted below.

13.6.1 Light Sources

13.6.1.1 Brief Overview

The need to possess light – other than the sun – began around 50,000 years before Christ, when there are records of a kind of candle. They were dishes with animal fat and a burning wick, only after Christ (A.C) that the rod-shaped candles we know today appeared. After the candle came the lamp. A container where whale oil was placed and a wick embedded in a piece of cork generated the flame. In 1807, a form of using gas to light was discovered become streets of London were lit for the first time with lamps. However, the light used in the houses and streets of the cities was based on kerosene, gas, or incandescent oil. These substances could cause poisoning and fire, in addition to not providing efficient lighting [21].

At the end of the 19th century, there was a decisive event for the evolution of electrical energy globally – the emergence of electric lamps. There were countless ideas and discoveries from the work of tireless minds that 'brought to light' – literally – one of the most important creations of humanity. Electric light is such a familiar phenomenon these days that we fail to realize the immense impact this invention has had on Earth over the past few centuries. And it all started over 200 years ago when the prototypes of what would become the light bulb we know today were created.

Currently, the evolution of light bulbs has been very fast. After incandescent lamps came neon, fluorescent, halogen, and sodium vapor, we are unveiling LED, OLED, and Quantum Dot Light-Emitting Diodes (QLED) lamps. The neon lamp was created in 1912 by the French chemist Georges Claude and was widely used in signs, in addition to being decorative. In 1938, the fluorescent lamp (made with argon and mercury gas, created by Nikola Tesla), arrived on the market. This model emits more energy in the form of light than heat, which also makes it more economical, savings of around 75% in its consumption. The halogen dichroic lamp appeared 20 years later – another type of incandescent lamp that contains cheap halogen elements such as bromine or iodine, being still more efficient and economical due to extended durability. However, it generates more heat than other lamps, and this increases energy consumption [21].

Traditional incandescent	Halogen incandescent	Compact fluorescent (CFL)	Light-emitting diode (LED)

Voltage (W): approximate voltage needed to produce 1,600 lumens

100	77	23	20

Energy saving (%)

0	30	75	83

Life span (h)

750	1,000	10,000	20,000

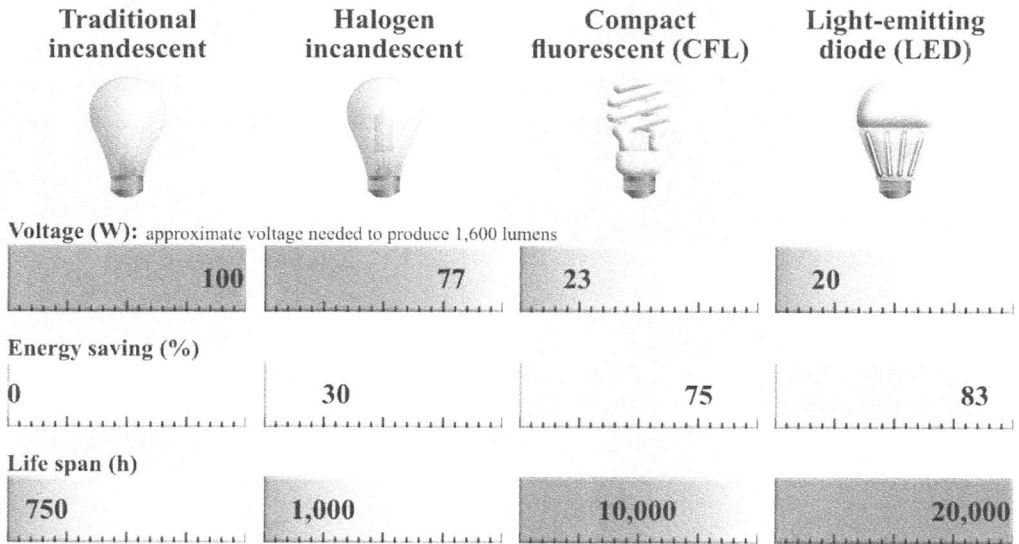

FIGURE 13.4 Characteristics of incandescent, halogen, fluorescent, and LED lamps.

For applications in public lighting and other outdoor areas that require high luminous flux, the technologies available on the market in the last 10 years have evolved significantly. High-intensity discharge lamps such as high-pressure mercury and sodium and metal halide dominated the market for many years. The first sodium vapor lamp appeared in 1930, but until 1962 that a model with high luminous efficiency and long durability was developed. It emits white and gold light and is a significantly economical source, widely used in avenues, industries, airports, and other large spaces [21].

However, LED technology was improved and tested for various applications reaching increasingly higher luminous fluxes until the manufacturers in the sector started their production in scale after evaluating that it was an economically viable technology. LED lamps promise to be the revolution for years to come, as they are rich in energy efficiency, durability, and economy. It is the ideal lamp for the current evolution of the world that seeks social sustainability and energy resources Figure 13.4 [22].

13.6.1.2 LEDs- Efficient Energy Sources against Light Pollution

LED is an electronic device that emits light through solid, semiconductor material such as gallium arsenide (GaAs) or gallium phosphide (GaP), with positive and negative polarities when energized by an electrical current. Figure 13.5a illustrates the basic LED conformation. LED is a semiconductor diode of the type P-N junction, which, when energized, emits visible light – hence LED (Light-Emitting Diode). Light is not monochromatic but consists of a relatively narrow spectral band produced by the electron's energetic interactions. The light-emitting process by applying an electrical source of energy is called electroluminescence. At any forward-biased P-N junction, within the structure, close to the junction, hole and electron recombination occurs. This recombination requires the energy possessed by the electrons to be released, which occurs in the form of heat or photons of light Figure 13.5b [23].

a) LED structural scheme

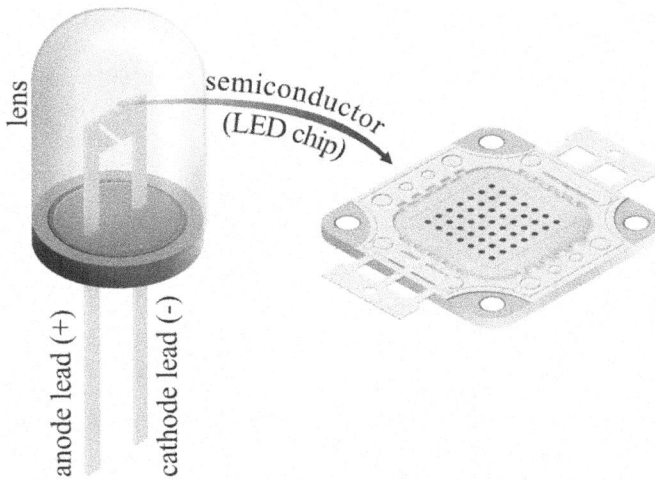

b) Principles of LED light operation

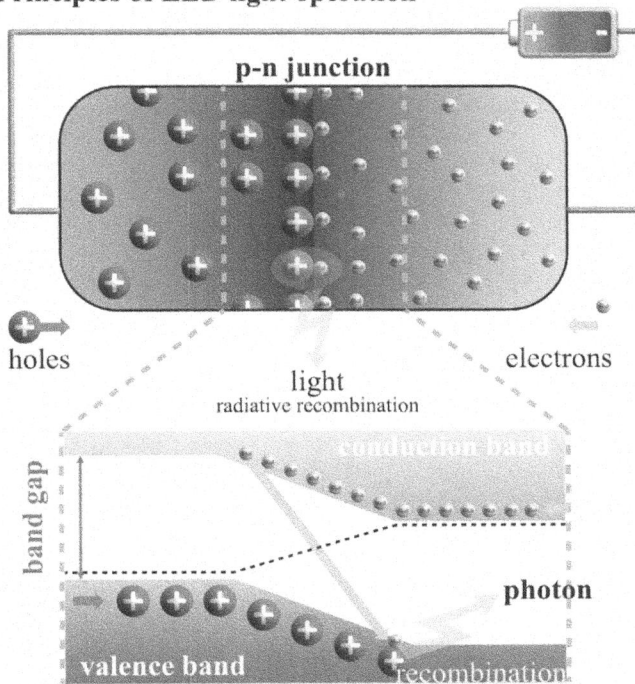

FIGURE 13.5   (a) LED structural scheme and (b) Principles of LED light operation.

The milestone for the use of LEDs for lighting purposes was the development of LEDs that emit white light. Either through the combination of the existing monochromatic LEDs (Red, Blue, and Green – RGB) or by the phosphor used in the Blue LED, which converts part of the blue radiation into yellow light – mixes with the rest of the blue light producing white light. As for its Spectral Energy Distribution (SPD), the cold white light LED (6,500 K) has

TABLE 13.1 Characteristics and Applications of Public Lighting Lamps

Lamps	Characteristics	Applications and Cost/ Benefit
High-Pressure Sodium Vapor	• High to very high-efficiency lamps; • Poor to moderate CRI; • Heating periods relatively long; • Luminous efficiency between 70–120 lm/W; • Lamp life from 5,000 to 30,000 h; • Low to medium CRIs and CCT (light in yellowish-orange tones)	• Economical public lighting and industrial lighting where high CRI is not required
High-Pressure Mercury Vapor	• Low to medium efficiency lamps; • Moderate CRI; • Heating periods relatively long. • uminous efficiency between 23 and 60 lm/W; • Lifespan from 6,000 to 28,000 h. • CCT low to medium	• Mainly used for street lighting, security, and industrial. • Low upfront costs, but not profitable for the long lamp life compared to equivalent alternatives.
Metallic Vapor Lamps	• Medium to high-efficiency lamps; • Several lumen pack options; • Warm-up time of a few minutes; • Difficult darkening; • Low lumen maintenance; • uminous efficiency between 47 and 105 lm/W • Lamp life from 6,000 to 20,000 h; • High CRI; • Wide range of CCT.	• Suitable for long hours of operation. • Most commonly used for industrial and street lighting
LED	• High luminous efficiency 70–160 lm/W; • Very High Lifespan 40,000–90,000 h or more; • High CRI (70–90+); • Wide range of CCT (3,000 K–6,500 K). • Greater controllability and adjustability or that is, instantly dimmable and with possible color temperature change.	• They are more expensive than other steam lamps. • High initial cost, but economical in consumption and in lifetime. • Growing application in Public Lighting.

Source: Data adapted from [24].

a predominant configuration of the blue spectrum (450 nm), while the neutral white light LED (4,200 K) has a more distributed spectral configuration. On the other hand, LEDs with low color temperatures (3,000 K and 2,400 K), with a warm light appearance, have a greater amount of red spectrum (630 nm).

Practically in the 2000s – LED light bulbs finally arrive in the lighting business with a lot of enthusiasm with the significant advantages that LED presents over others and is now considered the future of lighting. The LED lamp is more efficient than the others due to the difference in consumption. Table 13.1 shows the main differences between LED and other sources of light used in public illumination.

For public lighting, the light color typically varies between yellowish, neutral, and bluish-white corresponding to color temperatures between 2,500 and 5,000 K. The typical color specifications for the LED in outdoor areas at the beginning of its adoption were 5,000 K or more. However, there was a shift in the market, since this CCT has a predominant configuration of the blue spectrum. A high level of blue light in cold white light sources, on the other hand, may also be prejudicial to human health. Continuous exposure to these blue lights has been linked to cataract formation. So, turning them on at night can interrupt the production of sleep-inducing hormones such as melatonin, causing insomnia and fatigue [12]. Thus, more neutral CCTs (4,000 K) have been used, and currently, there is already a larger offer of products with CCTs around 3,000 K or lower. In this way, studies have shown that LED road light sources attract fewer insects than other technologies, with "warm white" LEDs (color temperature of 3,000 K) resulting in significantly lower numbers than with "cold white" LEDs (color temperature of 6,000 K) [4].

LED lighting, in contrast to various old lighting technologies, offers the opportunity to adjust or select color temperatures flexibly for various applications. This need can only be achieved thanks to the lamp efficiency in a branch of temperatures. Another optical functionality of LED is the greatest capacity to vary the emitted light cores and the presence of a current regulator (dimmer), which controls the light levels. Moreover, one way of reducing light pollution is to use luminaires which direct the light only on the areas to be illuminated. Directional light sources incorporating LEDs are especially suited for achieving optimized light distribution. Light emissions above the light source are generally not desired. In addition, the design of the LED lamp is easily adapted. Streetlight is recommended to have a design that avoids the effects like skyglow, glare, and light trespass and illuminate only angles and areas necessary for the safety and comfort of the population.

LED has greater advantages compared to other technologies that precede offer many possibilities to increase the feeling of comfort, safety, and well-being of human beings. In addition, the new lighting scenario possibilities enabled by the adjustability and controllability of LED luminaires can help to minimize the impact of light pollution on the urban environment and human beings and, subsequently, increase the community's well-being without affecting the particular characteristics of each location, historical and cultural preferences.

Recent developments in LED technology have added red-emitting phosphors to create a warmer white LED lamp for night use, but this only masks the blue hue without getting rid of it. Therefore, it is necessary to develop a material that would result in warm white light when used in a violet LED device, avoiding the problematic wavelength range [25]. Another evolution of LED is the OLEDs. Since the 1950s scientists studied the fundamental aspects of light generation from organic materials and proposed some devices. OLEDs are constructed by stacking carbon-based organic materials on a glass or plastic substrate. OLEDs have a larger area, are surface lights, and, due to the characteristics of the organic material, can be very thin and flexible. Although the new technology is advancing rapidly and it has many interesting applications, such as high-end flat TVs, OLED is not yet suitable for street lighting applications [23].

More recently, has emerged the QLED, which is the newest form of light-emitting technology and consists of nanoscale crystals that can provide an alternative for applications such as display technology. The structure of a QLED is very similar to OLED technology. But the difference is that the light-emitting centers are nanocrystals of cadmium selenide (CdSe), or quantum dots. QLEDs are a reliable, energy-efficient, adjustable color solution for display and lighting applications that reduce manufacturing costs while employing ultra-thin, transparent, or flexible materials, but like OLED is not available for public illumination [26].

13.6.2 Photodetectors Against Light Pollution

Photodetectors are one of the smart photonic devices used in intelligent cities against light pollution. The main class of photodetectors is photoconductive devices, which are generally made from semiconductor materials. Photoconductive materials can conduct electricity when exposed to light and are used to control light intensity and turn ON/OFF light. Among photoconductive devices, photoresistors, photodiodes, and phototransistors are the most used in light applications. In this way, we describe these photoconductive devices in more detail, showing their working principle, how their act against light pollution, and recent advances in nanomaterials for photodetectors applications.

13.6.2.1 Photoresistors

Photoresistors, also known as LDR (light-dependent resistors) or photoconductive cells, are light-controlled photodevices whose resistance changes with the amount of light they are exposed to. This type of sensor is composed of photoconductor material deposited onto a semi-insulating substrate. Usually, the conductor material is arranged in a zigzag pattern, separated by a pair of metal contact, to increase the area of the photoresistor that is exposed to light [27].

Cadmium sulfide (CdS), cadmium selenium (CdSe), aluminum sulfide (Al_2S_3), lead sulfide (PbS), and bismuth sulfide (Bi_2S_3) are the most common materials used as a photoconductor in photoresistors applications. The type of semiconductor used depends on the spectral characteristic of the desired application. In other words, photo devices made from different materials respond differently to light of different wavelengths. For example, according to the spectral characteristics of the photoresistor, it can be divided into three types of photoresistors, (i) Ultraviolet photoresistor; (ii) Infrared photoresistor; and (iii) Visible-light photoresistor [28]. For visible-light photoresistors, the semiconductor, commonly the CdS, presents sensibility curves similar to the human eye, which makes them widely used for street lights and other lighting systems.

Another division of photoresistors is based on characteristics of semiconductors used, intrinsic or extrinsic semiconductors. The intrinsic photoresistors are made from un-doped semiconductors, which its valence band is completely filled, while the conduction band is completely empty. Thus, when the photons reach the device, electrons are moved from the valence band to the conduction band conducting electricity. In another hand, extrinsic photoresistors are composed of semiconductors doped with impurities, which are added to increase the number of free electrons or holes to make it conductive. As a result, less energy is needed to move electrons from the valence band to the conduction band due to the smaller energy gap created by the impurities [29].

The working principle of the photoresistor is based on the photoelectric effect. The theory behind this effect is constructed in the fact that when electromagnetic radiation (light) reaches the semiconductor, the light photons with energy higher than the semiconductor bandgap are absorbed, their energy is transferred to the electrons from the valence band to conduct band generating electron-hole pairs. The results are free charge carriers in the system responsible for conducting electricity, thus, reducing the resistance [30], Figure 13.6.

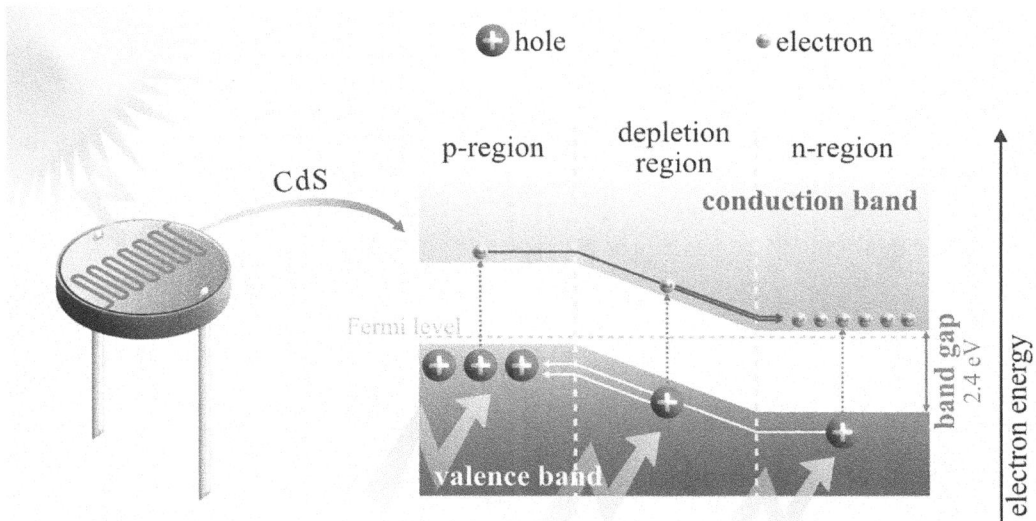

FIGURE 13.6   Schematic diagram of photoresistor principle working.

According to this principle, the resistance of photodevices is inversely proportional to light intensity, i.e., the higher the light intensity the lesser the resistance of the material. Besides streetlights control applications, photoresistors can be used in many different electronic circuit designs, including, fire or smoke alarms, switches, photographic light meters, and so on. One of the biggest applications is in the control system (switches) for lighting on street posts.

13.6.2.2 Photodiodes

Photodiodes are a type of diode susceptible to light/photons falls made from a P-type semiconductor (contains holes) diffused into the N-type (has free electrons) semiconductor. Due to the fact the electrons and holes repel each other, the PN-junction is formatted by a small depletion region or a "less conductive" region. The P-N junction is made up of a light-sensitive semiconductor.

Two specific modes of operation are possible for photodiodes, photovoltaic, and photoconductive modes. The photoconductive mode has to be seen as the only choice on the application. However, another way to take advantage of each mode is a PIN photodiode, Figure 13.7. In this case, an additional layer is called intrinsic semiconductor between P-type and N-type semiconductors. This layer act as a naturally occurring depletion region without applying a reverse-biased voltage, increasing the minority carrier current. A PIN photodiode in the photovoltaic mode can generate energy in the same range and sensitivity as a photodiode in the photoconductive mode [30].

Generally, photodiodes can be reverse-biased due the cathode operates at a higher voltage than the anode. Photodiodes can also be considered zero-biased, with cathode and anode at the same potential. Until now, most mainstream photodiodes adopt crystalline inorganic semiconductors such as Si, GaN, InGaAs, associated with small exciton binding

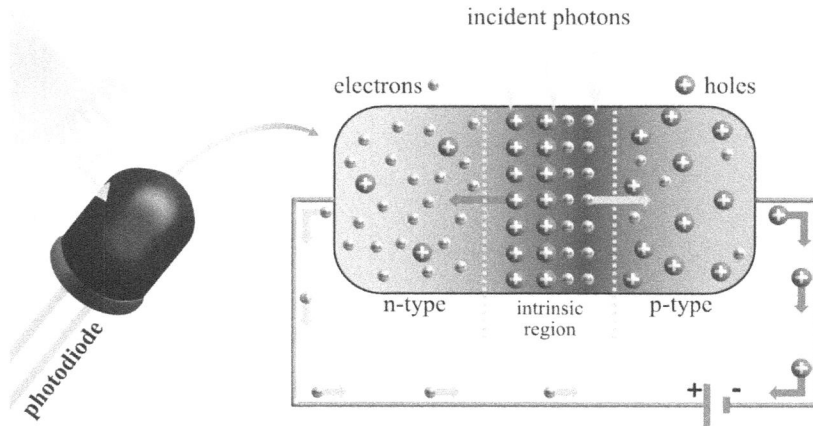

FIGURE 13.7   Photodiode principle working.

energy, charge-carrier mobility, and high stability [31]. It is commonly used in consumer electronic devices receiving infrared remote control devices, especially to control equipment like televisions and safety electronics like fire and smoke detectors. Any photosensor can be used for the accurate measurement of light intensity. Generally, photodiodes have a more linear response than photoconductors [32].

13.6.2.3 Phototransistors

Phototransistors are special design transistor that has a light-sensitive base region. When light hits a base of transistors made by semiconductors materials of the type PNP or NPN, base current develops. The magnitude of the current depends on the intensity of the light incident on it. The phototransistor is placed inside the opaque protection so that light particles or photons are easily reached on its surface. The collector region of the phototransistor is large compared to the common transistor because it is made of heavy and diffuse semiconductor material. Phototransistors have many applications, such as in encoders; card readers; security systems; infrared detectors; optocouplers; counting and lighting control systems [33].

A phototransistor is a terminal for two or three semiconductor devices of the PNP or NPN type. Like the conventional transistor, the phototransistor is a combination of two junction diodes, however, associated with the transistor effect, the photoelectric effect appears. Thanks to Albert Einstein for the discovery of the law of the photoelectric effect, in 1905. For this important contribution to science, Einstein received in 1922 the Nobel award (being the nomination done in 1921). After that, prodigious development in the understanding of such effect has been done, allowing multi- and widely technological applications.

When the base of the phototransistor absorbs light, they release the hole-electron pairs. Because of this pair of holes, the diode's depletion layer shrinks and the electron begins to move from the emitter region to the collector region. The small amount of light energy can be amplified by a large transistor collector current. In general, phototransistors have

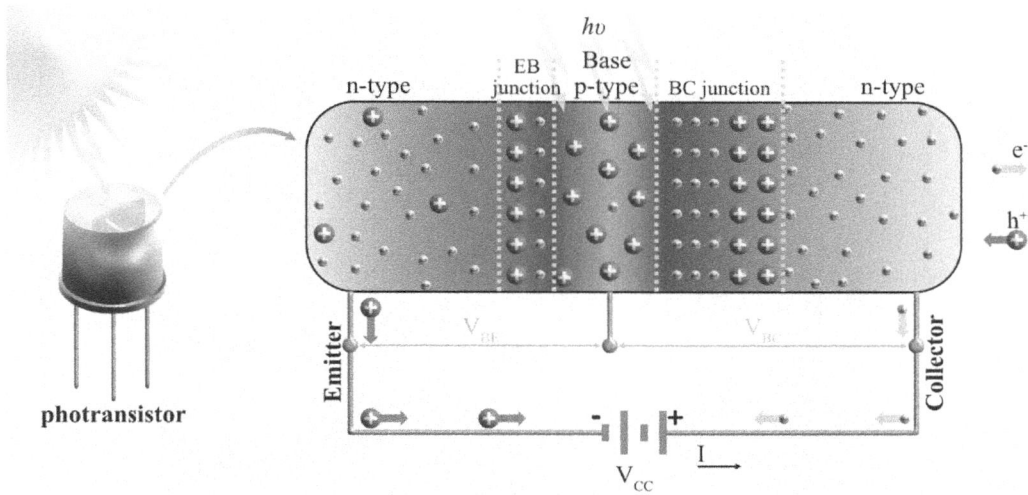

FIGURE 13.8   Working phototransistor of NPN type.

only two terminals, the collector (C) and the emitter (E), being a base (B) only for eventual polarization of electrical control [34].

As in other photoconductive cells, the incidence of light (photons) causes holes to appear in the vicinity of the base-collector junction (BC junction). This voltage (V_{BC}) drives the holes to the emitter, while electrons pass the emitter to a base (EB junction) this will cause an increase in the base current, which consequently will imply a variation in the collector current $-V_{CC}$ (Figure 13.8 demonstrates the working phototransistor of NPN type). This variation is proportional to the intensity of the incident light and this whole processes how the phototransistor works as a light sensor [34].

13.7 HOW PHOTODETECTORS ACT AGAINST LIGHT POLLUTION

Photodetectors like photoresistors, photodiodes, and phototransistors are smart photonic devices that can help to reduce waste light and save even more energy. They are the devices most used to control the public light system, which can be used in controlling the streetlights automatically or dimming the light intensity in smart cities. The manner that photodetectors act as nanodevices against the light pollution into the source of light (LED) or in the luminaries is the same.

In public illumination, they are typically localized in a lamppost or in the lamps, where there is local control. Thus, photodetectors can determine when the lights will be activated or deactivated. These photodevices act to detect the direct sunlight in their surroundings and determine if the brightness is enough to create a visible landscape without artificial light.

The streetlight system is composed of a microcircuit, which is commanded by a photodetector, in this case, a photoresistor as an example, is shown in Figure 13.9. The light incidence on the photoresistor surface induces a current change that can be used to control a relay (an electromagnetic switch). Basically, when the light falls on the photodetector, the current of the semiconductor increases dramatically, and the current flow is responsible

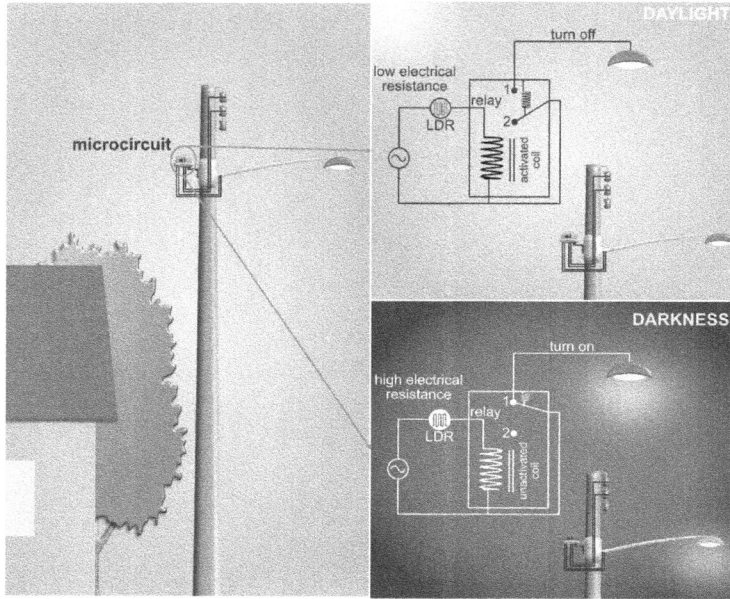

FIGURE 13.9   Working principle of photoresistor in daylight and darkness.

for opening the electromagnetic switch. Thus, the streetlight is maintained OFF. However, in the darkness, the current is so lower that there is no flow to the electromagnet and the switch closes allowing the streetlight to turn ON. Thus, energy is saved by ensuring the light is only on during hours of darkness [35]. Switching streetlights according to daylight levels and highway use, in the way that the streetlights are illuminated only when required is a manner to save energy and control the light pollution.

An innovative streetlight glowing system on vehicle detecting movement to reduce power consumption uses photodiode and photoresistors technology as an efficient sensor. An infrared sensor used on either side of the road sends a command for the LED output glowing for a way ahead. The idea is to detect vehicle movement on the street to switch on only a streetlight forth of the vehicle and switch off the trailing light to save energy. When all lights remain during the night on the highway, a lot of energy is wasted when there is no vehicle movement. However, this system consists of an achieved sensing-approaching vehicle. So, a block of streetlight is switched on ahead of the vehicle. After vehicles pass by, the light is switched off automatically. This system provides a solution to prevent all the street lights are turned on during the whole night and then contributes to reducing waste energy and light pollution [36,37].

Photodetectors can also be used to measure light pollution, detecting the sky glow caused by light excess. The photodiode is considered the most popular sensor type for light-based measurement applications like emission and absorption spectroscopy, and precision light measurement. Another application is the use in fiber optic communications systems, which generally operate in the infrared (IR), picking up high-speed signals [38].

Photodetectors used against light pollution can work in the same way, but there are some differences between them in the action's form. Photoresistors are made of a high-resistance semiconductor material, which resistance decreases with increasing light intensity. On another hand, the photodiode is made of semiconductors where the current flowing through it depends on the intensity of the light that falls at its junction. The junction will respond to light particularly longer wavelengths such as red and infrared rather than visible light. When light falls on the junction, more hole/electron pairs are formed, and then increase the output current. The main photodiodes advantage when used as light sensors is their quick answer to changes in light levels. However, the photodiode's disadvantage is the relatively small current flow, even when fully lit. Additionally, phototransistors work in a similar way to photoresistors, but they produce both current and voltage, while photoresistors produce only current due to changes in resistance [39].

Finally, the system of the phototransistors and photodiodes are practically the same, the difference is the photodiode uses the semiconductor of PN-type and phototransistor of PNP- or NPN type. Also, the phototransistor is more sensitive compared to the photodiode due to the use of a transistor. The transistor amplifies the base current that causes light absorption and therefore the large output current is obtained through the collector terminal. Nevertheless, photodiode response time is much faster than the phototransistor.

13.8 RECENT DEVELOPMENTS OF NANOMATERIALS FOR PHOTODETECTORS APPLICATIONS

As mentioned before, photoresistors are mainly made from CdS for light lamp applications. Cadmium-based materials have been a powerful material for photoresistors application, since it has a bandgap of 2.4 eV and excellent optoelectronic properties in the visible-light region, near to the eye human.

Many studies have focused in enhancing the sensitivity of CdS materials, making different nanostructures like nanorods, nanowires, and nanoflowers [40,41]. Similarly, other researchers have employed silica as nanopillars for CdS film to increase the surface ratio and improve the photosensitivity of photoresistors [42,43]. In another hand, the formation of heterojunctions with other materials, like ZnO, has been reported to enhance the photocurrent [44].

Despite this nanomaterial being widely employed, Cd is a poisonous element and environmentally unfriendly. So, CdS and CdSe as photoresistors are restricted in Europe due to the Restriction of Hazardous Substances Directive ban on cadmium [45]. To mitigate this disadvantage, numerous studies have been made, using new methodologies and materials to replace Cd.

The photoconductive property of a material, like efficient charge separation and charge transport, plays an important role in LDR application. The understanding of the nature of electronic excitations, distributions of traps, and recombination centers are essential for electronic and optoelectronic devices. Some metal oxides (ZnO, SnO_2, MoS_2, etc.) are considered to be good nanomaterials for photodetectors applications. Materials in nanoscale have a huge surface-to-volume ratio and completely different physicochemical properties in relation to the bulk form.

Nanotechnology has driven the synthesis of hybrid nanomaterials. In these cases, an organic conjugated polymer, such as polypyrrole (PPy), and inorganic metal oxide semiconductors, such as ZnO or SnO_2, was hybridized to produce a new class of visible LDRs [46,47]. Another interesting nanomaterial for LDR application is carbon. Carbon is plentiful, cheap, eco-friendly, and easy to prepare. Therefore, nanocrystalline carbon thin films deposited in amorphous carbon structure [48], carbon nanotube (CNTs) doped with boron (B), and nitrogen (N) [49] have shown excellent properties and are feasible for implementation as efficient and cost-effective components for use in LDRs.

Besides of materials cited above, other types of nanomaterials have been studied as an alternative for photoresistors in the last few years. For example, Bi_2S_3 [50], $TiO_2/SrRuO_3$ [51], SnSe [52], $Ni_{0.2}Zn_{0.8}Fe_2O_4$ (NZF) [53], and so forth. All of them have shown excellent photosensitive properties.

Nanomaterials for photodiodes are mainly made from Si, GaN, and InGaAs. Studies focused on illuminated InGaAs photodiodes grown on GaAs, for example, with short-wave infrared up to 2.4 μm has been demonstrated by Zimmermann et al. [54]. They studied the variation of dark current and photocurrent as a function of buffer thickness, buffer material, and temperature. Many other studies also have on developing InGaAs materials, preparing different semiconductor nanowires and quantum well [55,56]. Barladen et al. related an InGaAs/InP PIN photodiode to determine the suitability and optically

TABLE 13.2 Recent Nanostructured Materials and Their Performance as a Visible Photodetector

Material	Responsivity (A/W)	Detectivity (Jones)	Reference
rGO-CZS	$18.5.10^{-3}$	$2.08.10^{12}$	[61]
PQD-RGO	$1.07.10^3$	1.1013	[62]
GaN/N-SLG	4.10^2	$2.89.10^{12}$	[63]
GO-PbI$_2$	0.182	$1.79.10^{11}$	[64]
SnS/CdS	$10.4.10^{-3}$	$3.56.10^{11}$	[65]
MoSe$_2$-NCs	566.10^{-3}	$3.69.10^{11}$	[66]
WO$_3$	19	$1.06.10^{11}$	[67]
MAPbI$_3$	0.4	$1.1.10^{10}$	[68]
PVK/IGZO	$3.1.10^5$	$5.1.10^{16}$	[69]
CH3NH3PbI$_3$	7.8	$2.1.10^{13}$	[70]
P(VDF-TrFE)/ perovskite	0.02	$1.4.10^{13}$	[71]

Source: Photoresponsivity (A/W) and detectivity (Jones) are important parameters of the photodetectors to evaluate the ability to convert the light signals into electrical signals and measure the detector sensitivity under a certain wavelength.

rGO-CZS=Reduced graphene oxide-incorporated cadmium zinc sulfide; PQD-RGO=Perovskite quantum dots and Reduced graphene oxide; GaN/N-SLG=Gallium nitride and nitrogen doped single layer graphene; GO-PbI$_2$=nanorods/ graphene oxide- lead iodide; SnS/CdS=Tin monosulfide/Cadmium monosulfide; MoSe$_2$-NCs=molybdenum diselenide nanocrystals; WO$_3$=Tungsten trioxide; **MAPbI$_3$=Organic–inorganic hybrid perovskites; PVK/IGZO=Organic–inorganic hybrid perovskites/ indium gallium zinc oxide; CH3NH3PbI$_3$=Organic-inorganic hybrid perovskites; P(VDF-TrFE)/perovskite=ferroelectric poly (vinylidene-fluoride-trifluoroethylene)/perovskite.**

controlled high-speed current source at 4 K [57]. Other researchers have employed a high-efficiency InGaAs-based PIN photodiode with a resonant cavity enhanced structure. They observed a tuned resonance wavelength from 1,605 to 1,558 nm keeping the peak efficiencies above 60%. The max obtained was 66% at 1572 nm [58]. Lee and co-workers have reported studies based on PbS quantum dots infrared (IR) PIN photodetectors with IR sensitivity up to 1,500 nm to high gain and low dark current solution [59].

Phototransistors can be structured by homojunction or heterojunction. In the case of homojunction phototransistors, the entire device will be made of a single type of material; silicon or germanium. However, to increase their efficiency, phototransistors can be made of non-identical materials (Group III–V materials like GaAs) on both sides of the PN junction leading to heterojunction devices. Heterostructures that use different materials on either side of the PN junction are popular because they provide a high conversion efficiency. In addition, new classes of materials such as organic semiconductors, perovskites, and quantum dots were evaluated as active layers in phototransistors [60].

Other nanomaterials have been reported with high performance as visible-light photodetectors. Between them, different types of nanostructures as nanoplates, nanosheets, nanoflakes, nanorods, and so forth have been listed in Table 13.2. Although the extensive efforts, there still is a huge of opportunities to develop new nanomaterials for visible-light photodetectors. And also, improve parameters like responsivity, detectivity, etc., which are associated with the performance of a photodetector.

13.9 PHOTONICS MARKET VALUE

Photonics applications use the photon in the same way that electronic applications use the electron [72]. The photonics materials & component industry includes devices used for data conversion, sensors, amplifier systems, light modulation systems (modulators, switches, and routers), and fiber optic cables. The photonics market is segmented based on product type, such as LED; laser, sensor, detectors & imaging devices; optical communication systems & components; and others. Photonics products are used in a wide range of sectors such as production technology, industrial manufacturing; optical measurement & image processing, and based medical technology & life science. Nowadays, the photonics market is driven by the growing application of photonics-enabled products of healthcare in the pandemic situation. For example, demand for photonics UV-C LED light has increased as it is used to disinfect the virus [73].

Therefore, the photonic market is a great value market and it is growing expressly. The global photonics market was estimated at US$ 589.1 billion in 2019 and US$ 593.7 billion in 2020. This only 0.8% increase in 2020 was due to the pandemic harming various end-use industries due to the supply chain disruption [74]. Even in a difficult situation, the market of photonics is likely to recover soon, reaching an estimate of US$ 837.8 billion in 2025 with a CAGR (compound annual growth rate) of 7.1% from 2022 to 2025 [75]. China, the world's second-largest economy, is the global leading producer. The photonics market in the U.S. is the largest in America. At present, Europe is the second-largest producer in photonics, Germany is the major European producer in all photonics segments. The growth is due mainly to the 5G rollout by various countries that will increase the prices of fiber optics [76].

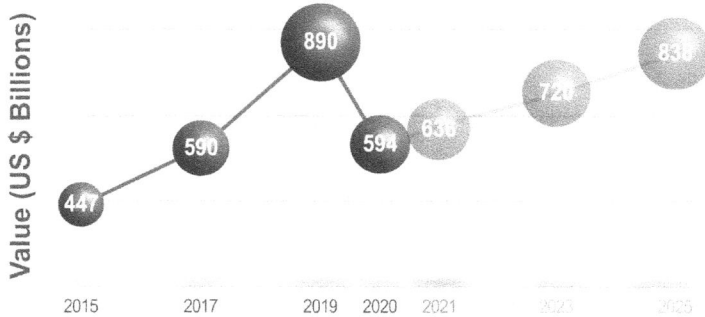

FIGURE 13.10 Graphic Trends and forecasts for the global photonics market from 2015 to 2025. (Box: The global photonics market can be too segmented by Product Type, Application, and End-use.)

The global photonics market can be too segmented by product type, application, and end-use, Figure 13.10. By Product Type, the photonics market is classified into five end-use applications: LED, Laser, Sensor, Detectors & Imaging Devices; Optical communication Systems & Components, and others. By application, the photonics market is classified into seven main end-use applications, namely Displays, Information & communication technology, Medical Technology & Life sciences (Biophotonics), Photovoltaic Measurement & automated vision, Lighting, and Production Technology. By End-use, the photonics market is classified into five main End-use: Medical industries, Industrial, Defense, Consumer Electronics, Media & Telecommunication, and others [77].

LED segment is responsible for the biggest piece in the global photonics market. The global LED lighting market is segmented into three bases: product, application, and end-user. By product, the market is forked into lamps and luminaries. Lamps are further sub-segmented as T-lamps, A-lamps, and others. The LED luminaires are sub-segmented as downlights, troffers, streetlights, and others that dominated the market in 2020 and accounted for the major of the global revenue share.

Luminaries are further based on application; the market is classified into indoor and outdoor. Outdoor applications include corporate campuses, industrial, government, airports, highways, roadways, healthcare infrastructure, public places, such as signage and traffic signals. The use of outdoor applications can be profoundly affected by public policy interventions on issues of light pollution reduction [1]. Increasing the adoption of photonics products has grown globally in the last decades, but has also generated problems related to light pollution. Stringent environmental regulations have required the reduction of light pollution that is environmentally hazardous [78].

The actual and projected market value has been driven by the emergence of smart cities. The quest for higher and widespread life quality is a call for researchers, to achieve results and innovations that can transform not only our cities in smart cities but smart states, smart countries, and why not a smart world.

13.10 CHALLENGES AND PROSPECTS

Man's invention of artificial light makes our lives safer and more comfortable in a night-time environment. With its development, new lights color emerged and has been used to make the city more attractive to tourists. However, the extensive use of light brings us a new type of pollution (commonly referred to as light pollution). Light pollution can be diverse forms such as skyglow, glare, light trespass, and clutter, among others. All of them present serious physiological and ecological problems. In addition, there is a waste of electricity, money, and emission of greenhouse unnecessarily.

To overcome these issues, a lot of effort has been made by associations that have a mission to protect the night skies for present and future generations. Progress and adjustments need to be made, but the perception of the stars as a projection of the past must not be lost, it must be considered that all their beauty reflects the radiance of humanity. The brightness and infinity of the starry sky and the life of ecosystems must be perceived in places, even by lighting control through the use of interconnected systems with intelligent light sources that can be modulated. Innumerous recommendations are made to reduce the obtrusive light. Among them, we can highlight the use of smart photonic devices to control the outdoor light in smart cities. Photonic devices as efficient light sources and photodetectors have an important role to lessen light pollution.

LEDs have emerged to replace traditional sources of light, due to their greatest capacity to vary the emitted light cores and the presence of a current regulator (dimmer), which controls the light levels. On other hand, photodetectors, like photoresistors, photodiodes, and phototransistors, have an important role to control the light intensity or turn ON/OFF the lamp according to the natural light levels and highway use. These important functions of the main smart photonic devices can help a lot the effect of obtrusive light responsible for pollution, as well as reducing the waste energy and the footprint.

Motivated by the impressive effect of the smart photonic devices against light pollution and by their biggest piece in the global photonics market, researchers of the whole world have explored new materials for photonic applications. The main goals are improving the performance of photodetectors and obtaining new sources of light, using eco-friendly, safer, and more efficient materials.

However, with the advances of smart cities pushed by political pressures, the development of smart photonic technologies needs more progress for solving urban management problems of the future. In addition to new technologies, more public policies and population awareness are needed since most people do not know anything about light pollution. We believe that this chapter can offer a valuable vision about how science can help our world against light pollution and also provide an overview of its main impacts on human health and the environment.

ACKNOWLEDGMENTS

The authors acknowledge Conselho Nacional de Desenvolvimento Científico e Tecnológico (CNPq) (Grant # 150866/2020–8, 307429/2018-0, 573636/2008-7, 435975/2018-8, and 421648/2018-0) for financial support and for the concession of a scholarship.

Special thanks to CEPID (2013/07296-2) and INCTMN (2008/57872-1) and Fundação de Amparo à Pesquisa do Estado de São Paulo (FAPESP) (grant numbers 2012/07067-0, 2013/23572-0, 2016/019405, 165734/2020-5 and 2013/07296) for technical support. This study was financed in part by the Coordenação de Aperfeiçoamento de Pessoal de Nível Superior – Brasil (CAPES) - Finance Code 001

REFERENCES

[1] Light Pollution, *Int. Darky-Sky Assoc.* (n.d.). https://www.darksky.org (accessed April 20, 2021).

[2] I.N. Da Silva, R.A. Flauzino, *Smart Cities Technologies*, BoD–Books on Demand, IntechOpen, London (2016).

[3] M. Bass, C. DeCusatis, J. Enoch, V. Lakshminarayanan, G. Li, C. Macdonald, V. Mahajan, E. Van Stryland, *Handbook of Optics, Third Edition Volume II: Design, Fabrication and Testing, Sources and Detectors, Radiometry and Photometry*, 3rd edn., McGraw-Hill, Inc., New York, NY (2009).

[4] A. Petrulis, L. Petkevičius, P. Vitta, R. Vaicekauskas, A. Žukauskas, Exploring preferred correlated color temperature in outdoor environments using a smart solid-state light engine, *LEUKOS*. 14 (2018) 95–106. https://doi.org/10.1080/15502724.2017.1377085.

[5] R.C. Lima, J. Pinto da Cunha, N. Peixinho, light pollution: Assessment of sky glow on two dark sky regions of Portugal, J. *Toxicol. Environ. Heal. Part A*. 79 (2016) 307–319. https://doi.org/10.1080/15287394.2016.1153446.

[6] F.M. Mims III, Monitor light pollution with photodiodes, MAKE Mag. (2016). https://makezine.com/projects/surveying-light-pollution/ (accessed May 20, 2021).

[7] M.I. Azman, M.N. Dalimin, M. Mohamed, M.F. Abu Bakar, A brief overview on light pollution, *IOP Conf. Ser. Earth Environ. Sci.* 269 (2019) 12014. https://doi.org/10.1088/1755-1315/269/1/012014.

[8] T. Longcore, C. Rich, Ecological light pollution, *Front. Ecol. Environ.* 2 (2004) 191–198. https://doi.org/https://doi.org/10.1890/1540-9295(2004)002[0191:ELP]2.0.CO;2.

[9] B. Fritz, G. Horváth, R. Hünig, Á. Pereszlényi, Á. Egri, M. Guttmann, M. Schneider, U. Lemmer, G. Kriska, G. Gomard, Bioreplicated coatings for photovoltaic solar panels nearly eliminate light pollution that harms polarotactic insects, *PLoS One*. 15 (2020) e0243296. https://doi.org/10.1371/journal.pone.0243296.

[10] C.C.M. Kyba, T. Ruhtz, J. Fischer, F. Hölker, Cloud coverage acts as an amplifier for ecological light pollution in urban ecosystems, *PLoS One*. 6 (2011) e17307. https://doi.org/10.1371/journal.pone.0017307.

[11] G. Zissis, Sustainable lighting and light pollution: A critical issue for the present generation, a challenge to the future, Sustainability. 12 (2020). https://doi.org/10.3390/su12114552.

[12] A. Haim, B.A. Portnov, Light at Night (LAN) Exposure and its Potential Effects on Daily Rhythms and Seasonal Disruptions BT – Light Pollution as a New Risk Factor for Human Breast and Prostate Cancers, in: A. Haim, B.A. Portnov (Eds.), Springer Netherlands, Dordrecht (2013): pp. 35–40. https://doi.org/10.1007/978-94-007-6220-6_5.

[13] Smart cities market – Growth, trends, Covid-19 impact, and forecasts (2021–2026), Mordor Intel. (2021). https://www.mordorintelligence.com/industry-reports/smart-cities-market (accessed April 10, 2021).

[14] M. Georgescu, The Importance of Internet of Things Security for Smart Cities, in: D.P.E.-I.N.D.S.E.-R.A. Flauzino (Ed.), *IntechOpen*, Rijeka (2016): p. Ch. 1. https://doi.org/10.5772/65206.

[15] O. Vermesan, P. Friess, *Internet of Things: Converging Technologies for Smart Environments and Integrated Ecosystems*, River Publishers (2013).

[16] S. Talari, M. Shafie-Khah, P. Siano, V. Loia, A. Tommasetti, J.P.S. Catalão, A review of smart cities based on the internet of things concept, *Energies*. 10 (2017) 421.

[17] B.N. Silva, M. Khan, K. Han, Towards sustainable smart cities: A review of trends, architectures, components, and open challenges in smart cities, *Sustain. Cities Soc.* 38 (2018) 697–713.

[18] H. Sun, C. Wang, B.I. Ahmad, *From Internet of Things to Smart Cities: Enabling Technologies*, CRC Press (2017).

[19] K. Chen, X. Zhou, X. Cheng, R. Qiao, Y. Cheng, C. Liu, Y. Xie, W. Yu, F. Yao, Z. Sun, F. Wang, K. Liu, Z. Liu, Graphene photonic crystal fibre with strong and tunable light–matter interaction, *Nat. Photonics*. 13 (2019) 754–759. https://doi.org/10.1038/s41566-019-0492-5.

[20] I.S. Mardikaningsih, W. Sutopo, M. Hisjam, R. Zakaria, Techno-economic feasibility analysis of a public street light with solar cell power, in: Proc. Int. MultiConference Eng. Comput. Sci., Hong Kong (2016): pp. 16–18.

[21] J. Brox, *Brilliant: The Evolution of Artificial Light*, Houghton Mifflin Harcourt, Boston (2010).

[22] S. Williams, *CFL's vs. Halogen vs. Fluorescent vs. Incandescent vs. LED*, HomElectrical (2017). https://www.homelectrical.com/cfls-vs-halogen-vs-fluorescent-vs-incandescent-vs-led.6.html (accessed May 10, 2021).

[23] V.C. Bender, T.B. Marchesan, J.M. Alonso, Solid-state lighting: A concise review of the state of the art on LED and OLED modeling, *IEEE Ind. Electron. Mag.* 9 (2015) 6–16. https://doi.org/10.1109/MIE.2014.2360324.

[24] P. Waide, S. Tanishima, I.E. Agency, *Light's Labour's Lost : Policies for Energy-efficient Lighting*, OECD/IEA, Paris (2006).

[25] S. Hariyani, J. Brgoch, Advancing human-centric LED lighting using Na2MgPO4F:Eu2+, *ACS Appl. Mater. Interfaces*. 13 (2021) 16669–16676. https://doi.org/10.1021/acsami.1c00909.

[26] A.M. Bagher, Quantum Dot Display Technology and Comparison with OLED Display Technology, *J. Am. Sci.* 17 (2021) 55–60.

[27] J. Fraden, Light Detectors BT – Handbook of Modern Sensors: Physics, Designs, and Applications, in: J. Fraden (Ed.), Springer New York, New York, NY (2010): pp. 461–501. https://doi.org/10.1007/978-1-4419-6466-3_14.

[28] N. Agnihotri, *Basic Electronics 10 – Dependent Resistors*, EngineersGarage (2021). https://www.engineersgarage.com/articles-basic-electronics-dependant-resistors-ldr-vdr-thermistor-ptc-ntc-magneto-resistor-strain-gauge (accessed June 16, 2021).

[29] L. Critchley, *The Different Types of Photoresistor*, AZO Mater (2021). https://www.azom.com/article.aspx?ArticleID=18355 (accessed June 16, 2021).

[30] J.-Y. Yoon, *Introduction to Biosensors: From Electric Circuits to Immunosensors*, Springer (2016).

[31] I. Sinclair, Chapter 8- Transducing components, in: I.B.T.-P.C. for C.D. Sinclair (Ed.), *Newnes*, Oxford (2001): pp. 214–240. https://doi.org/https://doi.org/10.1016/B978-075064933-9/50008-X.

[32] A. Mallik, Photodiode, *Himal. Phys.* 4 (2013). https://doi.org/10.3126/hj.v4i0.9437.

[33] H. Yoo, I.S. Lee, S. Jung, S.M. Rho, B.H. Kang, H.J. Kim, A review of phototransistors using metal oxide semiconductors: Research progress and future directions, *Adv. Mater.* n/a (2021) 2006091. https://doi.org/https://doi.org/10.1002/adma.202006091.

[34] M. Bansal, R.R. Maiya, Phototransistor: The story so far, *J. Electron.* 2 (2020) 202–210.

[35] M.A. Wazed, N. Nafis, M.T. Islam, A.S.M. Sayem, Design and fabrication of automatic street light control system, *Eng. e-Transaction.* 5 (2010) 27–34.

[36] M. Revathy, S. Ramya, R. Sathiyavathi, B. Bharathi, V.M. Anu, Automation of street light for smart city, in: 2017 Int. Conf. Commun. Signal Process., IEEE (2017): pp. 918–922.

[37] C. Badgaiyan, P. Sehgal, Smart street lighting system, *Int. J. Sci. Res.* 4 (2015) 271–274.

[38] H.D. Jabbar, M.A. Fakhri, M. Jalal AbdulRazzaq, Gallium nitride–based photodiode: A review, *Mater. Today Proc.* 42 (2021) 2829–2834. https://doi.org/https://doi.org/10.1016/j.matpr.2020.12.729.

[39] S. Donati, *Photodetectors: Devices, Circuits and Applications*, John Wiley & Sons (2021).

[40] B.H. Zhang, F.Q. Guo, L.H. Yang, J.J. Wang, Direct synthesis of highly pure CdS nanofilms via solvothermal method, *Mater. Res. Innov.* 19 (2015) 60–64.

[41] Y. Wang, X. Yang, Q. Ma, J. Kong, H. Jia, Z. Wang, M. Yu, Preparation of flower-like CdS with SDBS as surfactant by hydrothermal method and its optical properties, *Appl. Surf. Sci.* 340 (2015) 18–24.

[42] J. Liu, Y. Liang, L. Wang, B. Wang, T. Zhang, F. Yi, Fabrication and photosensitivity of CdS photoresistor on silica nanopillars substrate, *Mater. Sci. Semicond. Process.* 56 (2016) 217–221.

[43] J. Liu, X. Liang, Y. Wang, B. Wang, T. Zhang, F. Yi, Preparation of CdS nanorods on silicon nanopillars surface by hydrothermal method, *Mater. Res. Bull.* 120 (2019) 110591.

[44] J. Liu, Y. Xu, F. Yi, Fabrication and photosensitivity of ZnO/CdS/silica nanopillars based photoresistor, *J. Mater. Sci. Mater. Electron.* 32 (2021) 11326–11333.

[45] Restriction of Hazardous Substances in Electrical and Electronic Equipment (RoHS), Eur. Com. (2011). https://ec.europa.eu/environment/topics/waste-and-recycling/rohs-directive_en (accessed June 16, 2021).

[46] N.J.A. Cordeiro, C. Gaspar, M.J. de Oliveira, D. Nunes, P. Barquinha, L. Pereira, E. Fortunato, R. Martins, E. Laureto, S.A. Lourenço, Fast and low-cost synthesis of MoS2 nanostructures on paper substrates for near-infrared photodetectors, *Appl. Sci.* 11 (2021). https://doi.org/10.3390/app11031234.

[47] I. Rawal, R.K. Tripathi, O.S. Panwar, Easy synthesis of organic–inorganic hybrid nanomaterials: Study of DC conduction mechanism for light dependent resistors, *RSC Adv.* 6 (2016) 31540–31550. https://doi.org/10.1039/C5RA27774D.

[48] R.K. Tripathi, O.S. Panwar, I. Rawal, C.K. Dixit, A. Verma, P. Chaudhary, A.K. Srivastava, B.C. Yadav, Study of variable range hopping conduction mechanism in nanocrystalline carbon thin films deposited by modified anodic jet carbon arc technique: Application to light-dependent resistors, *J. Mater. Sci. Mater. Electron.* 32 (2021) 2535–2546. https://doi.org/10.1007/s10854-020-05020-z.

[49] M.S. Parvaiz, K.A. Shah, G.N. Dar, P. Misra, Computational modeling of carbon nanotubes for photoresistor applications, *Solid State Commun.* 309 (2020) 113831. https://doi.org/https://doi.org/10.1016/j.ssc.2020.113831.

[50] R. Nishikubo, A. Saeki, Solution-processed Bi2S3 photoresistor film to mitigate a trade-off between morphology and electronic properties, *J. Phys. Chem. Lett.* 9 (2018) 5392–5399. https://doi.org/10.1021/acs.jpclett.8b02218.

[51] H.-J. Liu, C.-H. Huang, C.-Y. Chen, S.-W. Hsiao, Y.-S. Chen, M.-H. Lee, Y.-C. Chen, P.-J. Wu, M.-W. Chu, J.G. Lin, Large photoresponsivity in the amorphous-TiO2/SrRuO3 Heterostructure, *Phys. Status Solidi Rapid Res. Lett.* 14 (2020) 2000273. https://doi.org/https://doi.org/10.1002/pssr.202000273.

[52] J. Cao, Z. Wang, X. Zhan, Q. Wang, M. Safdar, Y. Wang, J. He, Vertical SnSe nanorod arrays: From controlled synthesis and growth mechanism to thermistor and photoresistor, *Nanotechnology*. 25 (2014) 105705. https://doi.org/10.1088/0957–4484/25/10/105705.

[53] S. Singh, A. Bhaduri, R.K. Tripathi, K.B. Thapa, R. Kumar, B.C. Yadav, Improved sensing behaviour of self-healable solar light photodetector based on core-shell type Ni0.2Zn0.8Fe2O4@ poly (Urea-Formaldehyde), *Sol. Energy*. 188 (2019) 278–290. https://doi.org/https://doi.org/10.1016/j.solener.2019.06.003.

[54] L. Zimmermann, J. John, S. Degroote, G. Borghs, C. Van Hoof, S. Nemeth, Extended wavelength InGaAs on GaAs using InAlAs buffer for back-side-illuminated short-wave infrared detectors, *Appl. Phys. Lett*. 82 (2003) 2838–2840. https://doi.org/10.1063/1.1569042.

[55] B.M. Arora, M. Gokhale, A.P. Shah, K.S. Chandrasekharan, Synthesis of high-quality InGaAs/InGaAsP/InP quantum well structures by metlorganic vapor phase epitaxy, in: Proc. SPIE (1998). https://doi.org/10.1117/12.345357.

[56] D. Gustiono, E. Wibowo, Z. Othaman, Synthesis and characterization of InGaAs nanowires grown by MOCVD, *J. Phys. Conf. Ser*. 423 (2013) 12047. https://doi.org/10.1088/1742–6596/423/1/012047.

[57] E. Bardalen, B. Karlsen, H. Malmbekk, M.N. Akram, P. Ohlckers, Evaluation of InGaAs/InP photodiode for high-speed operation at 4 K, *Int. J. Metrol. Qual. Eng*. 9 (2018). https://doi.org/10.1051/ijmqe/2018015.

[58] S.Q. Liu, Q. Han, B. Zhu, X.H. Yang, H.Q. Ni, J.F. He, X. Wang, M.F. Li, Y. Zhu, J. Wang, X.P. Wang, Z.C. Niu, High-performance metamorphic InGaAs resonant cavity enhanced photodetector grown on GaAs substrate, *Appl. Phys. Lett*. 98 (2011) 201104. https://doi.org/10.1063/1.3592569.

[59] J.W. Lee, D.Y. Kim, F. So, Unraveling the gain mechanism in high performance solution-processed PbS infrared PIN photodiodes, *Adv. Funct. Mater*. 25 (2015) 1233–1238. https://doi.org/https://doi.org/10.1002/adfm.201403673.

[60] T. Yokota, K. Fukuda, T. Someya, Recent progress of flexible image sensors for biomedical applications, *Adv. Mater*. 33 (2021) 2004416. https://doi.org/https://doi.org/10.1002/adma.202004416.

[61] J. Mathew, V.G. Sreeja, P.S. Subin, E.I. Anila, Investigations on the effects of rGO incorporation on the photosensitivity of (Cd: Zn) S nanocrystalline thin film-based visible photodetectors by hydrothermal synthesis, *J. Mater. Sci. Mater. Electron*. 31 (2020) 2523–2529.

[62] F.A. Chowdhury, B. Pradhan, Y. Ding, A. Towers, A. Gesquiere, L. Tetard, J. Thomas, Perovskite quantum dot-reduced graphene oxide superstructure for efficient photodetection, *ACS Appl. Mater. Interfaces*. 12 (2020) 45165–45173.

[63] S. Sankaranarayanan, P. Kandasamy, R. Raju, B. Krishnan, Fabrication of gallium nitride and nitrogen doped single layer graphene hybrid heterostructures for high performance photodetectors, *Sci. Rep*. 10 (2020) 1–12.

[64] N. Sharma, I.M. Ashraf, M.T. Khan, M. Shkir, M.S. Hamdy, A. Singh, A. Almohammedi, F.B.M. Ahmed, I.S. Yahia, S. AlFaify, Enhancement in photodetection properties of PbI2 with graphene oxide doping for visible-light photodetectors, *Sens. Actuators A Phys*. 314 (2020) 112223.

[65] Y. Chang, J. Wang, F. Wu, W. Tian, W. Zhai, Structural design and pyroelectric property of SnS/CdS heterojunctions contrived for low-temperature visible photodetectors, *Adv. Funct. Mater*. 30 (2020) 2001450.

[66] A.B. Patel, P. Chauhan, H.K. Machhi, S. Narayan, C.K. Sumesh, K.D. Patel, S.S. Soni, P.K. Jha, G.K. Solanki, V.M. Pathak, Transferrable thin film of ultrasonically exfoliated MoSe2 nanocrystals for efficient visible-light photodetector, *Phys. E*. 119 (2020) 114019.

[67] H. Wang, J.L. Liu, X.X. Wu, S.Q. Zhang, Z.K. Zhang, W.W. Pan, G. Yuan, C.L. Yuan, Y.L. Ren, W. Lei, Ultra-long high quality catalyst-free WO3 nanowires for fabricating high-performance visible photodetectors, *Nanotechnology*. 31 (2020) 274003.

[68] D.H. Shin, J.S. Ko, S.K. Kang, S.-H. Choi, Enhanced flexibility and stability in perovskite photodiode–solar cell nanosystem using MoS2 electron-transport layer, *ACS Appl. Mater. Interfaces*. 12 (2020) 4586–4593. https://doi.org/10.1021/acsami.9b18501.

[69] S. Wei, F. Wang, X. Zou, L. Wang, C. Liu, X. Liu, W. Hu, Z. Fan, J.C. Ho, L. Liao, Flexible quasi-2D perovskite/IGZO phototransistors for ultrasensitive and broadband photodetection, *Adv. Mater*. 32 (2020) 1907527.

[70] J. Li, H. Li, D. Ding, Z. Li, F. Chen, Y. Wang, S. Liu, H. Yao, L. Liu, Y. Shi, High-performance photoresistors based on perovskite thin film with a high PbI2 DOPING LEVEL, *Nanomaterials*. 9 (2019) 505.

[71] F. Cao, W. Tian, L. Meng, M. Wang, L. Li, Ultrahigh-performance flexible and self-powered photodetectors with ferroelectric P (VDF-TrFE)/perovskite bulk heterojunction, *Adv. Funct. Mater*. 29 (2019) 1808415.

[72] What is photonics, Synopsys. (2021). https://www.synopsys.com/glossary/what-is-photonics.html (accessed June 15, 2021).

[73] COVID-19 Increases Demand for UV LEDs that Disinfect Spaces, Covid. (2021). https://www.photonics.com/Articles/COVID-19_Increases_Demand_for_UV_LEDs_that/a65629 (accessed June 15, 2021).

[74] Global Photosensor Market Size, Share, Value, And Competitive Landscape 2021–2026, MarketWatch. (2021). https://www.marketwatch.com/press-release/global-photosensor-market-size-share-value-and-competitive-landscape-2021-2026-2021-05-06 (accessed June 15, 2021).

[75] Photonics Market by Type (LED, Lasers, Detectors, Sensors and Imaging Devices, Optical Communication Systems & Networking Components, Consumer Electronics & Devices), Application End-Use Industry, and Region – Global Forecast to 2025, Marketsand. (2021). https://www.marketsandmarkets.com/Market-Reports/photonics-market-88194993.html (accessed June 15, 2021).

[76] Photonics Market by Type (LED, Lasers, Detectors, Sensors and Imaging Devices, Optical Communication Systems & Networking Components, Consumer Electronics & Devices), Application End-Use Industry, and Region – Global Forecast to 2025, Photonics. (2021). https://www.marketsandmarkets.com/Market-Reports/photonics-market (accessed June 15, 2021).

[77] The photonics market was valued at USD 773.64 billion in 2020, and is expected to reach a value of USD 1208.72 billion by 2026, at a CAGR of 7.96% over the forecast period (2021–2026), Globenewswire. (2021). https://www.globenewswire.com/news-release/2021/03/05/2188058/0/en/ (accessed June 15, 2021).

[78] Implementing the European Photonics 21 PPP strategy, EuroPho. (2017). https://ec.europa.eu/research/participants/documents/downloadPublic?documentIds=080166e5b1ee8cb9&apId=PPGMS (accessed June 15, 2021).

CHAPTER 14

Nanotechnology-Enabled Next-Generation LED Lights

Irfan Ayoub
National Institute of Technology Srinagar

Rishabh Sehgal
University of Texas at Austin

Vijay Kumar
National Institute of Technology Srinagar
University of the Free State

Rakesh Sehgal
National Institute of Technology Srinagar

Hendrik C. Swart
University of the Free State

CONTENTS

DOI: 10.1201/9781003185109-14

14.1 INTRODUCTION

With the ever-increasing demands and the significance of lighting and its different innovations in daily life, the year 2015 was celebrated as the international year of light along with its different innovations across the whole globe [1]. The light and different technologies based on it are considered to be vital for science from the past few centuries and possess a significant impact on numerous sectors of this technological age. Due to the ever-increasing demands of light, researchers believe that there is a need for the proper utilization of the available electrical energy available on the planet. Furthermore, to add on the research community believes there is a need for the development of energy-efficient devices, otherwise the time is not far off when the globe will face the scarcity of electrical energy [2]. Thus, the quest for ecologically safe, energy-efficient, cost-effective, easily approachable phosphor options for solid-state lighting systems has attained a lot of attention. The unique properties are shown by LEDs i.e., eco-friendliness, energy-efficient, low operative voltage, power utilization, compact size, and durability, and as a result, they are considered as the efficient alternative light sources for the future [3].

Due to the mentioned characteristics, the white LEDs (w-LEDs) are the most efficient sources for the general lighting and display devices, thus also referred to as fourth-generation lighting sources. Due to the remarkable characteristics of the LEDs, they have almost replaced a 40% and 80% market share for general and back lightning systems respectively. Due to a large number of advancements in this field large number of new materials are being utilized for the liquid-crystal display backlights such as cold cathode fluorescent light, W-LEDs, quantum dots along perovskite. Also, as the spectrum of the wavelengths covered by these LEDs possesses a constraint upconverting and downconverting, phosphors have been introduced to overcome the difficulties associated with these LEDs and thus covered the full electromagnetic spectrum [4]. The phosphor-converted LEDs (pc-LEDs) are the most promising outcomes of nanotechnology for next-generation lighting devices. The development of inorganic phosphors is considered a breakthrough relying on the luminescence phenomenon for solid-state lighting technology. The phosphors are the light-emitting materials compromised of a host matrix and an activator. It is the activator that assists in the absorption at its site followed by the relaxation and the emission processes. It is the phenomenon of relaxation that determines the efficiency of a particular phosphor. This phenomenon is defined as the minimization of the heat loss to the lattice for the attainment of better efficiency [5].

The ever-increasing demands for phosphors with better luminous characteristics in the diverse domain have brought together research activities in this sector intending to improve efficiency and applications. Usually, the transition metal ions or the rear-earth (RE) ions act as an activator. RE ions act as an activator for luminous centers in a variety of different hosts due to their partially filled 4f shells surrounded by 5s2 and 5p6 shells, resulting in the development of the phosphor with excellent efficiencies [6]. Because of the large number of allowed energy levels in comparison to the others, the RE ions possess outstanding luminescence properties and fine emission bands. It is because of these characteristics that they have garnered a lot of attention in different technical applications including optoelectronic devices and flat panel displays. Besides these, owing to the great technological pertinence of the RE doped phosphors such as solid-state lighting, medical imaging techniques, and detectors they too have attracted a lot of interest. Besides they also possess application in other fields due to their unique optical, electronic, and chemical properties [7,8]. Phosphor materials wherein the RE ions are used as activators have been classified into two types, the first one includes those that possess broad bands in their emission spectrums due to their d-f transitions while the latter includes those having narrow bands in their emission spectrum due to in-between transition among f levels. A few examples of the commonly used activators are Eu^{3+}/Eu^{2+}, Ce^{3+}, Tb^{3+}, Gd^{3+}, Yb^{3+}, Dy^{3+}, Sm^{3+}, and so on [9]. Keeping in view above all mentioned characteristics of the phosphors they are considered to be the novel outcome of the nanotechnological world. That is why advancements are being continuously made in this field. Also, the most critical and burning issue for the research community in the current scenario is developing the new phosphors materials for solid-state lighting (SSL) [5].

This book chapter focuses on the different aspects of lighting technology. A brief introduction about different materials is provided along with the principle of emission. Categorization of the pc-LEDs based on the light they emit and their efficiencies are also discussed. Besides their importance, a detailed discussion about the different synthetic methods for phosphors is also provided. Lastly, a brief description of different components and some conclusive remarks for the future perspective are given.

14.2 FLUORESCENT AND PHOSPHORESCENT MATERIALS

The phenomenon of luminescence is defined as the emission of light by any material. This word is has got its origin from the Latin Word Lumen, which means light [10]. The phenomenon of light emission is a distinguished property of any material that gets evolved by the interaction phenomenon and is governed by the principle of absorption and emissions. In this regard, the luminescence phenomenon is also termed as cold emission of light because this emission is not accompanied by any heat generation. The emission of light needs triggering for the initiation of the processes, the triggering can be done by some sort of reaction (chemical or biological), movement of the constituent particles, any radiation source, and so on. The materials that show the phenomenon of luminescence are called the luminophores or phosphors and are considered as the materials that change the absorbed energy into the emission while avoiding any kind of incandescence [11]. Till now most of the phosphor materials are found to be inorganic with only a handful being organic.

The usual size of the phosphor material lies in the range of 10^{-6} m–10^{-9} m i.e. (µm–nm). Structurally they are a composite of an inert host and an activator, which are the elements with 3d of 4f shell structure. Both the host and the dopant are equally responsible for the emission properties shown by a particular phosphor. Luminescence involves the absorption of the energy either by the host or dopant, a relaxation that is followed by the emission of light by a particular phosphor material [12]. These are the defects in the materials that are responsible for the emission phenomenon. Dopant ions are meant for this, however in some cases for the attainment of the efficient emission some other type of impurities are added that capture the energy supplied for the excitation which is then transferred to the host. Thus, by only altering the activator ions different emissive colors are obtained without changing the host. It is the phenomenon of relaxation that defines the effectiveness of the phosphor material.

For any phosphor materials, luminescence efficiency is defined as the ratio of the energy absorbed to the released. To date, the most used activator ions for the phosphor materials are the rear-earth ions due to their excellent properties which resulted in the efficient luminescence phenomenon. Out of the 17 rear-earth elements, the most important characteristics of the 13 out of them from Ce^{3+} to Yb^{3+} is the partially filled 4f shells overlying with the filled 5s and 5p that govern their optical investigations. Because of the unique energy level structure of these 13 elements, they exhibit the luminescence phenomenon in the visible range while the extra four exhibit this phenomenon near the visible range [13]. The rear-earth elements are being used for lighting applications since the nineteenth century. Trivalent lanthanide ions are considered to be unique for depicting the luminescence phenomenon because of their exceptional emission characteristics and high luminosity. Some of the parameters that are used for defining the efficiency of the phosphor materials are discussed below:

14.2.1 Quantum Efficiency (QE)

Phosphor materials that possess the value of QE of 80% or above are considered to be efficient. For phosphor to have higher values of the QE, the host and the activator ions must work together effectively such that the maximum amount of the incident energy should get absorbed. As most of the absorbed energy is lost during the relaxation process, thus all the absorbed energy is not being used for the emission phenomenon. The QE is defined as the number of photons emitted to the number of photons absorbed. The value of the QE determines the efficiency of a particular phosphor [14].

14.2.2 Color Rendering Index (CRI)

It is defined as the ability of the light source to depict an accurate color of the material that is being illuminated by the said source. The value of the Correlated Color Temperature (CCT) is the measure of this ability and is being denoted by R_a. The value of CRI for the sun is 100%, thus during the daylight, all the objects are being visualized in their ideal form. Thus, the higher the value that the CRI possessed by a light source the higher will be the accuracy/purity in reflecting the color of the illuminating substance. Besides the other parameters of a phosphor such as brightness, efficiency,

FIGURE 14.1 Color rendering index. ("Adapted with permission from reference [2], Copyright (2021), Elsevier".)

quantum yield, and so on, this too is very important for a phosphor substance. The difference in depicting the color of abject upon illuminating with the different light sources with varying CRI values is shown in Figure 14.1 [2]. The higher the value of the CRI possessed by an LED light source; the higher will be the cost of the source. It is clearly evident from this figure that as the CRI value increases the actual color of the image gets revealed more and more.

The achievement of color purity in the LEDs' luminaries is a challenging task for the researchers as an excellent LED light must have a high value of CRI. However, the Commission international de l'Eclairage (CIE) suggested the shortcomings of the CRI value as it is unable to define the color gamut and color fidelity index. Hence, the researcher's community has adopted a more precise way of measuring the color fidelity index (R_f). This is defined as how closely the color representation of the whole sample set is reproduced on an average by a test light in comparison to those under a reference illuminant [15]. But this newly defined color fidelity parameter does not act as a substitute for the already defined CCT parameter and is not used as a criterion.

14.2.3 Color Co-related Temperature

This is one of the parameters utilized for identifying the purity of the color in terms of the temperature by considering the temperature of the ideal black body as standard. The temperature of the color is expressed in the units of temperature i.e., Kelvin (K). This parameter is calculated only for those materials that exhibit the phenomenon of luminescence [2]. The lower values of the CCT depict the warmer light while the higher values depict the cooler light. There does not exist any fundamental white LED or illuminating source, it all depends upon the need of the light source which one requires. However, it has been observed that the light source whose CCT value lies in the range of 2,200–6,500 K is most readily used in every aspect. Light sources whose CCT value is less than 3,000 K are considered as warm light sources while those whose CCT value is greater than 4,000 K are considered as cold light sources [16]. The LED light sources that are being used mostly for illumination purposes possess the CCT value between 3,000 K and 5,000 K [17]. Depending on the value of the CCT industrialists have categorized the white light into different categories which are warm white (2,700 K), soft white (3,000 K), neutral white (3,500 K), cool white (4,100 K), bright white (5,000 K) and daylight white (6,500 K) [14].

14.2.4 CIE Chromaticity Coordinate

These coordinates are ideal metrics for the evaluation of the suitability of any color array for using them in different applications [18]. These metric variables have been developed by the international commission on illumination and are based on the tristimulus values obtained in three dimensions for any particular phosphor [19]. This concept has got its origin from the natural visualizing source (human eye) which also consists of three different color cones, each of them has been created by the combination of the three colors (RGB). As this value relies on the spectral power distribution of the emission thus these values are more exact and precise than others. Tristimulus values are functions of a wavelength corresponding to the red, green, and blue emissions. The precise position on the CIE chart is represented by a luminosity parameter 'Y' and CIE coordinates (x & y). Every specific color is obtained by the combination of the three primary colors (RGB), which is then displayed on a CIE chart by connecting all the three primary colors that result in the generation of the triangular symbol. Also, it has been observed that for all the non-chromatic sources the CIE coordinates lie inside the color gamut [14].

14.3 DIFFERENT TYPES OF LEDs

The evolution of the LEDs has altogether changed the solid-state light technology in every aspect. Because of their marvelous benefits over the other available lighting sources, they are being used in every field of science and technology. There are different ways of categorizing these LEDs i.e., on the basis of the host, wavelength, and emission color. Depending on the host there exist oxide, phosphate, silicate, aluminate, sulfide, and so on based on lattice phosphor. On the basis of the wavelength there exist only three LEDs i.e., white broadband LEDs, multi-wavelength LEDs, and single wavelength LEDs. Among them, the wavelength span for the first one ranges from near-ultraviolet to near-infrared. These closely resemble the mercury arc lamps as the light produced by them is white. In the case of the second one different LEDs (2–7) are joined or selected separately for the production of the white broadband. As many LEDs are involved in the configuration thus its lifetime is usually better in comparison to the others and possesses the capability to generate different wavelengths. The last category is devised in such a way so that it can produce only a specific wavelength. These LEDs are easier to install in comparison to the other two because of their compact size. Another and important way of classification of LEDs is based on their emission color. Depending on the emission color there exist various types of these LEDs such as red, green, blue, white, and so on emitting LEDs phosphors. Here a brief description of a few of them is provided.

14.3.1 Red-Emitting LED Phosphors

Among the different rear-earth dopants, europium is well known for the intense red emission for red LEDs phosphors. The Eu^{3+} ions possess intense red emission from in-between the wavelength ranges from 593 to 650 nm. This red emission arises due to the $^5D_0 - ^7F_2$ transition and has efficient color purity [20]. It is believed that if the Eu^{3+} ion is doped in the earth that too will emit their distinctive emission, but all the Eu^{3+} doped materials are not found suitable for lightning applications. For solid-state lighting, only those materials are considered efficient for it which possesses the broad excitation bands. Besides the fact that Eu^{3+} possesses broad excellent excitation and emission peaks due to the forbidden f-f transitions they also feature a charge transfer mechanism in the ultraviolet region of the electromagnet spectrum. This charge transfer phenomenon is being used for the sensitization of the red emission [21]. During the sensitization process, the Eu^{3+} ions are being converted to Eu^{2+} ions by transferring an appropriate amount of absorbed energy to it.

However, it has been observed that the mentioned mechanism does not show any significant advantage in the near-ultraviolet stimulated red-emitting phosphors. To resolve the issue only those host lattices are being selected in which the Eu^{3+} ions possess a position of lower symmetry. It is because the occurrence of Eu^{3+} at lower symmetry results in the relaxation of the forbidden f-f transitions along with the improvement in the efficiency in the excitation band [22]. The Eu^{2+} doped nitride phosphors are well-known for their red-emission as they not only possess remarkable stability (thermal and chemical) but also have superb luminous characteristics. Furthermore, these hosts offer an appropriate environment for the Eu^{2+} ions which results in the development of the crystal field effect thereby shifting the emission band to higher wavelengths [23]. Besides rear-earth ions nowadays transition metal ions are also used as dopants for the fabrication of the red-emitting LED phosphor materials [24]. As an example, an efficient $CaLa_2ZnO_5{:}Eu^{3+}$ red phosphor was developed by Vijay et al. the red emission of which is depicted in Figure 14.2 [25]. This figure shows the emission of the red color by $CaLa_2ZnO_5$ phosphor. In this figure, the researcher brought a blue LED torch. Then a paste of the fabricated phosphor powder has been poured on it which on aerating showed the red emission.

14.3.2 Blue-Emitting LED Phosphors

A large number of advancements have been made in the field of lighting technology since 1994 when the first InGaN LED chips were discovered. The earlier method of producing the white light i.e., by the combination of the blue and yellow phosphor has faced many problems such as low CRI and CCT values. To solve this issue, the researchers have proposed a new method called RGB, as this involves the combination of red, green, and blue for the production of the white light. It has been observed that the white light generated by this method possesses appropriate CRI and CCT values along with the efficient luminescence efficiency [26]. Thus, for the production of the white light using the RGB method, blue light being one component has achieved adequate importance. Besides this, blue light is also being used in different other applicative purposes such as plasma display panels, liquid crystal displays, tricolor lamps, and so on. In this regard, $BaMgAl_{10}O_{17}{:}Eu^{2+}$

FIGURE 14.2   Configuration of $CaLa_2ZnO_5:Eu^{3+}$ phosphor showing red emission. ("Adapted with permission from reference [25], Copyright (2021), Elsevier".)

is one of the most widely used blue phosphors [27,28]. This phosphor has attracted a lot of researchers for studying its optical and structural characteristics and also developing some novel fabrication methods [29]. Although its broad excitation spectrum is 225–410 nm, its relatively small absorption in the range of 360–410 nm has decreased its usage for LED manufacturing. Thus, different researchers are looking for some new blue-emitting phosphor that will meet the standards. Different studies have revealed that Eu^{2+} doped phosphor has shown outstanding blue luminescence. Among the different hosts doped with the Eu^{2+}, phosphates are well-known for showing blue luminescence. These hosts are favored by some important characteristic properties such as wide band-gap, low phonon energy along with excellent stability [2]. Few examples are $SrZnP_2O_7:Eu^{2+}$ [30], $NaMgPO_4:Eu^{2+}$ [31] and so on.

14.3.3 Green-Emitting LED Phosphors

Among the different types of tricolor LEDs, very little literature is present regarding green-emitting LED phosphor. However, the literature available is sufficient that one can easily visualize the advances achieved by the green-emitting LED phosphors in the LED technology. Tb^{3+} and Mn^{2+} ions are the readily used activator ions in different hosts that are responsible for the green emission [32]. However, for the enhancement of the luminescence efficiency Ce^{3+} and Eu^{2+} are used as sensitized ions for the hosts that possess Mn^{2+} activator ions. It has been observed that these sensitizer ions not only enhance the emission but occasionally they are also responsible for the green emission because of their 4f–5d transition. Although due to the crystal field phenomenon, the emission ranges of the Mn^{2+} ion can vary from red to green its forbidden transition 4T1–6A1 limits its application for LEDs [2]. The Tb^{3+} ions are exclusively used in different hosts and are being found most favorable for green emission. It has been observed that this ion only generates green emissions. Depending on the crystal field structure of the host in which it is doped, the yielded phosphor can either emit green light or nothing at all. The chance of not showing any luminescence phenomenon arises due to the probability of the terbium ion to attain a +4 charge state. Under these circumstances, the phosphor is being heated for the conversion of the Tb^{4+} to Tb^{3+} ions. Besides these ions, the Eu^{2+} ions also have been observed to produce green emissions in some hosts. Phosphor doped with Eu^{2+} ions possess a broad spectrum from the near ultra-violet to blue regions along with their strong emission bands. However, it has been observed that the silicates are the most favorable for the Eu^{2+} ions as they show emission from the green to the yellow range because of their excellent stability [2]. Few examples of the green-emitting phosphors are $SrSi_2O_2N_2:Eu^{2+}$ [33], $Ba_2LiSi_7AlN_{12}:Eu^{2+}$ [34] and so on.

14.4 ROLE AND IMPORTANCE OF LED TECHNOLOGY IN THE PRESENT WORLD

The different forms of lightning are being used for the last many decades and keep on changing to new forms with the development of technology. Because of the technological revolution different novel and potential LED phosphors have been developed, produced, and keep on changing along with their applicability in LED devices. The LED technology is considered to be the recent innovation that has completely replaced the old version of the lighting systems i.e., incandescent and compact fluorescent lamps. They offer a large number of benefits over the others which include compact size, efficient lifetime, durability, and so on. Because of their characteristic properties phosphor-converted LEDs (pc-LEDs) have opened new pathways in every aspect of the technological world. Researchers have made the following criteria that should be satisfied by the particular materials for using that in the LED-based solid-state lighting technology.

The phosphor materials should possess a coinciding excitation spectrum with that of the excitation source, should have an excellent luminescence and efficient quantum efficiency, the selected phosphor should have uniform morphology, non-hazardous, and should be cost-effective, and so on. All these characteristic features are found in pc-LEDs and are thus being used in every aspect of life such as for tri-color generation, in paints, in medical science, for the generation of the white light, for display technologies, and so on. Here a brief description of the generation of tri-color and white light is given. The notion of combining three separate thin emissive lines around 610 (red), 540 (green), and 450 (blue) led to the invention of the tri-color. This combination should be made in such a way that the resulting tri-color system has a high color rendering index [35]. In most cases, tri-color phosphors are found in the form of thin films. The phosphors are excited by the absorption of UV radiation emitted by mercury and electron interaction, which is then converted to visible light [36].

In general, the tri-color devices consist of 55%, 35%, and 15% of the red, green, and blue phosphors respectively. The most desirable emissions for these devices are presumably the red emission from Eu^{3+}, green emission from Tb^{3+}, and blue emission from Eu^{2+}. There exist a large number of hosts that can be used for the generation of the tri-color systems such as phosphates, aluminates, borates, silicates, and so on [37]. Among them the most readily used one is the aluminate, because of its characteristic features such as high thermal stability, strong luminous potency, and so on [38]. A few examples of the phosphor materials that have been used by the researchers for different colors are Y_2O_3:Eu^{3+} for red color, $BaMgAl_{10}O_{17}$:Eu^{2+} for the blue color, and $CeMgAl_{11}O_{19}$:Tb^{3+} for the green color [39].

White light-emitting devices (w-LEDs) with excellent electro-optical conversion efficiency, power effectiveness, long working life, and eco-friendliness can be used in a range of situations and can replace traditional lighting sources. The most widely used w-LEDs are made up of an InGaN-based blue LED and YAG: Ce^{3+} phosphor that emits yellow light [40]. But the lack of many characteristics such as poor color rendering index, large values of the correlated temperature, and significant quenching results in the poor quality of the white light generated by the complementary blue and yellow commercial phosphors [41–43]. This is due to the absence of the red-light producing part in the YAG: Ce^{3+}, which renders their utility for the w-LEDs. Thus, the quest for the new phosphors materials with red light as output is in need for the attainment of the quality white light. The different hosts that have been discovered to be proper for the generation of the red-light are oxides [44], silicates [45], borates [17], and so on.

Besides this, the most commonly used activators among the rare-earth ions for this are Eu^{3+}, Pr^{3+} and Sm^{3+}. Aside from this sometimes co-doping phenomenon is also used for improving the quality of the emission. A few examples of the different hosts that have been employed for improving the quality of the white light are Y_2SiO_5:Eu^{3+} [46], Sr_2CaMoO_6:Sm^{3+} [47]. Currently, they are being used as a light conversion agent (LCA) in the plastic films fabricated for covering the greenhouses [40,48]. The phosphor should have the activation wavelength values of the order of 290–350 nm and 510–580 nm for acting as an LCA agent. The emission of the LCA agent matches with the chlorophyll excitation distribution, which results in a considerable increment in photosynthesis [48]. It has been observed that the phosphor materials have a strong influence on improving the crop yield, dominance along matureness [49–50].

14.5 DIFFERENT SYNTHESIS METHODS

Different researchers have applied many different ways for the synthesis of different phosphor materials for the pc-LEDs. These phosphor materials consist of small activator ions and a host. There exist a large number of hosts which are being utilized for the fabrication of the different phosphor materials such as phosphates, oxides, aluminates, silicates, sulfides, halides, nitrites, etc. Sometimes these phosphor materials are also being coated for the attainment of optimal brightness and better efficiencies. Phosphors to be coated should have the characteristics such as; appropriate particles size, uniformity, high surface to volume ratio, and so on. The synthesis approach is selected depending on the specification of the phosphor to be fabricated. As most of the formulations of the phosphor contain many components thus obtaining the homogeneity, along with a single phase of the phosphors powder is challenging. The luminescence characteristics of a particular phosphor depend on its atomic structure, chemistry, contents of formation, texture, etc. All these phenomena are directly related to the thermodynamics and the kinetics of the synthesis method applied. Solid-state reaction, sol-gel, co-precipitation, solvothermal, combustion, spray pyrolysis, and microwave-assisted techniques are some of the methods employed for the different phosphor materials. Among the mentioned ones a brief introduction of some of the methods is presented here [2].

14.5.1 Solid-State Method

This synthesis method relies on the high-temperature processing of different components and thus possesses an impact on the physicochemical properties of the yielded phosphor materials. This approach is mostly used to synthesize solid luminous compounds which can't be produced at ambient temperature. Researchers found that the main elements that influence the likelihood along with the rate at which a particular reaction took place include the properties like reactant surface area, reaction environment, structural arrangement, and the reactivity of the different components [51]. For the initiation of the synthesis processes, different reagents such as metal oxides and dopants are taken in appropriate proportions followed by the proper grinding in a mortar. After the proper mixture of all the precursors, they are then added to distilled water and swirled at the best temperature. Many other reagents such as Li_2CO_3, BaF_2, NH_4F, etc. are utilized for the maintenance of temperature, enhancing the reaction pace, and for improvement of the crystallinity in the structure of the resulting phosphor. Besides this these reagents also enhance the other structural properties along with the luminescence phenomenon [52].

Most of the researchers also use acetone for ensuring that the chemicals are mixed uniformly. This reaction proceeds at a high temperature (1200–1600 °C) as the low temperature does not favor the occurrence of the reaction. Elevated temperature accelerates the mobility of the reactants which results in fast diffusion processes, lowers the activation energy, and speeds up the reaction. Following the mentioned steps results in the production of a luminous substance in powder form. This process has the advantage over others as the precursors needed remain readily available and the process does not produce any hazardous waste. However, the long reaction time and high-temperature requirements are some of the major drawbacks of the method. In addition, the contamination and nonuniformity among the obtained structures lower the efficiency of the method [53]. Furthermore, it has been observed that repetitive chemical washing results in damaging structural and luminescence properties. Some of the examples of the phosphor materials prepared by this method include $LiBaPO_4$: RE (RE=Eu^{2+}, Tb^{3+}, Sm^{3+}) [54], NaM_4 $(VO_4)_3$:Dy^{3+} (M=Ca, Ba, Sr) [55], etc.

14.5.2 Co-Precipitation Method

This technique is most favorable for the synthesis of silicate and fluoride-based phosphor materials. Among the different available techniques, this one is known to be one of the most cost-effective as it does not necessitate any use of expensive and complicated instrumentation [56]. In this technique, the desired materials (host & dopant) are taken in the stoichiometric ratios and then mixed completely by forming a solution using deionized as solvent. The solution is then kept on a stirrer with a hotplate for the attainment of the transparency of the solution. In this method, ammonium hydroxide is used as a precipitating agent. After the addition of the precipitating agent, the solution is left for some time, till the attainment of proper precipitation. The solution is then filtered for separation of the desired material which is then followed by extensive washing and drying. This method proves to be effective for the incorporation of the rear-earth ions as dopants in the developed nanophosphors. For improving the properties, a modified version of this technique has been devised wherein the reaction is carried out in an inert environment. Some of the examples of the different phosphor materials developed by this method are $Y_2O_2SO_4$:Eu^{3+} [57], K_2MF_6:Mn^{4+} (M=Ge, Si) [58], Ca_3La $(VO_4)_3$:RE^{3+} (RE=Eu^{3+}, Dy^{3+}) [59] etc.

14.5.3 Spray Pyrolysis Methods

These are considered to be the most accessible, fast, cost-effective, and easy for the production of nanoparticles with excellent morphological characteristics. Aerosol or flame spray pyrolysis (FSP) is the two different approaches of this method used for the fabrication of the phosphors [60]. FSP is an outcome of ultrasonic spray pyrolysis and is efficiently used for the fabrication of silicate nanophosphors. In these materials, the luminescence center is usually surrounded by the silica layer for the attainment of stability and amplification. This process entails four stages for the production of the required phosphor [61]. The first step is the formation of a drop from the precursors, followed by the shrinkage by the evaporation phenomenon. The third stage involves the conversion to the oxide, which results in the production of the desired phosphor and the completion of the last stage [60]. In the experimental setup, a burner ultrasonic generator is used for the production of the droplets. Distilled water is used as a solvent in which metal nitrates, activator ions, and silica are added for the formation of precursors. The obtained precursors are then placed in an ultrasonic vibrator for the formation of an aerosol [61]. The prepared aerosol is then oxidized by the passage of oxygen and fuel gas, which is then warmed at optimum temperature for the proper removal of the solvent through the evaporation process. Finally, the obtained materials are put in a furnace for breakdown. The desired materials obtained by this method possess pure chemical content and are of homogeneous shape. The main drawback of this method is that the products generated by this method are found to have a porous structure, which significantly impacts the morphology and thermal stability of the materials. To overcome this problem, the modified version of this method was employed wherein an artificial solution was used for the fabrication [62]. Few examples include Zn_2SiO_4: Mn [63], $(Ba, Sr)_2SiO_4$:Eu [64] etc.

14.5.4 Combustion Method

Among the different chemical processes, this approach is efficient in terms of time and efficiency gain. A wide range of phosphor materials has been synthesized by this method. Oxides, sulfides, aluminates, and so on are some of the well-known examples of the materials obtained by this approach of the synthesis [65–67]. On a comparative analysis with the solid-state method, it has been observed that the phosphor materials obtained by this method are highly homogeneous, possess small contaminants, and have a large surface-to-volume ratio. In this method, the solution of the metal nitrites and fuels is fired at a low temperature of the order of 500 °C. The commonly used fueling agents are urea and glycine. This reaction involved the oxidation reaction resulting in the release of an enormous amount of heat in the form of flame. With the evolution of a large amount of heat, the reaction temperature raises up to 1200 °C within a small instant of time (10 minutes) [68]. This increase in temperature for a brief time led to the development of the nanocrystalline phosphor powders. This powder is merged latterly into submicron-sized particles. However, researchers have developed an updated version of this method by employing hydrazine as a fueling agent, for enhancing the properties of the oxide powders. This updated version is more preferable because of its novel properties such as enhanced safety, complete recovery of the product, proper control over the reaction parameters and the environment [69]. Few examples of the materials fabricated by this method are $(Sr, Ca)_2Si_5N_8$:Eu^{2+} [70], $SrAl_4O_7$:Eu^{2+}, Dy^{3+} [71], $Y_3Al_5O_{12}$:Ce^{3+} [72].

14.6 DEVICE FABRICATION

Device fabrication is an important part of the LED luminaries before being released to the market. This process involves the manufacturing, packaging, and finally the fabrication of the desired lighting system. With the growing demands over the globe, attention is being diverted for cost reduction by choosing appropriate materials for synthesis. The selection of the appropriate materials required during the synthesis is the major concern in the development of LEDs. Therefore, before proceeding with the development, the first step is to select a proper substrate. There is a very limited number of options of substrates used for the development of efficient gallium nitride (GaN) based LEDs which include sapphire, silicon carbide (SiC), GaN, and silicon (Si). The direct bandgap of the order of 3.39 eV of the GaN is its outstanding property because of which it is being used for the development of multicolor lighting devices such as LEDs, lasers, and so on. The first single-crystal film of GaN was reported in 1969 which was found to be most suitable for the assessment of the electrical and optical properties. Researchers utilized the vapor-phase technique for the development of GaN crystals and reported a direct bandgap structure with a bandgap value of 3.39 eV [73].

However, it was challenging to create an efficient quality of GaN layer with a smooth, crack-free surface due to significant lattice distortion along with a difference in thermal expansion coefficients between the GaN and the Sapphire substrates. This mismatch led to many defects which in turn have a direct effect on the lifetime and the brightness of the developed luminaries. Due to the structural resemblance of the SiC with the GaN, it acts as an efficient and popular option for the replacement of sapphires. It not only reduces the defect concentration but also tends to enhance the effectiveness along with the endurance. Both the mentioned ones are unable to meet the demands because of the cost and complexity for their development. Thus, using the substrates with patterned surfaces proves to be efficient for increasing the light extraction, lowering the defect concentrations, and reducing the cost of production. The research in this field is continuously increasing for attaining better efficiencies in every aspect [2]. Before packaging the selected phosphor need to have the test for compatibility with the rest of the LEDs. Because despite early claims of high luminous characteristics, some of the phosphors failed to endure tests at the time of installation in LEDs. As a thumb rule, it has been observed that those host materials that possess a wide bandgap structure prove to be excellent phosphors material [74]. During the packing of the material mechanical support and the protection of the die remains the main concern. These are adjusted in such a way as to maximize the output. During the LED packaging exterior connection of the die is also provided. Usually, low-cost plastic packing in a variety of sizes is favored for LED packaging.

Nowadays the chip-on-board (COB) is being used, where the LED die is placed on the ceramic or the metal core printed circuit board (MCPCB). The usage of COB packaging proves to be an effective and efficient alternative to plastic ones. However, because of the lengthy connecting procedure, an extension of the COB was found in the form of a chip-in-flex (CIF). The CIF shortened the signal route down to the substrate thereby decreasing the connection length [75]. After packaging the LEDs, a proper encapsulation process is done for protecting against different contaminations and to separate the circuitry from different stresses. This encapsulation process also assists in the extraction of photons, as the phosphor materials are distributed throughout the encapsulate matrix before being placed on the top of the LED die. The most commonly utilized material for the encapsulation process is silicone as it possesses the property of being resistant to heat and blue emissions. However, it has been observed that this normal silicone does not stand at high thermal loads as there remains a danger of shattering, deterioration, and so on. However, using some high-grade silicone materials proves to be efficient but that is prohibitively costly which makes that unsuitable. Spin-on-glass composites have been offered as a substitute for silicone, and they provided impressive light attenuation in comparison to silicone over a lengthy illumination period, but at a lower brightness [76]. After passing all the mentioned stages the LEDs luminaries comes into the market. These luminaries are a stunning collection of a large number of LEDs, heat sinks, driving circuits, and lenses, all of which are combined into a single device [2]. These structures along with the different ways of packaging that have been applied for the LED phosphors. It is evident that it consists of an active layer, p-type and n-type layers along with cathode and anode. The PLCC, MCPCB, QFN, and others exhibit the different packaging approaches that have evolved with the development and progress made in this field.

14.7 CONCLUSION

This chapter focuses on the advancements that have been made in solid-state lighting technology. It has been observed that only those advancements meet the demands that possess characteristic properties like high luminescence efficiency and brightness, low energy consumption, eco-friendly nature, etc. The comparative studies have revealed that only nanotechnology-enabled phosphors LEDs meet all these demands. Keeping in view the increasing demands of these appliances and their usage in different areas it is important to focus on the development of novel and efficient phosphors. The investigation should be made for the phosphors materials that possess the characteristics like high-resolution profile, innovative fabrication route, persistency, quick-reaction, and regulated lifetime. This realm opens a plethora of possibilities for the researchers to focus on the development of new and creative phosphors along with their applications in different areas.

ACKNOWLEDGMENT

One of the authors (V. Kumar) acknowledges the financial support from DST-FIST (Grant No. SR/FST/PS-1/2021/169).

REFERENCES

[1] J. McKenna, A look ahead to the 2015 international year of light and light-based technologies, *J. Opt.* 16 (2014) 120201. https://doi.org/10.1088/2040–8978/16/12/120201.

[2] G.B. Nair, H.C. Swart, S.J. Dhoble, A review on the advancements in phosphor-converted light emitting diodes (pc-LEDs): Phosphor synthesis, device fabrication and characterization, *Prog. Mater. Sci.* (2019) 100622. https://doi.org/10.1016/j.pmatsci.2019.100622.

[3] M.M. Yan, Y. Li, Y.T. Zhou, L. Liu, Y. Zhang, B.G. You, Y. Li, Enhancing the performance of blue quantum-dot light-emitting diodes based on Mg-doped ZnO as an electron transport layer, *IEEE Photonics J.* 9 (2017). https://doi.org/10.1109/JPHOT.2017.2666423.

[4] G.B. Nair, H.C. Swart, S.J. Dhoble, A review on the advancements in phosphor-converted light emitting diodes (pc-LEDs): Phosphor synthesis, device fabrication and characterization, *Prog. Mater. Sci.* (2019). https://doi.org/10.1016/j.pmatsci.2019.100622.

[5] V.B. Pawade, H.C. Swart, S.J. Dhoble, Review of rare earth activated blue emission phosphors prepared by combustion synthesis, *Renew. Sustain. Energy Rev.* 52 (2015) 596–612. https://doi.org/10.1016/j.rser.2015.07.170.

[6] Y.-C. Li, Y.-H. Chang, Y.-F. Lin, Y.-S. Chang, Y.-J. Lin, Synthesis and luminescent properties of Ln3+ (Eu3+, Sm3+, Dy3+)-doped lanthanum aluminum germanate LaAlGe2O7 phosphors, *J. Alloys Compd.* 439 (2007) 367–375. https://doi.org/10.1016/j.jallcom.2006.08.269.

[7] N.G. Yeh, C.H. Wu, T.C. Cheng, Light-emitting diodes-their potential in biomedical applications, *Renew. Sustain. Energy Rev.* 14 (2010) 2161–2166. https://doi.org/10.1016/j.rser.2010.02.015.

[8] V.B. Pawade, S.J. Dhoble, Spin-orbit splitting difference and stokes shift in cerium 3+-activated aluminate phosphors, *Spectrosc. Lett.* 46 (2013) 472–475. https://doi.org/10.1080/00387010.2012.731025.

[9] V.B. Pawade, S.J. Dhoble, Novel blue-emitting SrMg 2Al 16O 27:Eu 2+ phosphor for solid-state lighting, *Luminescence.* 26 (2011) 722–727. https://doi.org/10.1002/bio.1304.

[10] B. Valeur, M.N. Berberan-Santos, A brief history of fluorescence and phosphorescence before the emergence of quantum theory, *J. Chem. Educ.* 88 (2011) 731–738. https://doi.org/10.1021/ed100182h.

[11] K. Shinde, S. Dhoble, H. Swart, K. Park, Phosphate phosphors for solid-state lighting (2012). https://books.google.com/books?hl=en&lr=&id=Wubt1IKnjEkC&oi=fnd&pg=PR5&ots=rfs jJciMIa&sig=FXXAEKOE586ecftFbujMqkljebc (accessed July 10, 2021).

[12] C. Li, H. Zheng, H. Wei, S. Qiu, L. Xu, X. Wang, H. Jiao, A color tunable and white light emitting Ca2Si5N8:Ce3+, Eu2+ phosphor via efficient energy transfer for near-UV white LEDs, *Dalt. Trans.* 47 (2018) 6860–6867. https://doi.org/10.1039/C8DT01430B.

[13] H. Dong, L.-D. Sun, C.-H. Yan, Energy transfer in lanthanide upconversion studies for extended optical applications, *Chem. Soc. Rev.* 44 (2015) 1608–1634. https://doi.org/10.1039/C4CS00188E.

[14] I. Gupta, S. Singh, S. Bhagwan, D. Singh, Rare earth (RE) doped phosphors and their emerging applications: A review, *Ceram. Int.* 47 (2021) 19282–19303. https://doi.org/10.1016/j.ceramint.2021.03.308.

[15] H. Yaguchi, A. David, T. Fuchida, … T. Yano, undefined 2017, CIE 2017 colour fidelity index for accurate scientific use, Lirias.Kuleuven.Be. (n.d.). https://lirias.kuleuven.be/1676302?limo=0 (accessed July 10, 2021).

[16] S. Singh, A.P. Simantilleke, D. Singh, Crystal structure and photoluminescence investigations of Y3Al5O12:Dy3+ nanocrystalline phosphors for WLEDs, *Chem. Phys. Lett.* 765 (2021) 138300. https://doi.org/10.1016/J.CPLETT.2020.138300.

[17] R. Shrivastava, J. Kaur, V. Dubey, White light emission by Dy3+ doped phosphor matrices: A short review, *J. Fluoresc.* 26 (2016) 105–111. https://doi.org/10.1007/s10895-015-1689-8.

[18] CIE Draft Standard DS 008.2/E-2000, Lighting of indoor work places, *Color Res. Appl.* 25 (2000) 385–386. https://doi.org/10.1002/1520-6378(200010)25:5<385::AID-COL16>3.0.CO;2-D.

[19] Y.-C. Lin, M. Karlsson, M. Bettinelli, *Inorganic Phosphor Materials for Lighting* (2017), pp. 309–355. https://doi.org/10.1007/978-3-319-59304-3_10.

[20] P. Niu, X. Liu, Y. Wang, W. Zhao, Photoluminescence properties of a novel red-emitting phosphor Ba2LaV3O11:Eu3+, *J. Mater. Sci. Mater. Electron.* 29 (2018) 124–129. https://doi.org/10.1007/s10854-017-7895-1.

[21] S. Ray, G.B. Nair, P. Tadge, N. Malvia, V. Rajput, V. Chopra, S.J. Dhoble, Size and shape-tailored hydrothermal synthesis and characterization of nanocrystalline LaPO4:Eu3+ phosphor, *J. Lumin.* 194 (2018) 64–71. https://doi.org/10.1016/j.jlumin.2017.10.015.

[22] P. Dorenbos, The Eu3+ charge transfer energy and the relation with the band gap of compounds, *J. Lumin.* 111 (2005) 89–104. https://doi.org/10.1016/j.jlumin.2004.07.003.

[23] R.J. Xie, N. Hirosaki, Silicon-based oxynitride and nitride phosphors for white LEDs-A review, *Sci. Technol. Adv. Mater.* 8 (2007) 588–600. https://doi.org/10.1016/j.stam.2007.08.005.

[24] A. Fu, L. Zhou, S. Wang, Y. Li, Preparation, structural and optical characteristics of a deep red-emitting Mg2Al4Si5O18: Mn4+ phosphor for warm w-LEDs, *Dye. Pigment.* 148 (2018) 9–15. https://doi.org/10.1016/j.dyepig.2017.08.050.

[25] V. Kumar, S. Som, S. Dutta, S. Das, H.C. Swart, Red-light-emitting inorganic La2CaZnO5 frameworks with high photoluminescence quantum efficiency: Theoretical approach, *Mater. Des.* 93 (2016) 203–215. https://doi.org/10.1016/j.matdes.2015.12.153.

[26] S. Ye, F. Xiao, Y.X. Pan, Y.Y. Ma, Q.Y. Zhang, Phosphors in phosphor-converted white light-emitting diodes: Recent advances in materials, techniques and properties, *Mater. Sci. Eng. R Rep.* 71 (2010) 1–34. https://doi.org/10.1016/j.mser.2010.07.001.

[27] S.S. Lee, H.J. Kim, S.H. Byeon, J.C. Park, D.K. Kim, Thermal-shock-assisted solid-state process for the production of BaMgAl10O17:Eu phosphor, *Ind. Eng. Chem. Res.* 44 (2005) 4300–4303. https://doi.org/10.1021/ie048953j.

[28] K.B. Kim, Y. Il Kim, H.G. Chun, T.Y. Cho, J.S. Jung, J.G. Kang, Structural and optical properties of BaMgAl10O17:Eu2+ phosphor, *Chem. Mater.* 14 (2002) 5045–5052. https://doi.org/10.1021/cm020592f.

[29] Z.H. Zhang, Y.H. Wang, X.X. Li, Y.K. Du, W.J. Liu, Photoluminescence degradation and color shift studies of annealed BaMgAl10O17:Eu2+ phosphor, *J. Lumin.* 122–123 (2007) 1003–1005. https://doi.org/10.1016/j.jlumin.2006.01.351.

[30] J.L. Yuan, X.Y. Zeng, J.T. Zhao, Z.J. Zhang, H.H. Chen, G. Bin Zhang, Rietveld refinement and photoluminescent properties of a new blue-emitting material: Eu2+ activated SrZnP2O7, *J. Solid State Chem.* 180 (2007) 3310–3316. https://doi.org/10.1016/J.JSSC.2007.09.023.

[31] S.W. Kim, T. Hasegawa, T. Ishigaki, K. Uematsu, K. Toda, M. Sato, Efficient red emission of blue-light excitable new structure type NaMgPO4:Eu2+ phosphor, *ECS Solid State Lett.* 2 (2013) R49. https://doi.org/10.1149/2.004312SSL.

[32] V. Singh, R.P.S. Chakradhar, J.L. Rao, D.K. Kim, Mn2+ activated MgSrAl10O17 green-emitting phosphor-A luminescence and EPR study, *J. Lumin.* 128 (2008) 1474–1478. https://doi.org/10.1016/j.jlumin.2008.02.001.

[33] C.-Y. Wang, R.-J. Xie, F. Li, X. Xu, Thermal degradation of the green-emitting SrSi2O2N2:Eu2+ phosphor for solid state lighting, *J. Mater. Chem. C.* 2 (2014) 2735–2742. https://doi.org/10.1039/C3TC32582B.

[34] T. Takeda, N. Hirosaki, S. Funahshi, R.-J. Xie, Narrow-band green-emitting phosphor Ba2LiSi7AlN12:Eu2+ with high thermal stability discovered by a single particle diagnosis approach, *Chem. Mater.* 27 (2015) 5892–5898. https://doi.org/10.1021/ACS.CHEMMATER.5B01464.

[35] Y. Zhao, X. Wang, Y. Zhang, Y. Li, X. Yao, Optical temperature sensing of up-conversion luminescent materials: Fundamentals and progress, *J. Alloys Compd.* 817 (2020) 152691. https://doi.org/10.1016/j.jallcom.2019.152691.

[36] C. Tunsu, T. Retegan, C. Ekberg, Sustainable processes development for recycling of fluorescent phosphorous powders-rare earths and mercury separation A literature report (n.d).

[37] D. Singh, Europium doped silicate phosphors: Synthetic and characterization techniques, *Adv. Mater. Lett.* 8 (2017) 656–672. https://doi.org/10.5185/amlett.2017.7011.

[38] W. Bin Hung, T.M. Chen, Efficiency enhancement of silicon solar cells through a downshifting and antireflective oxysulfide phosphor layer, *Sol. Energy Mater. Sol. Cells.* 133 (2015) 39–47. https://doi.org/10.1016/j.solmat.2014.11.011.

[39] C.R. Ronda, T. Jüstel, H. Nikol, Rare earth phosphors: Fundamentals and applications, *J. Alloys Compd.* 275–277 (1998) 669–676. https://doi.org/10.1016/S0925-8388(98)00416-2.

[40] E.F. Schubert, J.K. Kim, Solid-state light sources getting smart, *Science (80-.).* 308 (2005) 1274–1278. https://doi.org/10.1126/science.1108712.

[41] H.S. Jang, H. Yang, S.W. Kim, J.Y. Han, S.G. Lee, D.Y. Jeon, White light-emitting diodes with excellent color rendering based on organically capped CdSe quantum dots and Sr3SiO5:Ce 3+, Li+ phosphors, *Adv. Mater.* 20 (2008) 2696–2702. https://doi.org/10.1002/adma.200702846.

[42] K.A. Denault, J. Brgoch, M.W. Gaultois, A. Mikhailovsky, R. Petry, H. Winkler, S.P. Denbaars, R. Seshadri, Consequences of optimal bond valence on structural rigidity and improved luminescence properties in SrxBa2-xSiO4:Eu 2+ orthosilicate phosphors, *Chem. Mater.* 26 (2014) 2275–2282. https://doi.org/10.1021/cm500116u.

[43] Q. Wang, D. Deng, Y. Hua, L. Huang, H. Wang, S. Zhao, G. Jia, C. Li, S. Xu, Potential tunable white-emitting phosphor LiSr4(BO 3)3:Ce3, Eu2 for ultraviolet light-emitting diodes, *J. Lumin.* 132 (2012) 434–438. https://doi.org/10.1016/j.jlumin.2011.09.003.

[44] J.G. Li, X. Li, X. Sun, T. Ishigaki, Monodispersed colloidal spheres for uniform y2O 3:Eu3+ red-phosphor particles and greatly enhanced luminescence by simultaneous Gd3+ doping, *J. Phys. Chem. C.* 112 (2008) 11707–11716. https://doi.org/10.1021/jp802383a.

[45] S. Singh, D. Singh, Synthesis and spectroscopic investigations of trivalent europium-doped M2SiO5 (M=Y and Gd) nanophosphor for display applications, *J. Mater. Sci. Mater. Electron.* 31 (2020) 5165–5175. https://doi.org/10.1007/s10854-020-03076-5.

[46] B. Lauritzen, N. Timoney, N. Gisin, M. Afzelius, H. De Riedmatten, Y. Sun, R.M. MacFarlane, R.L. Cone, Spectroscopic investigations of Eu3+:Y2SiO 5 for quantum memory applications, *Phys. Rev. B Condens. Matter Mater. Phys.* 85 (2012) 115111. https://doi.org/10.1103/PhysRevB.85.115111.

[47] L. Wang, H.M. Noh, B.K. Moon, S.H. Park, K.H. Kim, J. Shi, J.H. Jeong, Dual-mode luminescence with broad near UV and blue excitation band from Sr2CaMoO6:Sm3+ phosphor for white LEDs, *J. Phys. Chem. C.* 119 (2015) 15517–15525. https://doi.org/10.1021/acs.jpcc.5b02828.

[48] X. Liu, B. Lei, Y. Liu, The application of phosphor in agricultural field, in: *Phosphors, Up Convers. Nano Part. Quantum Dots Their Appl.* Springer, Singapore, (2016): pp. 119–137. https://doi.org/10.1007/978-981-10-1590-8_4.

[49] J. Wu, D. Newman, I.V.F. Viney, Study on relationship of luminescence in CaS:Eu, Sm and dopants concentration, *J. Lumin.* 99 (2002) 237–245. https://doi.org/10.1016/S0022-2313(02)00342-3.

[50] M. Suchea, S. Christoulakis, M. Androulidaki, E. Koudoumas, CaS:Eu, Sm and CaS:Ce, Sm films grown by embedding active powder into an inert matrix, *Mater. Sci. Eng. B Solid-State Mater. Adv. Technol.* 150 (2008) 130–134. https://doi.org/10.1016/j.mseb.2008.03.017.

[51] A. R. West. Solid state chemistry and its applications. John Wiley and Sons, 1985 Paper back, Price £ 14.95, US $ 19.95, *Cryst. Res. Technol.* 21 (1986) 166–166. https://doi.org/10.1002/crat.2170210140.

[52] N. Taghavinia, G. Lerondel, H. Makino, T. Yao, Europium-doped yttrium silicate nanoparticles embedded in a porous SiO 2 matrix, *Nanotechnology.* 15 (2004) 1549–1553. https://doi.org/10.1088/0957-4484/15/11/031.

[53] L.Y. Bao, W. Gao, Y.F. Su, Z. Wang, N. Li, S. Chen, F. Wu, Progression of the silicate cathode materials used in lithium ion batteries, *Chinese Sci. Bull.* 58 (2013) 575–584. https://doi.org/10.1007/s11434-012-5583-3.

[54] J. Sun, X. Zhang, Z. Xia, H. Du, Luminescent properties of LiBaPO 4:RE (RE Eu 2, Tb 3, Sm 3) phosphors for white light-emitting diodes, *J. Appl. Phys.* 111 (2012) 013101. https://doi.org/10.1063/1.3673331.

[55] R. Singh, S.J. Dhoble, Dy 3+-activated NaM 4 (VO 4) 3 (M=Ca, Ba, Sr) phosphor for near-UV solid-state lighting, *Luminescence.* 26 (2011) 728–733. https://doi.org/10.1002/bio.1305.

[56] W.S. Peternele, V. Monge Fuentes, M.L. Fascineli, J. Rodrigues Da Silva, R.C. Silva, C.M. Lucci, R. Bentes De Azevedo, Experimental investigation of the coprecipitation method: An approach to obtain magnetite and maghemite nanoparticles with improved properties, *J. Nanomater.* 2014 (2014). https://doi.org/10.1155/2014/682985.

[57] J. Lian, H. Qin, P. Liang, F. Liu, Co-precipitation synthesis of Y2O2SO4:Eu3+ nanophosphor and comparison of photoluminescence properties with Y2O3:Eu3+ and Y2O2S:Eu3+ nanophosphors, *Solid State Sci.* 48 (2015) 147–154. https://doi.org/10.1016/j.solidstatesciences.2015.08.004.

[58] L.L. Wei, C.C. Lin, M.H. Fang, M.G. Brik, S.F. Hu, H. Jiao, R.S. Liu, A low-temperature co-precipitation approach to synthesize fluoride phosphors K2MF6:Mn4+ (M=Ge, Si) for white LED applications, *J. Mater. Chem. C.* 3 (2015) 1655–1660. https://doi.org/10.1039/c4tc02551b.

[59] B.V. Rao, K. Jang, H.S. Lee, S.S. Yi, J.H. Jeong, Synthesis and photoluminescence characterization of RE3+ (=Eu3+, Dy3+)-activated Ca3La(VO4)3 phosphors for white light-emitting diodes, *J. Alloys Compd.* 496 (2010) 251–255. https://doi.org/10.1016/j.jallcom.2009.12.175.

[60] P.S. Patil, Versatility of chemical spray pyrolysis technique, *Mater. Chem. Phys.* 59 (1999) 185–198. https://doi.org/10.1016/S0254-0584(99)00049-8.

[61] H. Hasegawa, T. Ueda, T. Yokomori, Y2Si2O7:Eu/SiO2 core shell phosphor particles prepared by flame spray pyrolysis, *Proc. Combust. Inst.* 34 (2013) 2155–2162. https://doi.org/10.1016/j.proci.2012.07.067.

[62] H.S. Kang, Y.C. Kang, H.D. Park, Y.G. Shul, Green-emitting yttrium silicate phosphor particles prepared by large scale ultrasonic spray pyrolysis, *Korean J. Chem. Eng.* 20 (2003) 930–933. https://doi.org/10.1007/BF02697301.

[63] Y.C. Kang, S.B. Park, Zn2SiO4:Mn phosphor particles prepared by spray pyrolysis using a filter expansion aerosol generator, *Mater. Res. Bull.* 35 (2000) 1143–1151. https://doi.org/10.1016/S0025-5408(00)00306-8.

[64] H.S. Kang, Y.C. Kang, K.Y. Jung, S. Bin Park, Eu-doped barium strontium silicate phosphor particles prepared from spray solution containing NH4Cl flux by spray pyrolysis, *Mater. Sci. Eng. B Solid State Mater. Adv. Technol.* 121 (2005) 81–85. https://doi.org/10.1016/j.mseb.2005.03.013.

[65] F. Deganello, G. Marcì, G. Deganello, Citrate-nitrate auto-combustion synthesis of perovskite-type nanopowders: A systematic approach, *J. Eur. Ceram. Soc.* 29 (2009) 439–450. https://doi.org/10.1016/j.jeurceramsoc.2008.06.012.

[66] S.L. González-Cortés, F.E. Imbert, Fundamentals, properties and applications of solid catalysts prepared by solution combustion synthesis (SCS), *Appl. Catal. A Gen.* 452 (2013) 117–131. https://doi.org/10.1016/j.apcata.2012.11.024.

[67] H.H. Nersisyan, J.H. Lee, J.R. Ding, K.S. Kim, K. V. Manukyan, A.S. Mukasyan, Combustion synthesis of zero-, one-, two- and three-dimensional nanostructures: Current trends and future perspectives, *Prog. Energy Combust. Sci.* 63 (2017) 79–118. https://doi.org/10.1016/j.pecs.2017.07.002.

[68] L.E. Shea, J. McKittrick, O.A. Lopez, E. Sluzky, Synthesis of red-emitting, small particle size luminescent oxides using an optimized combustion process, *J. Am. Ceram. Soc.* 79 (1996) 3257–3265. https://doi.org/10.1111/j.1151-2916.1996.tb08103.x.

[69] R. García, G.A. Hirata, J. McKittrick, New combustion synthesis technique for the production of (InxGa1-x)2O3 powders: Hydrazine/metal nitrate method, *J. Mater. Res.* 16 (2001) 1059–1065. https://doi.org/10.1557/JMR.2001.0147.

[70] S.L. Chung, W.C. Chou, Combustion synthesis of Ca 2 Si 5 N 8: Eu 2+ phosphors and their luminescent properties, *J. Am. Ceram. Soc.* 96 (2013) 2086–2092. https://doi.org/10.1111/jace.12369.

[71] T. Peng, H. Yang, X. Pu, B. Hu, Z. Jiang, C. Yan, Combustion synthesis and photoluminescence of SrAl2O 4:Eu, Dy phosphor nanoparticles, *Mater. Lett.* 58 (2004) 352–356. https://doi.org/10.1016/S0167-577X(03)00499-3.

[72] M. Upasani, B. Butey, S. V. Moharil, Synthesis, characterization and optical properties of Y3Al5O12:Ce phosphor by mixed fuel combustion synthesis, *J. Alloys Compd.* 650 (2015) 858–862. https://doi.org/10.1016/j.jallcom.2015.08.076.

[73] H.P. Maruska, J.J. Tietjen, The preparation and properties of vapor-deposited single-crystalline GaN, *Appl. Phys. Lett.* 15 (1969) 327–329. https://doi.org/10.1063/1.1652845.

[74] J. Brgoch, S.P. Denbaars, R. Seshadri, Proxies from Ab initio calculations for screening efficient Ce3+ phosphor hosts, *J. Phys. Chem. C.* 117 (2013) 17955–17959. https://doi.org/10.1021/jp405858e.

[75] K.L. Suk, H.Y. Son, C.K. Chung, J. Do Kim, J.W. Lee, K.W. Paik, Flexible Chip-on-Flex (COF) and embedded Chip-in-Flex (CIF) packages by applying wafer level package (WLP) technology using anisotropic conductive films (ACFs), in: *Microelectron.* Reliab., Pergamon, (2012): pp. 225–234. https://doi.org/10.1016/j.microrel.2011.08.003.

[76] L.B. Chang, K.W. Pan, C.Y. Yen, M.J. Jeng, C. Te Wu, S.C. Hu, Y.K, Kuo, Comparison of silicone and spin-on glass packaging materials for light-emitting diode encapsulation, *Thin Solid Films.* 570 (2014) 496–499. https://doi.org/10.1016/j.tsf.2014.03.033.

CHAPTER 15

Layer Structured Materials for Photonics

Felipe M. de Souza and Ram K. Gupta

Pittsburg State University

CONTENTS

15.1 INTRODUCTION

The concept of photonics lies in the interaction of materials with photons in a way that they can detect, emit, control, or manipulate light. Within that broad concept, photonic materials have been drastically increasing their relevance in several technological sectors after the start of the digital era. This process is occurring due to the strong necessity of fast-flux of information, data transmission, electronic devices with non-harmful lights in their displays i.e., blue light, better resolution for imaging in medical instruments for more accurate diagnosis, among many other applications. In that sense, the photonics research led to an optimized set of novel materials that can deliver better performance, while presenting a lower cost, longer lifetime, and less energy wastage when compared to traditional devices used to produce light. One example of efficient photonic materials is the light-emitting diodes (LEDs), which could emit lights with higher intensity and television displays with better resolution. Another example was the optic fibers, which promoted a great enhancement in internet speed and data transport for telecommunication and yet can be manufactured at a relatively lower cost due to the use of abundant starting materials such as silica. Hence, even though photonic materials are usually within electronic devices and pieces of equipment, that can make them go unnoticed, they play an important role to improve overall performance while functioning as a feasible and ever-growing market. Such materials that can be used for photonic-related applications are usually layered structured.

Two-dimensional (2D) materials used for that purpose are likely to present van der Waals interactions in between its layers, whereas the layers themselves are covalently bonded. This type of structure leads to anisotropic properties which can be a valuable feature for electronic as well as optic applications. Metal monodichalcogens with the general formula of MX such as SnS, GeS, SnSe, and GeSe are some of the few examples. Also, metal dichalcogenides with the general formula of MX_2, which include MoS_2, WSe_2, WS_2, $MoSe_2$, $TiSe_2$, $ZrSe_2$, TiS_2, ZrS_2 along with several other layered materials can be employed as photonic materials. In addition, these materials are widely researched in other areas such as energy storage and fuel cells, for example. Yet, their employment for photonic-related applications is due to their satisfactory properties in terms of light absorption, which occurs mostly by its interaction with photons in an energy level below its bandgap. Through that, materials with a bandgap above 1.5 eV and optical properties in the near infra-red (NIR) such as MX and MX_2, can be used for light interaction-related properties.

An example for this case is the bulky 2H-phase MoS_2 which has an energy bandgap (E_g) of 1.3 eV, whereas when obtained as a 2D 2H-MoS_2 monolayer it can reach an $E_g = 1.8$ eV. The layered structure leads to a decrease in the dielectric screening effect for the electric fields, which is one of the factors that cause a change in the E_g. In addition, there are other ways to vary 2D layered materials through chemical doping, use of different substrates, surface modification, and external stimuli such as mechanical strain or an applied electric field. Through that, 2D materials can reach an E_g that roams around 1–2 eV, which enables their use for applications related to NIR refraction [1]. In that sense, the experimental band gaps for materials such as metals, semimetals, TiS_2, or graphene can be around 0 eV, whereas for hexagonal boron nitride (hBN) it can be as high as 6 eV. Within that line, some factors have been observed such as the increase of E_g when a transition metal is bonded with Te, Se, or S, at which there is a larger increase of E_g by going from Te, Se to S. Also, there is often a change in E_g from indirect to direct concerning a decrease in the number of layers. Another possible tuning of E_g can be performed by chemical alloying of the metal dichalcogens with a different chalcogen such as for the cases of ZrS_2–$ZrSe_2$, $HfSe_2$–HfS_2, or SnS_2–$SnSe_2$ [2–4]. Figure 15.1 shows the schematics for the E_g measured experimentally for widely studied 2D materials.

Another material that receives considerable attention within the photonics field is Ge–Sb–Te due to its drastic change in optical properties during phase transition, which in the case of this material can be induced by optical heating. Its optical applications could be attributed to four factors. The first is related to its capability to drastically change its refraction index (n) as well as extinction coefficient. Second, its non-volatile phase change allows low-energy processes to maintain the switched phase. Third, ultrafast switching is possible due to the fast phase change process, which can occur in the order of nanoseconds. Fourth, both electrical and optical pathways are available to perform the phase change at the interface between the photonic and electric areas, as well as within the optical components.

An example of Ge–Sb–Te use has been performed by Wu et al. [5] developed an efficient optical switch that presented less than 1 dB of insertion loss for the output ports along with up to 20 dB of switching extinction ratio. The satisfactory performance was attributed to the fast n and extinction coefficient among the two Ge–Sb–Te phases. Another factor

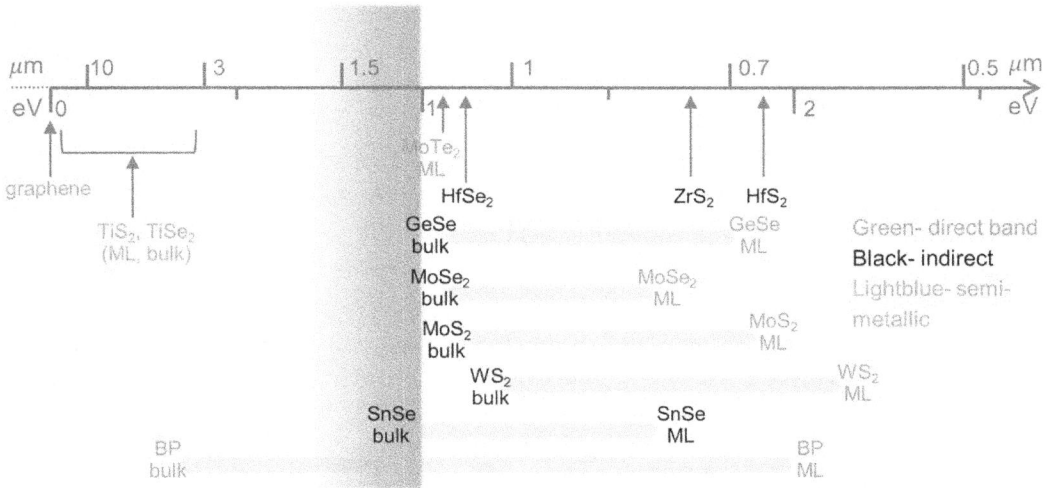

FIGURE 15.1 The bandgap of several 2D layered materials suitable for photonics due to their tunable properties. (Adapted with permission from reference [1]. Copyright (2020), American Chemical Society.)

for that was due to the patterning of Ge–Sb–Te in a nanoscale below wavelengths, likely improving optical confinement and reduction of mode leakage. On top of that, Ge–Sb–Te was placed at the field maximum of the optical mode. Through that, phase transition was improved, which led to a better switching process and avoidance of optical loss. Within those lines, a 1×2 optical switch was assembled based on the phase transition of a Ge–Sb–Te film that was incorporated into a microring resonator. In that sense, when Ge–Sb–Te was in its amorphous phase, there was a resonant signal that coupled with the microring leading to outputs from the drop port. On the other hand, when Ge–Sb–Te was in its crystalline phase, there was a decoupling with the microring leading to outputs from the through port. The scheme for that system is provided in Figure 15.2.

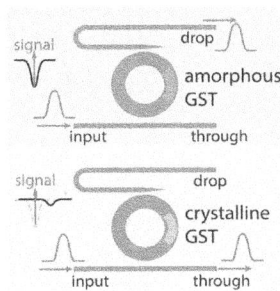

FIGURE 15.2 Scheme for a 1×2 optical switch operated based on the fast amorphous and crystalline phase transitions of a Ge–Se–Te film incorporated into a microring resonator. (Adapted with permission from reference [5]. Copyright (2019), American Chemical Society.)

One of the reasons for the use of such materials is based on its large n, which improves the optical confinement as well as diminishes mode leakage. Another important phenomenon related to photonic materials is the electron-hole (e^-–h^+) and exciton screening. In that sense, excitons are observed through the Coulombic interaction between e^-–h^+ which are screened by the material's dielectric constant. In that sense, layered materials with van der Waals interaction in between its layers leads to confinement of e^-–h^+ wave functions, which forms excitons in the intralayer that present binding energy that can reach up to 1 eV. Larger binding energies related to more stability for the excitons close to room temperature while usually being predominant on the optical spectra. In that sense, excitons present optical absorption resonances at an energy level that is below the E_g. This effect leads to an increase in the interaction between light and matter enhancing the photonic material's optical and electronic properties.

15.2 SYNTHESIS OF LAYERED MATERIALS

One of the main aspects regarding the synthesis of photonic materials is the processing temperature. In the case of SiO_2 or Si_3N_4 temperatures up to 800°C can be employed for the synthesis of layered materials, which is sufficient to perform their synthesis. However, when complementary metal-oxide semiconductors are present in the circuit, then the processing temperature should be below 600°C. In that sense, there are many approaches available for the synthesis of layered materials for photonics. Chemical vapor deposition (CVD) is a widely used one as it consists of the vaporization of chemical precursors that react over a substrate leading to the formation of a film. One common example is the growth of MoS_2 through MoO_3 and S powders. Yet, from the scalability standpoint, it is more desirable to use starting materials in the gas phase as it usually implies better deposition over the substrate. In that sense, solid-source CVD is a widely available method as it can be used to synthesize a plethora of MX_2 layered structures based on several transition metals such as Mo, Fe, Ti, V, Ta, Nb, Hf, Zr, W, Pd, Re, and Pt along with chalcogens such as Te, Se, and S. However, it imposes a challenge in terms of scalability.

Sutter et al. [6] obtained a single-layered crystalline structure of GeS using the vapor transport method. It consisted of the vaporization of bulky GeS at 430–450°C which could be grown over mica or highly oriented pyrolytic graphite as both function as van der Waals-based substrates. The convenience of this approach was based on the facile deposition of the powdered material over the substrate which took around 10 min to take place after the temperature was reached, leading to a relatively robust layered structure suitable for photonic applications. Other materials such as GeSe or black phosphorus have also been synthesized in a similar manner [7,8].

The gas source CVD and metal-organic CVD (MOCVD) are also valuable approaches for the synthesis of high-performance 2D materials for photonics as it consists of a gaseous precursor that grows over a substrate. This method allows some degree of control for the reaction system such as a number of layers, uniform film formation, and scalability. However, the necessity of a gas-phase precursor diminishes the number of starting materials available for this process. In that sense, MoS_2, SnS, WS_2, TiS_2, and WSe_2 have been obtained by these methods. Based on the MOCVD approach Eichfeld et al. [9] synthesized

WSe₂, which is a promising material for optoelectronic applications due to its direct $E_g = 1.65$ eV when obtained as a monolayer. In that line, WSe₂ was synthesized by using $(CH_3)_2Se$ and $W(CO)_6$. This approach is not only scalable but also provides proper control of the reaction system, facilitating phase control. In addition, the authors observed that several parameters such as Se:W ratio, substrate, pressure, and temperature, also influence the 2D film's properties, as under an optimized scenario domains up to 8 μm were obtained, as demonstrated in Figure 15.3.

Another variation from these methods is the atomic layer deposition (ALD), which is usually commanded by the reaction kinetics of film formation which also enables control over the number of layers and morphology. Usually, the annealing process is required to improve the film's quality. This approach was employed by Basuvalingam et al. [10] to obtain a controlled phase of metallic TiS₂ and semiconducting TiS₃ nanolayers. The phase was controlled based on the optimization of temperature and reactants ratio. The range of temperatures went from 100 to 200°C which is considerably lower compared to other methods. Yet, a post-deposition annealing process was performed at 400°C under an atmosphere containing a high concentration of S. This step was required as it improved the crystallinity of the film as well as photoluminescence at 0.9 eV, which is related to the semiconducting behavior of TiS₃. Throughout the study, it was concluded that the utilization of H₂S plasma to generate S₂ led to the formation of amorphous TiS₃ which was obtained at 100°C. Above that temperature, there was the formation of TiS₂.

Physical vapor deposition (PVD) is another technique available for the synthesis of 2D materials, even though it is a lesser-used method when compared to the ones previously mentioned. Within this technique, there are some approaches such as sputtering, electron-beam evaporation, pulsed laser deposition, and thermal evaporation. One aspect of this technique is that during the formation of transition metal chalcogenides there is often a lower concentration of chalcogens, which leads to a deficit in anions due to the different evaporation

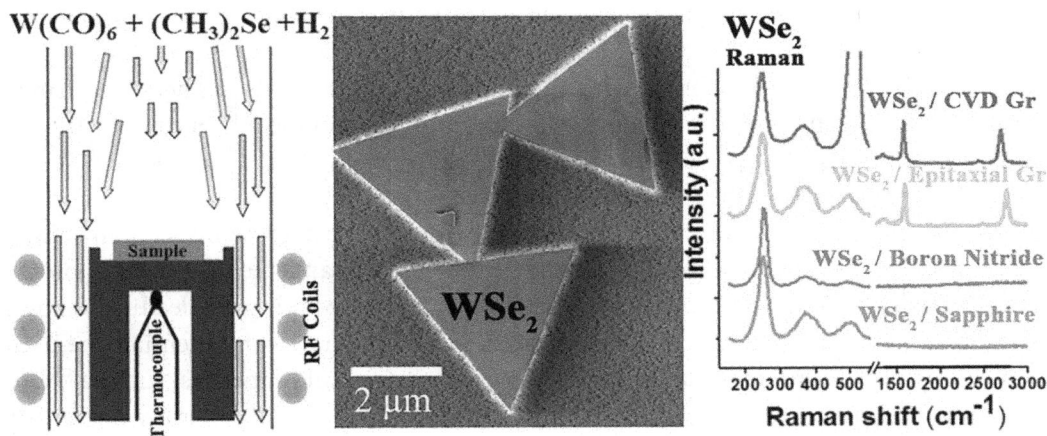

FIGURE 15.3   (Left) Schematics for the synthesis of WSe₂ through MOCVD. (Right) Transmission electron microscopy (TEM) image of WSe₂ domains. (Adapted with permission from reference [9]. Copyright (2015), American Chemical Society.)

rates between metals and chalcogenides. For that, an extra source of chalcogenides can be added to the reaction system. Yet, there are quite a few metal chalcogenides that have been synthesized by this method such as GeSe, SnSe, GeS, and SnS. Another variation from PVD is molecular beam epitaxy which allows precise control over morphology and thickness of the grown films. However, it is applicable to limit the number of transition metal dichalcogens or chalcogens-based materials. One example of the limitation of this method can be seen for the synthesis of MoS_x-based materials since the melting point of Mo is 2623°C. In the case of S, even at around 800°C, its vapor pressures are higher than 1 atm, which leads to a shorter reaction time between S and the metallic substrate. Despite that, other transition metal dichalcogens such as $MoSe_2$ have been effectively grown over hBN substrate [11].

Another method for the synthesis of layered materials is the layer transfer method, which consists of the mechanical exfoliation of a thicker layered material that is later transferred into a different substrate. It is an option to obtain heterostructures whereas enabling coupling and decoupling with an electronic device, for instance. In that sense, a layered material can be first synthesized under harsh conditions to be later transferred to a substrate that functions as proper support. However, the current challenges for this reside in properly transferring the layered material into a substrate that can adhere properly along with avoiding contamination with precursors or solvents [12].

The possibility to synthesize the same photonic layered material through different methods allows a better understanding of the method's influence on the properties of the material. Dang et al. [13] proposed the synthesis of WO_3 through the electrodeposition method. The electrodeposition approach is usually performed by dissolving the precursor in an electrolyte and applying an electrical potential to create a driving force for its deposition over a substrate. It can promote the formation of a relatively thin and evenly distributed film which can be formed over as well as within the substrate's gaps. In that sense, polystyrene spheres with a diameter of 267 nm were deposited over a SnO_2 coated glass doped with F. Then the substrate was immersed in a solution containing dissolved W powder. Through the same method, sodium tungstate was used as a precursor to being electrodeposited over a titanium substrate.

A relatively novel method for the growth of opal photonic crystals has been performed by using the sol-gel method over natural templates which could include butterfly wings or beetle scales, for instance [14,15]. The reason for that is related to the presence of specific nanostructures present in the exoskeleton of these insects that possess a layered structure containing relatively different refractive indices. Because of that, these insects can show a variety of colors as their exoskeleton can form unique 3D arrangements named photonic crystals, at which when light hits their structure it scatters in specific ways, leading to novel optical behaviors. This is a valuable tool since these bio templates can potentially be used for the growth of photonic crystals that could possess its optical properties, such as bandgap, completely within the visible wavelength. Yet, such types of properties are not widely available since the current technology for processing these photonics enables the optical properties on the microwave, near-infrared and infrared regions [15].

However, the biomaterials cannot withstand high temperatures, radiation, or harsh chemical environments, often required for the synthesis of inorganic photonics, which imposes a hindrance to their straight use. For that, some methods have been performed to replicate the structure of butterfly wings using inorganic compounds as was the case for Huang's study [16]. In their study, a butterfly wing scale was used as a bio-template to grow Al_2O_3 coating, performed through ALD. An inverted polycrystalline shell structure of Al_2O_3 with controlled morphology was obtained after the bio-template removal through thermal treatment. Due to that, the Al_2O_3 presented similar optical properties to those of the butterfly wing such as colors in between the range of blue and violet. Yet, the Al_2O_3 promoted a shift in the reflection peak toward longer wavelengths which was related to the variation of the structure's periodicity and refraction index. On top of that, the Al_2O_3 butterfly wing replica presented similar beam splitter and waveguide properties as the original wing. In that sense, this approach provides a relatively cheaper and simpler option to obtain complex structures that can function as photonic materials. The morphology and nanostructure of the Al_2O_3 coated butterfly wing can be seen in Figure 15.4.

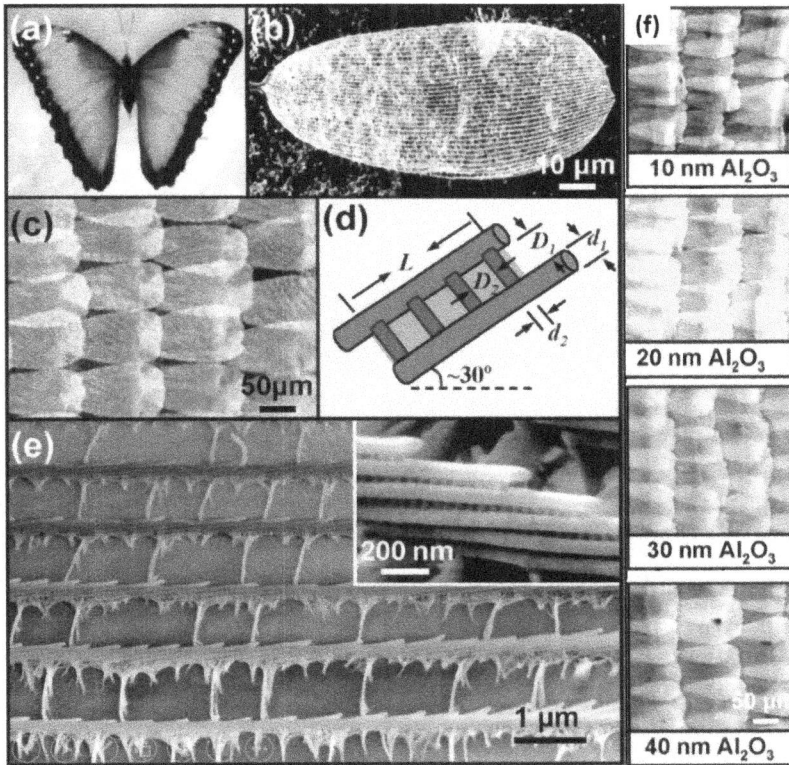

FIGURE 15.4　(a) Butterfly photocopy. (b) Scanning electron microscopy (SEM) for a butterfly wing scale over a silicon substrate. (c) Blue area of a butterfly wing seen from optical microscopy. (d) Scheme for a lamellae unit cell that can promote light scattering. (e) SEM of the scale's surface, displaying the lamellar structure (inset with higher resolution). (f) Al_2O_3 replica of the butterfly scale shows a color change from blue to pink as the film's thickness increases. (Adapted with permission from reference [16]. Copyright (2006), American Chemical Society.)

In another study, Galusha et al. [15] proposed the fabrication of a photonic inorganic framework by using a sol-gel bio-templating method. In this case, the exoskeleton scales of different beetles were used as bio templates. The process was performed under low temperature while using a hybrid silica sol-gel containing inorganic-organic structure along with acid or thermal treatment to remove the bio-template leading to an inorganic replica of the colored beetle scales used in the procedure.

15.3 OPTICAL PROPERTIES OF LAYERED MATERIALS

A consistent way that scientists can measure the optical properties of certain materials is through the refractive index, which is a unitless number that defines the velocity that light travels through a material. Tuning this property allows the photonic materials to refract or disperse light in different angles, which makes them applicable in prisms, lenses, displays, and so on. In that sense, there is also the complex refractive index ($\underline{n}$) defined in Equation 15.1 which is related to the dielectric permittivity ($\mathcal{E}$) through Equation 15.2. Also, the imaginary part of the refractive index (k) can be related to the absorption coefficient (α) through Equation 15.3.

$$N = n - ik \tag{15.1}$$

$$n = \mathrm{Re}\left(\sqrt{\varepsilon}\right) \tag{15.2}$$

$$\alpha = \frac{4\pi k}{\lambda} \tag{15.3}$$

Based on these concepts it has been noted that layered chalcogenides, for instance, usually present a large refractive index which is required for proper interaction in a confined geometry for integrated devices and photonic waveguides. Such properties can be further enhanced by an external stimulus which can lead to an increase in refractive index along with a shift in the optical phase without leading to large optical loss which can be defined as large Δn with a relatively smaller k. Within the idea of varying the refractive index of monolayer dichalcogenides, Yu et al. [17] proposed the use of a metal-oxide semiconductor as a complementary component that was incorporated into monolayered MoS_2, WSe_2, and WS_2 which led to an increment in around 20% of real refractive index along with around 60% for the imaginary. Such an increase in property was likely due to inherently strong excitonic effects present in monolayered materials along with the influence of free charge carriers as they were able to broaden the spectral width to promote the exciting transitions. It is worth noting that, injected charge carrier can also promote renormalization of E_g and change the binding energy of excitons, which for this case was negligible. One of the analyses of this study was performed through the spectral reflectance of the transition metal dichalcogens monolayered materials under electrical gating. It consists of measuring the intensity of the reflected light ratio from the layered material and the reflected light from a mirror. Figure 15.5 shows the variation of the refractive index in function of carrier density (1/cm²) for the WS_2. It was observed that both imaginary and real refraction index

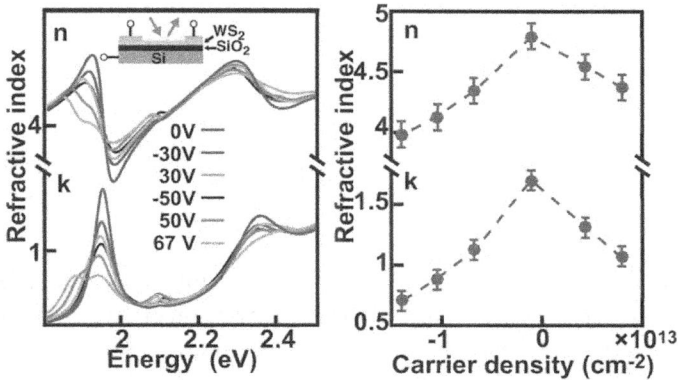

FIGURE 15.5   Real (n) and imaginary (k) refractive indexes in the function of the carrier density. (Adapted with permission from reference [17]. Copyright (2017), American Chemical Society.)

reached their maximum at the point with neutral charge carriers and decreased with a higher density of either e^- or h^+, which allowed a considerable tunability of the refraction index of WS_2. This variation in property is mostly notable for monolayered materials, as this case was tested also for MoS_2 and WSe_2, which displayed similar behavior.

Another relevant property that has been explored in monolayered structures is photo-elasticity, which consists of a variation in optical properties when a material is exposed to a mechanical strain. This effect has been explored on WS_2, WSe_2, MoS_2, and $MoSe_2$ by Mennel et al. [18] which determined their respective second-order nonlinear photoelastic tensors. It was noted that even when small strains are applied there was a considerable change in properties for each material, hence enabling a wider range of optical properties that enable different sectors of use according to the necessity. The optical and electronic properties of layered materials arise from their nearly atomic thick structure, as it facilitates the formation of e^-–h^+ pairs (excitons), which leads to the quantum confinement effect. Hence, reducing the number of layers from bulky materials to a 2D structure can grant them novel optic-electronic properties. The exploration of this phenomenon is relatively new for the scientific community which can lead to exciting discoveries. Within that line, thin-layered transition metal dichalcogenides have demonstrated interesting properties that are yet not fully unraveled.

A recent example of this came from the study of a few atomic layered WSe_2 that presented some defects in its structure that enabled it to move to a photo-excited state in the band around 1.63 and 1.72 eV, 760 and 721 nm, respectively. Such values of energy facilitated the formation of e^-–h^+ which led to an enhancement in its photocatalysis properties, conductivity and eased the formation of excitons within the visible range of wavelength. Aside from reaching mono or few-layered structures, there is also the possibility to create nanopatterns that can further enhance the optoelectronic properties. This effect was explored by Wei et al. [19] that synthesized MoS_2 patterned nanodots that contained excitons that were laterally confined. It was observed that the valley polarizations were the same for either MoS_2 monolayer or nanodots. However, the exciton energies were dependent on the MoS_2 nanodot's size. This study demonstrated a way to predict the optical

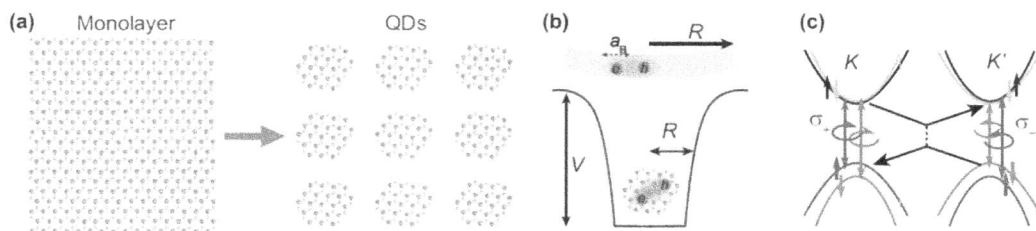

FIGURE 15.6   (a) MoS$_2$ monolayer structure converted into MoS$_2$ nanodots. (b) Scheme for quantum confinement. (c) Coupling of spin-orbit along with inversion of asymmetry for the monolayer MoS$_2$. (Adapted from reference [19]. Copyright (2017) The Authors, Some right reserved. Springer Nature. This article is licensed under a Creative Commons Attribution 4.0 International License.)

and electronic properties of semiconductors which can allow them to be integrated into optoelectronic systems as the nanodots of MoS$_2$ acquired appreciable photoluminescence properties when compared to the pristine MoS$_2$. The schematics for the variation in optical properties for the MoS$_2$ monolayer to the nanodots are presented in Figure 15.6.

Another optical property that is often desired in photonics is photochromism is the phenomenon of color change when a material absorbs a photon. The rate at which this phenomenon occurs in a material is a factor that can determine its applications. In that sense, fast photochromism can be employed for optical switches whereas slow photochromism can be employed for optical storage of data. Other applications of this phenomenon include the manufacture of privacy shields, coatings able to conserve energy, eye protection, among others. Some organic and organometallic materials possess these properties which are likely to contain spiro-pyrans and/or spiro-oxazines which can go through a change in their chemical structure after the absorption of a photon. One of the approaches to increase the concentration of these photoactive materials consists of chemically crafting them into a matrix structure, which allows higher loading without causing agglomeration while also decreasing photo fatigue [20]. In that sense, there are a considerable number of matrices based on silica such as SiO$_2$, Me$_2$SiO, Me$_2$SiO/ZrO$_2$ or organic materials such as poly(methyl methacrylate), poly(phenyl glycidyl ether)-co-formaldehyde that can be covalently bonded with photoactive functional groups or compounds. Another class of photoactive materials is rare metals such as Eu^{2+}, Eu^{3+}, Er^{3+}, Nd^{3+}, Ce^{3+}, and Tr^{3+} which can be doped on glassy substrates enabling their use in tunable lasers in both visible and UV range, and scintillator fields.

Photonic material may also display electroluminescence properties which is the emission of light during the passage of an electrical current in their structures. This property is the core aspect for the functioning of LEDs and organic LEDs which already compose a strong market of displays worldwide. However, there are several requirements that materials should meet to be suitable for practical applications in this regard such as reasonable cost of starting materials and manufacture, high quantum efficiency, stability through long-term, and robustness. Within that line to introduce competitive materials Gutiérrez et al. [21] proposed the fabrication of a metal-organic framework (MOF) based on silver with electroluminescence properties (Ag-MOF). For that, the synthesized material was

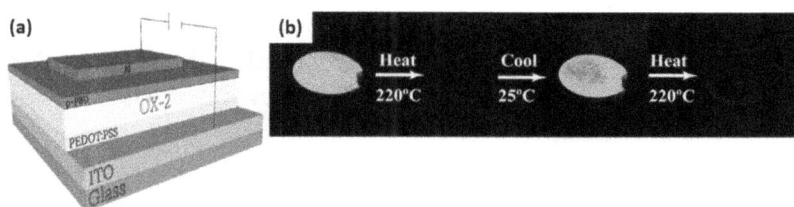

FIGURE 15.7 (a) Schematic of the LED device. (b) The thermochromic effect on the Ag-MOF after heating and cooling cycles. (Adapted with permission from reference [21]. Copyright (2020), Elsevier.)

built into a LED device for testing. That was performed by assembling an indium thin oxide as an anode, along with an aluminum cathode. Yet, the further incorporation of PEDOT:PSS for conductivity and 2-(4-tert-butyl phenyl)-5-(4-biphenylyl)-1,3,4-oxadiazole (p-PBD) as a fluorescent organic agent. By adding these two components there was a proper improvement in electroluminescence stability there were likely due to a balance in the mobility of the carriers and proper matching of the electrode's energy levels and electroluminescence properties. The scheme for the device is presented in Figure 15.7a.

To improve the emission resolution is important that the device can emit in a short range of wavelength. In this case, two bands were observed at 520 and 620 nm, whereas the former corresponded to the maximum of the photoluminescence and the latter to the e^-–h^+ pair recombination. In that sense, it is known that this red emission process (~620 nm) is mostly related to the e^-–h^+ recombination. This phenomenon occurs in an electronic trap state, this factor is also attributed to grain boundaries between the nanocrystals or charge carriers with unbalanced mobility. Hence, to diminish the grain boundaries as well as improve charge carrier mobility the Ag-MOF was incorporated into a semiconductor substrate to form an inorganic-organic LED. For this case, polyvinyl carbazole (PVK) was employed as its function as an h^+ transport polymer. Through that, there was a considerable decrease in the turn-on voltage from around 4.5 to 1.5 V along with the emission of a more intense and homogeneous white light. This suggested that incorporating the Ag-MOF into an h^+ conducting polymeric matrix facilitated the carrier mobility. Based on all these factors, it was notable that the electroluminescent phenomenon can be a relatively complex process.

For that, the authors proposed a mechanism at which the e^- were introduced into the LUMO state of either the Ag cluster and/or the PVK. At the same time, the h^+ was more likely to be formed at the HOMO state of PVK due to its more accessible energy levels when compared to the Ag cluster. Also, the presence of the Ag cluster provides a conducting pathway that may allow a longer lifetime for an e^-–h^+. Yet, the recombination of this e^-–h^+ can occur from the conducting and valence bands (VBs) of either PVK or the Ag-MOF emitting at 420 and 520 nm, respectively. Aside from this phenomenon the Ag-MOF also presented thermochromic properties which presented a consistently linear behavior as the increase in temperature led to a decrease in emission intensity with a coefficient of determination (R^2) of around 0.993. Such precise response against temperature allows this material to be used in luminescent thermometers as it also displayed good cycle reversibility and stability as the Ag-MOF was submitted to 10 cycles of heating and cooling. Through that, it was notable that its emission diminished at 100°C and recovered when it was cooled back to room temperature. Further analysis to explain the reason for the occurrence of the thermochromism in luminescent MOFs is still necessary. Yet, this phenomenon is likely attributed to the enlargement of vibrational modes in the structure as temperature increases, which culminates in the emission of non-radiative recombination. Another factor might be related to the distance of the ligands from the metal cluster as it can be also altered by temperature. The visual aspect of the thermochromic effect is presented in Figure 15.7b. Finally, the authors also observed the influence of mechanical stress over the Ag-MOF as they observe a mechanic chromic response. Likewise, the thermochromic behavior, the Ag-MOF also presented considerable reversibility for this process as it was compressed into a pellet and ground into a powder in cycling processes. In that sense, the phenomenon of changing the radiation intensity that is emitted according to the mechanical stress applied upon the Ag-MOF was attributed to the distortion of its 2D wave-like layered structure. That process altered the distance between the Ag cluster and organic ligands which were likely to decrease the charge transfer process.

15.4 TECHNIQUES TO STUDY OPTICAL PROPERTIES

There are a considerable number of analyses that can be employed to determine the optical properties of layered materials, yet the challenges, in this case, are related to the difficulties based on the birefringence outside the plane and the preparation of crystals with smooth surfaces along with a plane orientation different than basal. The summary of these techniques is presented in Table 15.1.

TABLE 15.1   Summary of the Measurement of a Material's Optical Properties Based on Different Techniques. Adapted with Permission from Reference [1]. Copyright (2020), American Chemical Society

Method	Expreimental Schematic	Pros	Cons
Ellisometry		Independent measurement of n & k	Highly – sensitive to surface conditions
Normal Reflection		Simple geomenty, high spatial resolution	n & k not independently determined
Transmission		Good for measuring k below band gap	Requires flat, parallel surfaces, sample thickness must be known; sample thinning may be required
Polarized Light Transmission		Measure birefringence of LM	Same as transmission
Waveguide/Device Integration		prediction of performance in indended devices, including processing – dependent effect	results may not reflect intrinsinc material properties
Photo-thermal Defelction Spectroscopy		Good for measuring very small k	Knowledge of thermal properties needed for data analysis

Ellipsometry is a technique used to determine the dielectric properties of a photonic material as the ratio of reflection for the Fresnel coefficients is measured through Equation 15.4.

$$\rho = \frac{r_p}{r_s}$$

(15.4)

where r_p and r_s are the parallel and perpendicular field components, respectively, for the incident light beam over the plane of the sample. In the case of an interface between a thin film with air both refringence index and imaginary part of refringence index can be obtained. However, it is worth noting that ellipsometry is sensitive to surface layers which can lead to inaccurate information if an inadequate model is utilized to harvest the data. An example for the use of spectroscopic ellipsometry has been demonstrated by Kim et al. [22] which studied the optical properties of $Zn_{(1-x)}Co_xO$ alloys. The measurements were performed by using a rotating-analyzer ellipsometer at room temperature within the photon energy region of 1.5–5 eV. Through that, the pseudo electric constant can be calculated through Equation 15.5, where φ is the incident angle. For that, a few considerations had to be made such as assuming a flat surface in between sample and air and a φ of 70° along with a 30° for a fixed polarizer from the incident plane.

$$= \sin^2\varphi + \sin^2\varphi \tan^2\varphi \frac{(1-\rho)^2}{(1+\rho)^2}$$

(15.5)

Based on that, a comparison between $Zn_{(1-x)}Co_xO$ with increasing concentration of Co^{3+} along with ZnO was made with its real and imaginary parts of pseudo electric constants presented in Figure 15.8a and b, respectively. It was notable that as x increased there was

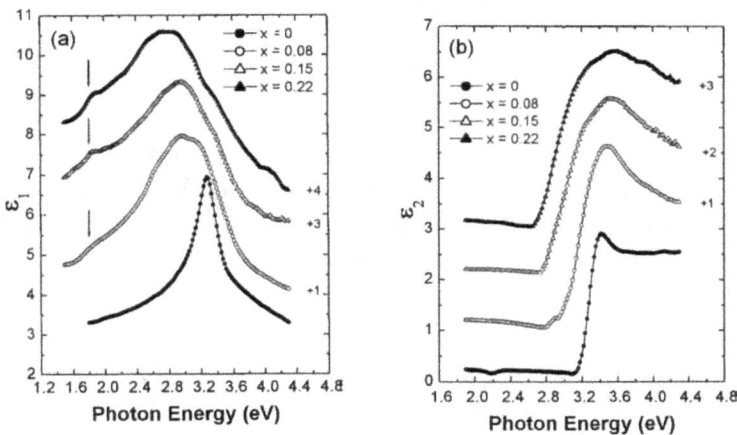

FIGURE 15.8 (a) Real and (b) Imaginary pseudo electric parts for the $Zn_{(1-x)}Co_xO$ films with increasing concentrations of x where ZnO is $x=0$. (Reproduced with permission from reference [22]. Copyright (2002), AIP Publishing.)

a decrease in the fundamental bandgap (E_0) transition from the highest VB to the lowest conducting band. This phenomenon was attributed to the increase in carrier scattering effect as the crystal was more disorganized due to the insertion of Co^{3+} into the crystal lattices.

Normal reflectivity is a method that can be employed for several geometries. In the case of normal incidence, the Fresnel coefficient can be written as in Equation 15.6. This measurement yields one value of R for each wavelength (λ). Hence, to obtain the values for both n and k, it is required to use the transformations of Kramers-Kronig (KK).

$$R = \frac{(n-1)^2 + k^2}{(n+1)^2 + k^2} \tag{15.6}$$

Both ellipsometry, as well as normal reflectivity, can be adjusted to measure the aniso-tropic optical properties. For the case of reflectivity from normal incidence, the optical anisotropy in-plane can be obtained by plotting polar reflectivity along with using linearly polarized light. However, for the case of layered materials, there is usually strong refrac-tion of incident light that angles toward the normal surface, which is due to their high n of layered materials. Because of that, measuring out-of-plane optical properties for this case is often a challenging process. For another case, the transmission-geometry optical techniques are suitable to measure k values that are below-band gap whereas also enabling to quantify adsorption that is defect-assisted. Also, birefringence in layered materials can be quantified by measuring the interference fringes during the transmission of polarized and highly convergent light. Through this methodology both ordinary and extraordinary refraction indexes, n_o and n_e, respectively of MoS_2 have been determined [23]. Yet, one of the current hindrances for the analysis of materials is the requirement of polishing and thinning of the specimen, which diminishes the broad use of transmission-geometry opti-cal methods. In addition to that, accurately knowing the material's thickness is a require-ment for the proper measurement of k [23].

Within that line, these methods such as ellipsometry, normal reflection, and transmis-sion are mostly employed for the optical properties' measurement for relatively bulkier materials, which leads to analyses that are often based on averaged thickness values. Now, for the characterization of photonic integrated devices that are usually a nonuniformity in the structure which makes the optical properties, particularly the optical loss, of a mate-rial depend on the device's geometry, interfacial regions, and material's microstructure. This case was explored by Wei et al. [24] that studied the optical loss of MoS_2 by utilizing a microring resonator. Through that, the authors obtained a transmission spectrum col-lected before and after the MoS_2 to analyze its optical loss. The measured value of opti-cal loss for the simulated process was 1390 dB/cm whereas through the resonator it was 850 dB/cm at 633 nm. This analysis is schematized in Figure 15.9.

Another parameter for the characterization of photonics is the measurement of electro-optic coefficients. For that, a crystal should be processed in a cell of Pockels geometry. Then, the analysis is conducted based on the variation of the crystal's position concerning the optical field as well as the electric field applied in the Pockels cell. Through that, the

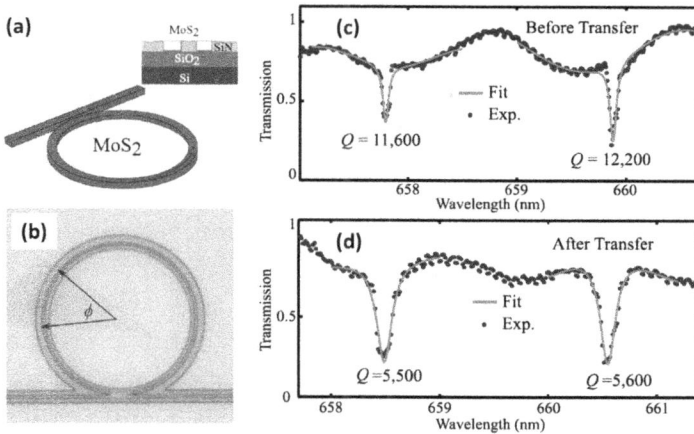

FIGURE 15.9 A (a) schematic of the experimental set-up. (b) Optical image of a device. Transmission spectrum of a resonator before (c) and after (d) Transferring MoS$_2$. (Adapted with permission from reference [24]. Copyright (2015), AIP Publishing.)

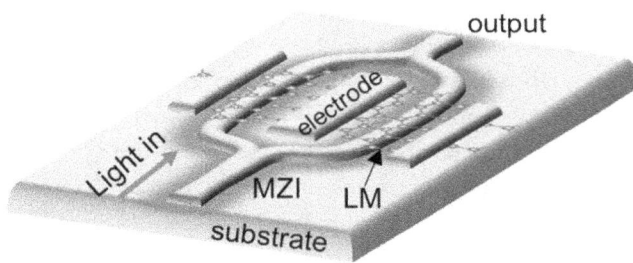

FIGURE 15.10 Scheme for an MZI device used to measure the electric optic coefficients. (Adapted with permission from reference [1]. Copyright (2020), American Chemical Society.)

electro-optic tensor can be obtained. An interferometer device can be used to measure the electro-optic coefficient in the case of a given layered material integrated photonic device placed on a chip. The device consists of a Mach-Zehnder interferometer (MZI), which has a geometry that allows the electric modulation field to propagate in both lines as its schematics are presented in Figure 15.10. It is worth noting that the Pockels effect must be measured over either bulky crystals or transparent substrates for thin films. Also, MZI requires the fabrication of a device. In that sense, there is a necessity to measure electro-optic coefficients without the need for a device framework which would make the analysis more feasible.

15.5 CONCLUSION

The optimization of optic properties for integrated photonics is an important piece of technological advancement for electronic devices. In that sense, it can provide a faster flux of information, data transmission, emission of wavelengths toward the infra-red instead of ultra-violet, which tend to be less harmful to eyes, among many other features. The layered materials have demonstrated appreciable optic properties that can be tuned which allow their use as optical switches and modulators. For that, such properties of these 2D materials should be controlled most likely by focusing on the low-loss and below-band gap spectral regions, which is an inherently challenging process in terms of synthesizing materials with appreciable optical properties along with properly characterizing them. In that regard, understanding the relationship between the processing and refractive properties of layered materials still requires further studies. An example of that is the optical loss that is observed due to the grain boundaries effect, which is not yet clear. Also, the synthetical approaches for the development of layered semiconductor materials such as MOCVD can be a potential way to scale up their fabrication. Another method is layer transfer which allows the synthesis of many-layered materials that can be incorporated into several substrates, facilitating the fabrication of integrated photonics. Within that line, introducing novel layered materials may allow them to achieve higher performances when compared to established materials that can perform phase change. In that sense, these novel layered materials may be capable of performing order-order transformations which can be performed through non-thermal processes. Yet, there are several requirements to enable novel materials to be incorporated into mass production such as low and fast power switching, strong modulation for the optical phase, long usage for the device, and low insertion loss as well as material fatigue. Based on these observations, the use of layered materials for the development of integrated photonics is a considerably challenging area as it has room for growing based on the development of properly controlling the optical properties. On top of that, there is the need to establish consistent and efficient synthetical approaches along with more feasible characterization techniques to facilitate the study of these materials' properties. Despite these challenges, the use of semiconductor layered materials for photonic applications has demonstrated an interesting set of applications for high technological ends.

REFERENCES

[1]   A. Singh, S.S. Jo, Y. Li, C. Wu, M. Li, R. Jaramillo, Refractive uses of layered and two-dimensional materials for integrated photonics, *ACS Photonics.* 7 (2020) 3270–3285.

[2]   Y. Wang, L. Huang, B. Li, J. Shang, C. Xia, C. Fan, H.-X. Deng, Z. Wei, J. Li, Composition-tunable 2D SnSe2(1−x)S2x alloys towards efficient bandgap engineering and high performance (opto)electronics, *J. Mater. Chem. C.* 5 (2017) 84–90.

[3]   D.L. Greenaway, R. Nitsche, Preparation and optical properties of group IV–VI2 chalcogenides having the CdI2 structure, *J. Phys. Chem. Solids.* 26 (1965) 1445–1458.

[4]   S.M. Oliver, J.J. Fox, A. Hashemi, A. Singh, R.L. Cavalero, S. Yee, D.W. Snyder, R. Jaramillo, H.-P. Komsa, P.M. Vora, Phonons and excitons in ZrSe2–ZrS2 alloys, *J. Mater. Chem. C.* 8 (2020) 5732–5743.

[5]   C. Wu, H. Yu, H. Li, X. Zhang, I. Takeuchi, M. Li, Low-loss integrated photonic switch using subwavelength patterned phase change material, *ACS Photonics.* 6 (2019) 87–92.

[6]   E. Sutter, B. Zhang, M. Sun, P. Sutter, Few-layer to multilayer germanium(II) sulfide: synthesis, structure, stability, and optoelectronics, *ACS Nano.* 13 (2019) 9352–9362.

[7]   Y. Xu, X. Shi, Y. Zhang, H. Zhang, Q. Zhang, Z. Huang, X. Xu, J. Guo, H. Zhang, L. Sun, Z. Zeng, A. Pan, K. Zhang, Epitaxial nucleation and lateral growth of high-crystalline black phosphorus films on silicon, *Nat. Commun.* 11 (2020) 1330.

[8] D.-J. Xue, S.-C. Liu, C.-M. Dai, S. Chen, C. He, L. Zhao, J.-S. Hu, L.-J. Wan, GeSe thin-film solar cells fabricated by self-regulated rapid thermal sublimation, *J. Am. Chem. Soc.* 139 (2017) 958–965.

[9] S.M. Eichfeld, L. Hossain, Y.-C. Lin, A.F. Piasecki, B. Kupp, A.G. Birdwell, R.A. Burke, N. Lu, X. Peng, J. Li, A. Azcatl, S. McDonnell, R.M. Wallace, M.J. Kim, T.S. Mayer, J.M. Redwing, J.A. Robinson, Highly scalable, atomically thin WSe2 grown via metal–organic chemical vapor deposition, *ACS Nano.* 9 (2015) 2080–2087.

[10] S.B. Basuvalingam, Y. Zhang, M.A. Bloodgood, R.H. Godiksen, A.G. Curto, J.P. Hofmann, M.A. Verheijen, W.M.M. Kessels, A.A. Bol, Low-temperature phase-controlled synthesis of Titanium Di- and Tri-sulfide by atomic layer deposition, *Chem. Mater.* 31 (2019) 9354–9362.

[11] R. Yue, Y. Nie, L.A. Walsh, R. Addou, C. Liang, N. Lu, A.T. Barton, H. Zhu, Z. Che, D. Barrera, L. Cheng, P.-R. Cha, Y.J. Chabal, J.W.P. Hsu, J. Kim, M.J. Kim, L. Colombo, R.M. Wallace, K. Cho, C.L. Hinkle, Nucleation and growth of WSe 2 : enabling large grain transition metal dichalcogenides, *2D Mater.* 4 (2017) 45019.

[12] K. Kang, K.-H. Lee, Y. Han, H. Gao, S. Xie, D.A. Muller, J. Park, Layer-by-layer assembly of two-dimensional materials into wafer-scale heterostructures, *Nature.* 550 (2017) 229–233.

[13] X. Dang, X. Jiang, T. Zhang, H. Zhao, WO3 inversce opal photonic crystals: unique property, synthetic methods and extensive application, *Chinese J. Chem.* 39 (2021) 1706–1715.

[14] S. Zhu, X. Liu, Z. Chen, C. Liu, C. Feng, J. Gu, Q. Liu, D. Zhang, Synthesis of Cu-doped WO3 materials with photonic structures for high performance sensors, *J. Mater. Chem.* 20 (2010) 9126–9132.

[15] J.W. Galusha, L.R. Richey, M.R. Jorgensen, J.S. Gardner, M.H. Bartl, Study of natural photonic crystals in beetle scales and their conversion into inorganic structures via a sol–gel biotemplating route, *J. Mater. Chem.* 20 (2010) 1277–1284.

[16] J. Huang, X. Wang, Z.L. Wang, Controlled replication of butterfly wings for achieving tunable photonic properties, *Nano Lett.* 6 (2006) 2325–2331.

[17] Y. Yu, Y. Yu, L. Huang, H. Peng, L. Xiong, L. Cao, Giant gating tunability of optical refractive index in transition metal dichalcogenide monolayers, *Nano Lett.* 17 (2017) 3613–3618.

[18] L. Mennel, M. Paur, T. Mueller, Second harmonic generation in strained transition metal dichalcogenide monolayers: MoS2, MoSe2, WS2, and WSe2, *APL Photonics.* 4 (2018) 34404.

[19] G. Wei, D.A. Czaplewski, E.J. Lenferink, T.K. Stanev, I.W. Jung, N.P. Stern, Size-tunable lateral confinement in monolayer semiconductors, *Sci. Rep.* 7 (2017) 3324.

[20] C. Sanchez, B. Lebeau, Design and properties of hybrid organic–inorganic nanocomposites for photonics, *MRS Bull.* 26 (2001) 377–387.

[21] M. Gutiérrez, C. Martín, B.E. Souza, M. Van der Auweraer, J. Hofkens, J.-C. Tan, Highly luminescent silver-based MOFs: scalable eco-friendly synthesis paving the way for photonics sensors and electroluminescent devices, *Appl. Mater. Today.* 21 (2020) 100817.

[22] K.J. Kim, Y.R. Park, Spectroscopic ellipsometry study of optical transitions in Zn1−xCoxO alloys, *Appl. Phys. Lett.* 81 (2002) 1420–1422.

[23] B.L. Evans, P.A. Young, R.W. Ditchburn, Optical absorption and dispersion in molybdenum disulphide, *Proc. R. Soc. London. Ser. A. Math. Phys. Sci.* 284 (1965) 402–422.

[24] G. Wei, T.K. Stanev, D.A. Czaplewski, I.W. Jung, N.P. Stern, Silicon-nitride photonic circuits interfaced with monolayer MoS2, *Appl. Phys. Lett.* 107 (2015) 91112.

CHAPTER 16

Layered Structured Materials and Nanotechnology for Photodetection

Felipe M. de Souza, Magdalene Asare, and Ram K. Gupta

Pittsburg State University

CONTENTS

16.1 INTRODUCTION

The efforts of the scientific community that led to the development of particles that reached the nanoscale opened a vast field with many novel materials and applications as it has been stated in Feynman's lecture back in 1959, "There is plenty of room in the bottom". Following that line, researchers focused on the synthesis of several types of nanomaterials such as nanolayered structures. Graphene is a known example that has high conductibility, mechanical, and thermal properties [1,2]. Yet, one of the current limitations lies in its application as a switching device due to its high conductivity that differs from those of semiconductors. Based on that, one of the most researched classes of semiconductors is the transition metal dichalcogenides (TMDs) which can be composed of metallic elements such as Mo, Ti, W, V, Zr, Hf, Ni, Co, Zn, Fe among others which are intercalated with chalcogens such as Te, Se, or S. These structures are covalently bonded within the plane and have van der Waals (vdW) interactions in between the planes. The vdW strength can vary depending on the elements, yet it also allows them to be exfoliated likewise graphene. In addition to that, TMDs present a large bandgap range that can go from the infrared to

DOI: 10.1201/9781003185109-16

visible light which allows their use as optical detectors, optical switching devices, micro-electronic, prims, nano pigments for inks, reflective coating for lenses, mirrored surfaces, waveguides along with other applications that require photoactive properties. On top of that, TMD presents a broad range of varied and tunable properties such as ferroelectrics, magnetism, tunable conductivity, among others.

However, one of the challenges lies in effectively synthesizing 2D materials that can maintain such properties as when compared to their 3D bulky counterparts. These layered materials can be potentially used as optical sensors since their photonic band gaps (E_g), as well as their refractive indexes, can be tuned, which enables them to diffract, scatter or filter the desired wavelength. Based on that, these materials can be synthesized through bottom-up or top-down methods, such as chemical vapor deposition (CVD), layer-by-layer stacking, electron-beam lithography, electrochemical etching, and self-assembly of crystalline colloidal arrays, among others. One of the aims of employing such synthetical approaches is to enable control over the layered material's refractive index, crystal lattice constants, and symmetry. Another aspect is that particles with a size range of between 1,000 and 100 nm may perform visible-light Bragg diffraction which is a useful phenomenon for the sensors [3]. Because of these and other properties, these layered materials can be potentially employed in photodetectors. However, to be competitive against a well-established market the photosensors derived from these materials are required to deliver a fast response toward external stimuli and in a reversible way, which allows them to be used in several sectors such as for medical and veterinary analysis, for the detection and possibly quantification of cells, antibodies, antigens, and cells.

In addition, one major advantage that photodetectors hold when compared to electrochemical detectors is their immunity against electromagnetic interference, which can extend their field of application. Another field of analysis for optical sensors includes the detection of drugs, that enable the identification of adulterated pills without sufficient active principle, for instance. In another sector, photodetectors can be used in smart windows or displays which can vary their color, become translucent or opaque when exposed to a certain wavelength or an electric field [4]. Based on the same principle, these optically active materials can also be used to develop camouflage due to their dynamic color transition that is influenced by the environment. Within that line $BaTiO_3$, WO_3, and VO_2 are some examples of layered materials that can be used to tune the refractive index in a given photo active-matrix along with acting as temperature and electric field sensors based on their optical properties. Alongside these applications, there are also holographic sensors which are composed of materials that can change their optical features when in contact with an analyte. It means that after interacting with the analyte the material can change its spectral response as well as its diffraction efficiency which can be, in some cases, empirically observed as a change in color and/or brightness, as schematized in Figure 16.1.

The diffraction efficiency can be defined as the ratio between the diffracted beam's intensity by the incident beam. Through that, a quantitative measurement from either brightness or color change can be obtained, which correlates with the analyte's concentration.

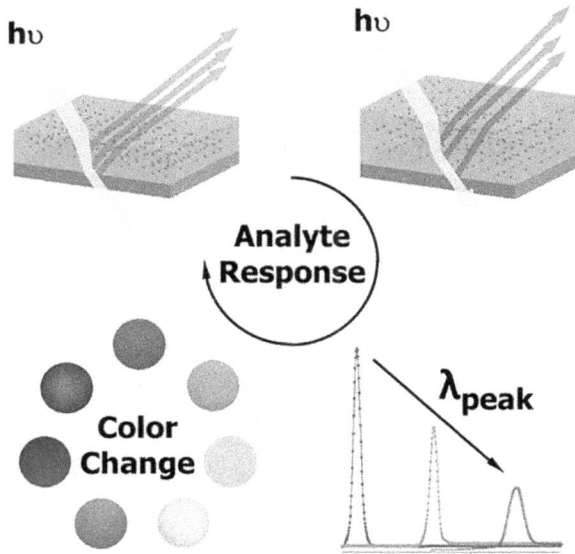

FIGURE 16.1   Schematics for the functioning of a photosensor. (Reproduced with permission from [5]. Copyright (2014), American Chemical Society.)

Based on that, different optic sensors can be designed according to the property that changes during the recording such as surface or volume relief, reflection or transmission, and amplitude or phase change [6]. Usually, there could be at least three components to design a photosensor: substrate, recording media, and embedded nanoparticles. The substrate should be transparent; hence it should be neither birefringent, optically active nor anisotropic as the latter could lead to a decrease in diffraction efficiency. In addition, it should present enough mechanical strength to support the photosensitive material that can be deposited through several variations of CVD methods, for instance. In that sense, when a photosensor is used in liquid-state systems the substrate should be stable against shrinking and swelling cycles to avoid detachment of photoactive material. To prevent this process a unimolecular sublayer can be placed underlying the substrate before the coating of active material.

Some of these techniques are silane coupling which consists of the bonding between organic and inorganic moieties over the substrate's surface by forming hydrogen bonds between the substrate and a silane coupling agent with a structure of X_3–Si–$(CH_2)_n$–R. Also, gelating coatings can be used which are generally based on the adhesion of (3-aminopropyl)-triethoxysilane which is dried over the substrate and promotes hydrogen bonding with it along with exposed amino groups that can be later used for functionalization with active material. The recording media consists of a photosensitive material that can capture and store information from the radiation. An example of that is AgX where X are halides that can be incorporated into a polymeric matrix that is coated over a substrate. Some of the polymers available for that are poly(vinyl alcohol) (PVA), poly(2-hydroxy methacrylate), and poly(acrylamide). Nanoparticles can also be added to enhance

TABLE 16.1   Analytes that can be Identified through Holographic Sensors Based on Layered Materials

Type	Analyte
Gases	O_2, N_2, NH_3, alkynes, alkenes, and alkanes
Ions	H^+, monovalent ions (Na^+ and K^+), divalent ions (Mg^{2+}, Cu^{2+}, Fe^{2+}, Ca^{2+}, Ni^{2+}, Co^{2+}, Zn^{2+}, Pb^{2+}), Co^{3+}, IO_6^{5-}
Enzymes	Amylase, trypsin
Metabolites	Lactose, glucose, acetylcholine, urea, testosterone
Drugs	Penicillin G
Physical Parameters	Pressure and temperature
Others	Organic solvents, humidity, alcohol, ionic strength, magnetic field, and light
Gases	O_2, N_2, NH_3, alkynes, alkenes, and alkanes
Ions	H^+, monovalent ions (Na^+ and K^+), divalent ions (Mg^{2+}, Cu^{2+}, Fe^{2+}, Ca^{2+}, Ni^{2+}, Co^{2+}, Zn^{2+}, Pb^{2+}), Co^{3+}, IO_6^{5-}
Enzymes	Amylase, trypsin
Metabolites	Lactose, glucose, acetylcholine, urea, testosterone
Drugs	Penicillin G
Physical Parameters	Pressure and temperature
Others	Organic solvents, humidity, alcohol, ionic strength, magnetic field, and light

the detection and recording properties of the sensor. Yet, the nanoparticles must be stable over the polymer matrix, show good affinity toward the analyte, proper refractive index, and size. For the case of the latter, nanoparticle sizes within the range of 10–30 nm can be used to diminish light scattering, which is directly related to a higher diffraction efficiency and higher contrast. In addition, nanoparticles that present a refractive index that differs from the recording material can be used to create a larger dynamic range enabling it to record multiple holograms. Through that, these holographic sensors can be used for a myriad of detection systems. A summary of some of the analytes that these photoactive sensors can identify is provided in Table 16.1 [5].

16.2 MECHANISM OF PHOTODETECTION

Photodetectors have become an increasingly important piece of technology that is used on communication devices, imaging, displays, and more sophisticated applications such as space observer telescopes. Based on that, photodetection consists of the conversion of specific light radiation into an electric signal. Such a process can take place through three sequential mechanisms which are harvesting of light, separation of excitons, and charge carrier transport to the desired electrodes. The equation that describes the light efficiency can be expressed by external quantum efficiency which is shown in Equation 16.1. Where I_{ph} is defined as photocurrent which represents the variation among current under light and in the dark; Φ represents photon flux, which is the number of incident photons; q represents the unit of an electron charge; $h\nu$ represents the energy of one photon, and P represents the power of the incident light [7].

$$\text{EQE} = \frac{I_{ph}}{q\Phi} = \frac{h\nu}{q} \frac{I_{ph}}{P} \tag{16.1}$$

The other two parameters that are used within the field of photodetection are the responsivity and detectivity defined in Equations 16.2 and 16.3, respectively. Within that line, detectivity represents the capability of a photosensor to acquire a signal while distinguishing it from noise. Based on these expressions p and I_{dark} represent the light intensity and dark current, respectively. Another parameter for photo detective properties is the response time (t) which is defined as the time required for a material shined by light to reach photocurrent saturation.

$$R = \frac{I_{ph}}{P} = EQE\frac{q}{h\nu} \tag{16.2}$$

$$D = \frac{I_{ph}}{P\sqrt{2qI_{dark}}} \tag{16.3}$$

Within that line, the photodetectors can be classified as photoconductors or photodiodes. The first relates to materials that go through a change in their conductivity when shined by a light whereas the second relates to materials that possess semiconductors based on p–n or Schottky junction behavior [8,9]. Based on that, a photodiode has a faster response due to the involvement of both electrons (e⁻) and holes (h⁺) in the generation of a photocurrent. However, they usually present a shorter carrier lifetime due to the tendency of the e⁻–h⁺ pair to recombine. Because of that, photodiodes can present an External Quantum Efficiency (EQE) smaller than 1. Oppositely, photoconductors present only one type of charge carrier as the other one is unmovable. Because of that, the charge carrier has a longer lifetime which provides values larger than 1 for the EQE along with high responsivity. Yet, that is accompanied by a longer response time. Based on these concepts, one of the challenges within the development of optical sensors as well as optic electronic devices is to develop materials that possess a specific E_g to promote photo excitement under room temperature. Within that line, silicon-based photosensors present an E_g of 1.4 eV, which corresponds to the visible wavelength of around 900 nm. The commercial photosensors based on the crystalline structure of InGeAs/HgCdTe are used for long-wavelength detection processes. However, these materials present a costly production process through molecular beam epitaxy and can sometimes require low temperatures to be operational. The understanding of the photodetection process can be defined through different mechanistic processes which are (i) photoelectric, (ii) photo-bolometric, (iii) photo-thermoelectric, along with other phenomena such as photogate, photoinduced charge transfer (tunneling), and field-effect doping.

The photoelectric effect occurs when incident photons can bring e⁻ from the valence band (VB) to the conducting band (CB) leading to the formation of excitons. Then, a potential can be applied to propel the photoexcited electrons to flow promoting the photocurrent process, which can be described by Equation 16.4.

$$I_{ph} = AVq\mu\Delta n \tag{16.4}$$

where A is the area of the active layer, V is the applied potential (V), q is the charge, μ is the charge carrier mobility, and Δn is the density of photoinduced carriers. Based on that, ideally, a material should present high carrier mobility, absorb in a broad wavelength range and maintain the excitons for a proper time before they recombine. Yet, it is a difficult condition to achieve as such materials like graphene, for instance, present high carrier mobility with the drawback of low light absorption, which leads to a fast recombination process for the excitons. On the other hand, traditional semiconductors may display the opposite behavior.

The photo-thermoelectric effect can be a disadvantageous phenomenon for photodetection as it occurs through successive energy conversion processes as the light absorbed by the material is converted to thermal and then to electrical energy. Because of that, there can be a delay in the overall photo excitement, leading to relatively lower efficiency in some cases. Yet, the occurrence of this effect can provide important information to understand the mechanistic process as an incident light can promote a considerable variation in electron temperature, which might provide an insight into the development of photodetection as well as optoelectronic applications [10,11]. In that line, the description of the Seebeck coefficient is a measurement of the thermoelectric power of a material, which is related to both densities of states Density of States (DOS) as well s the Fermi level. Based on that, the Seebeck coefficient can be expressed by Equation 16.5. From that, k_b is the Boltzmann constant; T is the temperature of the sample; q is the value of a single electron charge; G is the conductance, and V_{GS} is the gate bias.

$$S = \frac{\pi^2 k_b^2 T}{3q} \frac{1}{G} \frac{dG}{dV_{GS}} \frac{dV_{GS}}{dE} \tag{16.5}$$

Following up on that, the photo-thermoelectric current (I_{PTE}) can be calculated through Equation 16.6.

$$I_{PTE} = \frac{S_2 - S_1}{R} \Delta T \tag{16.6}$$

As S_1 and S_2 are the Seebeck coefficents obtained from two regions; T is the temperature and R is the resistance. The temperature gradient leads to the diffusion of DOS as carriers at higher temperatures tend to diffuse from low to high DOS sites. The variance, in principle, is what leads to the occurrence of the photo-thermoelectric effect. Based on that, a device that operates under this phenomenon can be exposed to more than one vertical gate bias on different regions of the material which can induce thermoelectric power discrepancies [12]. The occurrence of this effect is related to the two ways that the energy from photon-excited e^-–h^+ pairs can be released. The first way occurs through energy transport from the charge carrier which can promote photocurrent. The second way consists of the energy transfer to the lattice that is converted to heat [13]. This type of mechanism can be induced in materials such as graphene as well as TMDs.

An example of promoting the photo-thermoelectric effect on the former has been performed by Stutzel et al. [14] which synthesized a graphene nanoribbon presenting a width from 10 to 35 nm that was intercalated with two metallic electrodes over a SiO$_2$/Si substrate. A laser of 633 nm was used to measure the photocurrent. The main aspect of this work was that the heat generated under illumination generated higher temperature differences when graphene was in contact with the metal electrodes rather than graphene p–n junctions. That was attributed to the large difference between the Seebeck coefficient among graphene and the metallic electrode. This led to a photoresponse mechanism dominated by the photo-thermoelectric effect when compared to the photovoltaic effect. Within that line, the TMDs are another class of nanomaterials that can function as photosensors through the same mechanism which has been explored by Buscema et al. [15] that analyzed the behavior of a MoS$_2$ single layer interaction with light using sub-band-gap illumination. The authors observed that the photo-thermoelectric effect was predominant for the generation of photocurrent through both sub-bandgap radiation as well as above-band gap radiation of 750 and 532 nm, respectively. This study provided a potential insight for the use of single-layered MoS$_2$ as a material suitable to power devices through harvesting their thermal waste energy.

The photovoltaic effect is another important mechanism that is often described in research involving solar cells at which there is a concerted process of a photosensitive material that absorbs a photon, followed by the formation of excitons (e$^-$–h$^+$), then these charges are separated and collected by free charge carriers at the electrode. However, in the field of layered and thin films, the photovoltaic effect is described as the generation of photoexcited carriers in the presence of an electric field [10,16]. These photogenerated charges can be extracted either through diffusion in a short-circuit system or through an electric field that accelerates them to the electrodes. Based on that phenomenon, the photovoltaic effect can be described by Equation 16.7.

$$I_{\text{ph}} = \frac{\eta_i P_{\text{opt}}}{\hbar w_0} \frac{W}{L} \tau_c q \left(\mu_e + \mu_h \right) V_{\text{sd}} \tag{16.7}$$

As the terms related to the expression are internal quantum efficiency (η_i), the energy of a single photon ($\hbar w_0$), device's width (W) and length (L), carrier lifetime (τ_c), electron (μ_e) and hole (μ_h) mobilities and the applied potential (V_{sd}).

Compositing materials to improve their light adsorption properties is a known process that can be performed with graphene and a semiconductor, for instance. Through that, upon illumination, a photoexcited charge can be generated from the semiconductor and transferred to graphene. This phenomenon promotes a gradient of carrier density in the graphene, which is notable through the measurement of electron flow in the charge neutrality point (V_{CNP}). In that way, it can be considered as a gate that is activated through light photogates named as photogating effect. Based on these concepts it is worth noting that the photon that hit the semiconductor's surface is absorbed if the photon's energy is higher than the E_g, hence $h\nu \geq E_g$. When that occurs, then the e$^-$–h$^+$ pairs are formed, which leads to excitons that possess an intrinsic efficiency (η_{gen}) which is a parameter that depends on the material's absorption coefficient. To free charges to be created there should be a driving force that can be in form of an electrical field or thermal energy that is higher than the Coulombic force between the e^-–h^+ this process can be associated with the efficiency term (η_{diss}). Taking a semiconductor-graphene composite the charge transfer is bound to a thermodynamic barrier that emerges either from the interface between them or by charge entrapment in the semiconductor. Thus, smooth and clean interfaces are necessary to enhance the charge transfer efficiency (η_{trans}). By combining these three terms the quantum efficiency can be calculated based on Equation 16.8.

$$\eta = \eta_{gen}\eta_{diss}\eta_{trans} \tag{16.8}$$

Through that, the charge can be extracted by being exposed to a voltage that directs it to the drain contact. Also, electrical neutrality must be preserved, for that, another charge must be injected into the system simultaneously. This can be achieved by increasing the trapped carrier lifetime (τ). However, higher τ diminishes the response time of the photodetector. Hence, to optimize that the transit time should decrease there should be a short space between the electrode along with large electric fields. Within that line, graphene synthesized through electron-beam lithography is a material that can offer high carrier mobility [17]. Based on these concepts Figure 16.2 shows the schematics for the mechanism for photoresponse in terms of photothermal electric, photovoltaic and photogating effects for graphene-based materials.

The photo-bolometric effect is another type of mechanism by which photoactive materials can deliver a response after interaction with light. It is based on a relation between the material's conductivity and temperature. The occurrence of this phenomenon was studied by Yan et al. [19] that fabricated a hot-electron bolometer based on bilayer graphene dual gate type of device. For that, graphene–metal junction was avoided along with a uniform graphene structure. In that sense, both photo-thermoelectric and photovoltaic effects were minimal due to the lack of an output DC voltage when the device was illuminated. Based

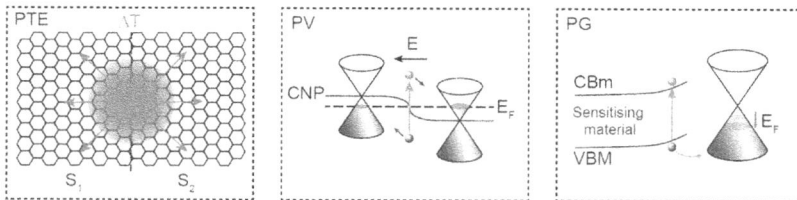

FIGURE 16.2 Schematics for the mechanisms of photo-sensing for graphene-based materials through photothermal electric (PTE), photovoltaic (PV), and photogating (PG). From that, Fermi level (E_F), electric field (E), valence band maximum (VBM), and conduction band minimum (CBm) are presented. (Adapted with permission from [18]. Copyright (2018), The Authors, Some right reserved. Licensee MDPI, Basel, Switzerland. This article is licensed under a Creative Commons Attribution 4.0 International License.)

on that, when there was an input of DC, the output DC voltage was varied as the light when off and on. This effect can be calculated through Equation 16.9 as I_{DC} represents the DC applied; ΔR is the variation in resistance, and ΔT is the variation of temperature that occurred due to the light absorption. It was observed that the system's temperature was another factor that changed the way that heating happened, since under lower temperatures, around $-263°C$, both electrical and optical heating were considerable. On the other hand, when exposed to high temperatures, around $-243°C$, then optical heating was the main contributor.

$$\Delta V = I_{DC}\Delta R = I_{DC}\frac{dR}{dT}\Delta T \tag{16.9}$$

16.3 PHOTO DEVICES

16.3.1 Graphene-Based Photo Devices

Along with TMDs, polymeric matrices, and nanoparticles, graphene is a keystone nanomaterial that holds great potential for optoelectronic applications. Such field of use is possible due to the inherent properties of graphene-like the linear behavior for electrons in terms of energy and momentum, which is likely attributed to its highly conjugated structure. Because of that, graphene possesses high charge carrier mobility which can reach 10^5 cm^2/Vs at room temperature and increase up to values 10 times higher when exposed to lower temperatures. On top of that, graphene presents a high absorption coefficient of 7×10^5 cm^{-1} that can cover a range from 300 to 2,500 nm that can overcome the corresponding values for traditional semiconductors. Hence, because of these semi-metallic and tunable properties, graphene can be incorporated into optoelectronic devices such as photosensors. Yet, despite these unique properties, there are still some challenges for the proper implementation of graphene into commercial devices based on photodetection. Some of these drawbacks are related to its short linear dynamic range (LDR) and low efficiency to generate and extract excitons (e^-–h^+). To address those issues, graphene can be composited with other photosensitive materials to enhance its performance. An example of that is the incorporation of FeCl$_3$ into graphene's structure to improve stability and light-harvesting properties by increasing its LDR. Based on that, it deems necessary to functionalize graphene to improve its properties. One way lies in chemically inserting oxygen groups in its surface to obtain graphene oxide (GO) which can work as a photodetector over a wider range of the spectrum that goes from UV up to frequencies of the order of THz. Another approach such as embedding graphene's surface or in between its layers with nanoparticles, quantum dots, or TMDs. Another approach consists of saturating the graphene's π bonds through chemisorption with F atoms. These functionalization approaches can be performed through solution methods i.e., solvent/ion exchange, sol-gel, hydrolysis, or hydrothermal as well as dry techniques i.e., CVD, physical vapor deposition, plasma sputtering, thermal evaporation, and annealing to name some. In the sense of photodetection, graphene is usually placed in contact with metallic connections, either through shadow masks or lithography. Along with that, an encapsulation process may be necessary to improve the stability of some light-sensitive materials against radiation. This process can be performed by depositing a polymeric layer of materials such as poly-methyl-methacrylate or lithium-perchlorate-PEO (LiClO$_4$-PEO). The schematics of these processes are shown in Figure 16.3.

Based on these approaches, graphene's photo-sensing properties can be tuned by incorporating FeCl$_3$ in between its layers with the use of a laser beam. Through this process, a p–p' junction can be obtained. This process provides some advantages such as considerable improvement in LDR along with broad wavelengths operational ranges that can go from the ultra-violet (UV) to mid-infrared. Also, there is a considerable diminishment of the photo-thermoelectric effect that is overcome by the photovoltaic effect after the intercalation with FeCl$_3$. The convenience is that there is a smaller population of hot-carriers in graphene's structure that even through are species that aid in the photo-sensing process can lead to a decrease in resolution because of smearing of the photoactive regions [16]. Another graphene-based material is GO which consists of chemically bonded oxygen into its structure. This leads to a loss of conjugation making it an insulator. Despite that, its E$_g$ and conductibility can be tuned by reducing the oxygenated groups to form reduced GO (rGO). This effect has been studied in separate reports from Chitara et al. which synthesized an rGO that could operate at a wavelength of 360 nm with a responsivity of 120 mA/W [20].

In another work, Chitara et al. [21] fabricated a photosensor that could operate at a wavelength of 1,550 nm with the drawback of lower responsivity of 4 mA/W. Through these studies, it was notable that rGO can cover a wide range of wavelengths. It is worth noting that rGO usually possesses a low conductivity, which can hinder its application in certain types of photodetectors. To address that issue, Yang et al. [22] proposed the synthesis of rGO through a wet process followed by a drop-cast over an etched

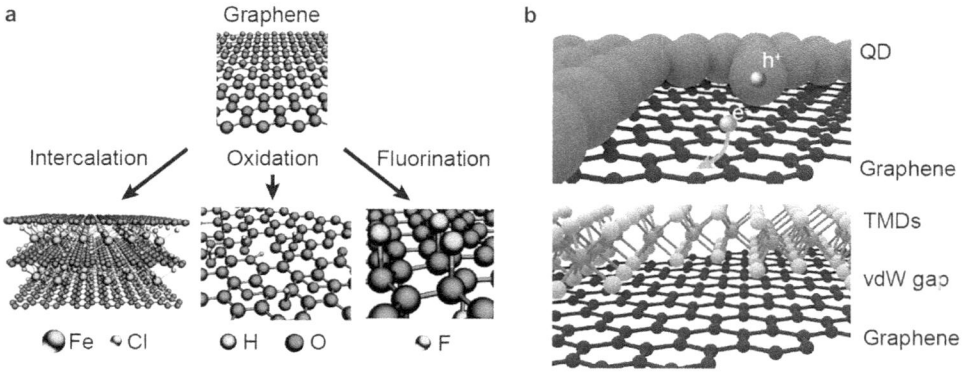

FIGURE 16.3 (a) Schematics for some of the functionalization processes for graphene. (b) Graphene's surface-functionalized with quantum dots and TMDs. (Adapted with permission from [18]. Copyright (2018), The Authors, Some right reserved. Licensee MDPI, Basel, Switzerland. This article is licensed under a Creative Commons Attribution 4.0 International License.)

FIGURE 16.4 Scheme for the fabrication of a highly conducting rGO-based free-standing photodetector obtained through wet synthesis and drop-cast over a Si wafer followed by annealing process. (Reproduced with permission from [22]. Copyright (2017), Elsevier.)

silicon wafer, which provided a high yield along with a larger area to coat the film. The rGO had superior properties such as a conductivity of 87,100 S/m and a charge carrier mobility of around 16.7 cm^2/V·s. Through that, the authors fabricated a device based on a free-standing rGO thin-film that presented a fast and broad photoresponse by reaching 100 ms and a spectral covering that went from UV to terahertz. Such speed response could match those of CVD and mechanically exfoliated graphene-based photodetectors. The schematic for the free-standing rGO photodetector is presented in Figure 16.4.

The compositing of graphene with other semiconducting materials such as TMDs can lead to an appreciable improvement in properties. To focus on that, Singh et al. [23] fabricated a UV photodetector that consisted of a composite of layered graphene structure with MoS$_2$ quantum dots that were synthesized through CVD followed by the hydrothermal method. Then, SiO$_2$/Si was used as a substrate. The purpose of combining these two materials was to take advantage of the UV absorption properties of WS$_2$ quantum dots combined with graphene's high charge mobility. Following that, satisfactory results were obtained such as high photo detectivity and responsivity of around 7.47×10^{12} Jones (Hz$^{1/2}$cm/W) and 1,814 A/W, respectively over a UV light of 365 nm and power density of 50.74 μW/cm^2. Based on these results an operational principle was proposed to the graphene-WS$_2$ composite at which graphene presented a lower work function (Φ_G) than the one for the WS$_2$ (Φ_{WS2}). This condition led to the formation of a considerable amount of e$^-$–h$^+$ and since graphene has higher carrier mobility the electrons move toward it whereas the holes remained at the WS$_2$ structure. That process is also possible due to the applied external potential which produces the photocurrent. The graphene-WS$_2$ photosensor along with its energy diagrams are presented in Figure 16.5.

Another approach that can be performed to enhance the photodetection properties of graphene lies in combining it with polymers which can, in simple terms, enhance the photothermal electric effect considering that the polymer can create a charge inhomogeneity and provide a large Seebeck coefficient. Based on these premises Miseikis et al. [24] fabricated a photosensor device based on an intercalated structure of graphene/polymer/graphene. In that sense, PVA was used as a dielectric to yield a tunable p–n junction. Through that, the PVA provided a Seebeck coefficient of 140 μV/K along with a relatively low charge inhomogeneity of 8×10^{10} cm^2. The scheme for the graphene-PVA photodetector is presented in Figure 16.6.

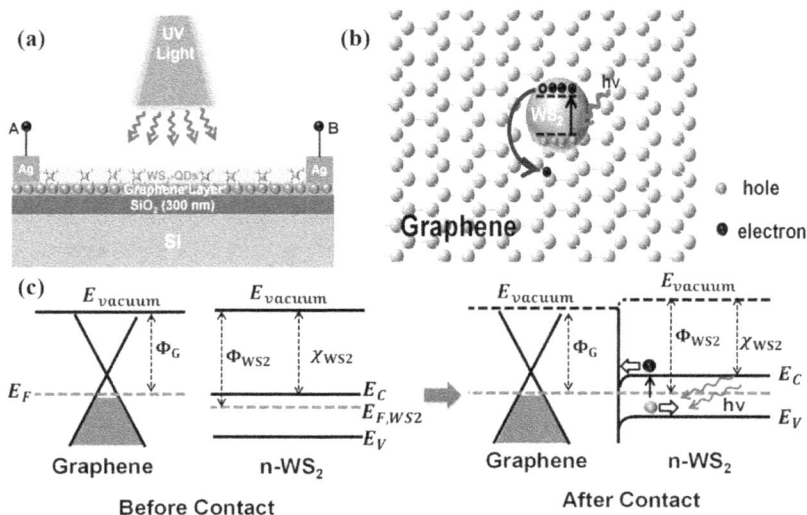

FIGURE 16.5 (a) Scheme for the graphene (2D) composited with WS_2 quantum dots (0D) photodetector device. (b) Scheme for the photoresponse of WS_2 quantum dots distributed over graphene's surface. (c) Band structure diagrams for graphene and WS_2 before and after contact, at which Φ_G and Φ_{WS2} are the work functions of graphene and WS_2, respectively. E_F and $E_{F, WS2}$ are the Fermi Levels for graphene and WS_2, respectively. E_c and E_v are the conduction and valence energy levels. (Adapted with permission from [23]. Copyright (2019), American Chemical Society.)

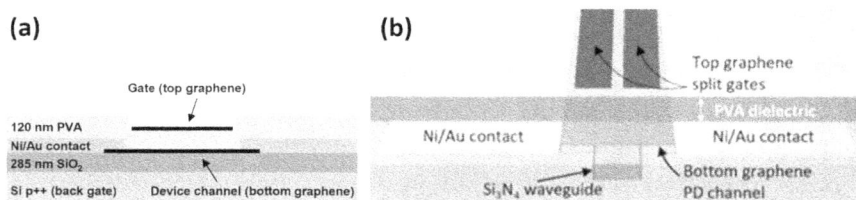

FIGURE 16.6 (a) Scheme for the cross-section of PVA intercalated with graphene photodetector device. (b) Different perspectives for the graphene-PVA-based photosensor. (Adapted with permission from [24]. Copyright (2020), American Chemical Society.)

16.3.2 Other Layered Materials-Based Photo Devices

Ideally, semiconductor-based photodetectors should present some required properties to be implemented commercially such as absorbing light efficiently over a broad spectral range, presenting a high ratio between photo and dark current, generating a charge carrier with high mobility during exposure to light, presenting a very low charge trapping, high stability, and simple fabrication while using low cost and abundant resources. However, the market is currently dominated by relatively expensive semiconductor materials such as InGaAs and GaN along with Si which is considerably cheaper and abundant. Yet, complex techniques are used for their manufacture which consists mostly of CVD processes under high temperature and sputtering techniques in high-vacuum systems. In addition to that, electronic pre-amplifiers are often required to increase the device's electric signal.

To counter those conditions, a class of materials such as metal halide perovskites (MHPs) demonstrates remarkable properties to be implemented as a photosensor due to its properties like direct and facile tunability of E_g, charge carriers with LDR, tolerance to defects, and high absorption coefficients. On top of these properties, these semiconductor materials can be processed through wet techniques under low temperatures, making their use more feasible. Their general formula can be expressed as ABX_3 where X are halides (I, Br, and Cl), B are divalent cations such as Sn^{2+} or Pb^{2+}, and A are monovalent cations like $CH_3NH_3^+$ or Cs^+, for instance. In that sense, it is worth noting that MHPs that present can obtain considerably high photoresponse with values up to around 2.6×10^6 A/W. Despite that, the presence of organic cations which are volatile can be associated with lower stability and increased deterioration when exposed to air.

On the other hand, all-inorganic MHPs such as $CsPbX_3$ present considerably higher stability under air which can last more than 60 days. Among the different halides that can be incorporated into the MHP's structure, the use of Br to obtain $CsPbBr_3$ presents an E_g of 2.3 eV and stability at room temperature, which can make it a promising material for photonic devices. However, the main challenges regarding its large-scale use are related to environmental concerns as the recycling of Pb is required to avoid contamination with heavy metals, along with some improvements in its performance such as its slow and limited photoresponse. The latter is likely related to phase impurities that are rich in 0D Cs on Cs_4PbBr_6 or the 2D Pb phase of $CsPb_2Br_5$ that lead to a considerable decrease in its optoelectronic properties. In that sense, one of the core aspects of the research of these materials focuses on diminishing the phase populations of $CsPb_2Br_5$ and Cs_4PbBr_6 and directing to the phase containing $CsPbBr_3$ as it is more likely to deliver a better performance [25]. Along with these factors $CsPbBr_3$ can be synthesized in a variety of nanostructures such as thin films, single, micro, and nanocrystals, nanowires, along hybrid photodetectors. Based on the fabrication of $CsPbBr_3$ thin-film photosensors, one of the main challenges lies in obtaining a regular shaped and compact film that contains large grain sizes to diminish the grain boundaries effect. This is a relevant factor in the sense that grain boundary can decrease non-radiative recombination and lead to low-efficiency charge transport.

To avoid that, Li et al. [26] employed a dissolution followed by a recrystallization technique that led to thin films of $CsPbBr_3$ with low roughness, large area, and without cracks on the structure. The thin films were capped with ligands over their surface which were slowly washed by using a mixture of ethanol and toluene. This approach provided a proper control of the particle size which could range from 10 nm to 1 μm. This phenomenon was possible due to the ligand's high dynamic binding as well as surface diffusion coefficient. This promoted a fast trade of the atomic and surfactant species over the thin-film's surface facilitating the washing process. In that sense, toluene enhanced the diffusion process due to its low polarity, and ethanol enhanced the surface atom's dissolution. After that, the solvents are evaporated leading to the crystallization process, leading to the formation of a compact film with large grains. Such an approach led to a drastic increment instability, EQE which was around 700% along with a fast response speed of 1.0 and 1.8 ms for the rise and decay times, respectively. Another possibility for the uses of $CsPbBr_3$, aside from the different morphologies previously mentioned, is through composting.

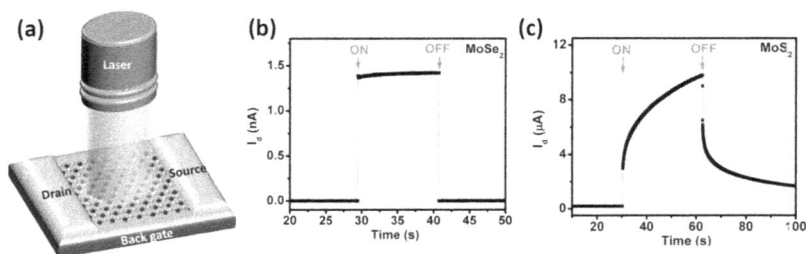

FIGURE 16.7 (a) Scheme for the photosensor device fabricated through CVD process to obtain single-layered $MoSe_2$ or MoS_2. Photocurrent time responses for (b) $MoSe_2$ and (c) McS_2 photo sensors both at $P = 0.59$ and 0.31 W/cm^2, respectively. (Adapted with permission from [28]. Copyright (2014), American Chemical Society.)

Ji et al. [27] fabricated a photosensor based on $CsPbBr_3$ along with NiO_x and TiO_2. The perovskite layer was fabricated through the spin-coating technique, whereas the charge transport layers were grown through atomic layer deposition. In that sense, the NiO_x aid in the formation of a perovskite layer by optimizing the interfacial interaction between the substrate and perovskite as it acted as an h$^+$ transport layer. Alongside that, TiO_2 functioned as a passivation agent while extracting the photogenerated electrons. By incorporating these layered materials for charge transport the device yielded a low dark current of around 10^{-11} A, 0.056, and 0.25 ms for rising and fall times, respectively, an LDR of 186.7 dB along with a detection limit of 857 pW/cm^2, which is considered as ultralow.

As previously discussed TMDs are another group of feasible candidates for photosensors along with other areas that include energy storage, electrocatalysis, and so on. To explore their use within the photosensors area, Chang et al. [28] studied the optoelectronic properties of monolayer $MoSe_2$ synthesized through CVD which achieved high crystallinity. The device's schematics are presented in Figure 16.7. When the comparison between $MoSe_2$ and MoS_2 was made it was notable that $MoSe_2$ presented a weaker bound exciton peak which led to a response time of around 25 ms (Figure 16.7b) which was much faster than the 30 s response for the MoS_2 (Figure 16.7c). Such variation in response was attributed to the defect-free morphology of $MoSe_2$ which enabled a better charge transport for the generation of photocurrent. Also, it was notable that $MoSe_2$ had a behavior of ambipolar n-type, whereas MoS_2 displayed n-doped electrical features.

16.4 CONCLUSION AND OUTLOOK

The technological development of photosensors based on the principle of conversion of light into an electrical signal holds great importance in recent times as it is related to applications such as high-resolution imaging for medical devices along with scanners that also function for security purposes, cameras, along with the detection of over other analytes. In that sense, photo-sensing can extend to the identification or response toward specific wavelengths that can be correlated with the presence of biomolecules or for use in integrated circuits, microelectronics, and other similar electronic devices. Within that line, there is the constant push to use novel semiconductor materials that can deliver a similar performance when compared to the commercially available ones, which are obtained through relatively expensive synthetical approaches such as laser beam epitaxy along with the use of scarce materials such as Ga, In, Ge, among other semi-metallic elements. To work around this situation came the need for semiconductor materials that can offer a considerable degree of optical-electronic properties tunability, in terms of conductivity, bandgap, dynamic linear range, quantum efficiency, absorption coefficients among other parameters.

One of the feasible options for the next generation of photosensors is graphene and its derivatives as it can attend to most of the requirements for photo-sensing properties such as high conductibility and an absorption coefficient that can cover a broad wavelength range. However, by itself, graphene cannot be used as a photosensor as it is not able to properly absorb photons to induce the generation of e^--h^+. Hence, to take advantage of its inherent optical properties, graphene can be composited with TMDs which can promote the e^--h^+ due to their optical band gap values. Considerable progress has been made in this regard yet is still challenging to manufacture a decisive photoactive composite material based on graphene and TMD to function as a suitable alternative to the commercially available materials. Another promising area lies in the use of MHPs as photo-sensing materials as they also present a vast range of bandgap tunability, high tolerance to defects along with high charge carrier and absorption coefficients. Along with that, this class of materials can be synthesized with a variety of methods including wet and dry techniques, which leads to a myriad of morphologies that can be applied in several ways. In addition, current research has been devoting some effort into fabricating composites of MHPs which can function as promising materials that can photo induce the generation of e^- and h^+, promoting a fast response and efficiency that can enable their large-scale application. Yet, it is still deemed important to develop composites with satisfactory environmental credentials along with satisfactory performances.

REFERENCES

[1] R.K. Gupta, Z.A. Alahmed, H.A. Albrithen, F. Yakuphanoglu, Highly efficient photosensor based on reduced graphene oxide, *J. Nanoelectron. Optoelectron.* 9 (2014) 474–478.

[2] A. Mekki, A. Dere, K. Mensah-Darkwa, A. Al-Ghamdi, R.K. Gupta, K. Harrabi, W.A. Farooq, F. El-Tantawy, F. Yakuphanoglu, Graphene controlled organic photodetectors, *Synth. Met.* 217 (2016) 43–56.

[3] Y. Yin, Y. Lu, Y. Xia, Assembly of monodispersed spherical colloids into one-dimensional aggregates characterized by well-controlled structures and lengths, *J. Mater. Chem.* 11 (2001) 987–989.

[4] A. Llordés, G. Garcia, J. Gazquez, D.J. Milliron, Tunable near-infrared and visible-light transmittance in nanocrystal-in-glass composites, *Nature.* 500 (2013) 323–326.

[5] A.K. Yetisen, I. Naydenova, F. da Cruz Vasconcellos, J. Blyth, C.R. Lowe, Holographic sensors: Three-dimensional analyte-sensitive nanostructures and their applications, *Chem. Rev.* 114 (2014) 10654–10696.

[6] S.I. Stepanov, Applications of photorefractive crystals, *Rep. Prog. Phys.* 57 (1994) 39–116.

[7] G. Konstantatos, E.H. Sargent, Nanostructured materials for photon detection, *Nat. Nanotechnol.* 5 (2010) 391–400.

[8] D.T. Phan, R.K. Gupta, G.S. Chung, A.A. Al-Ghamdi, O.A. Al-Hartomy, F. El-Tantawy, F. Yakuphanoglu, Photodiodes based on graphene oxide-silicon junctions, *Sol. Energy.* 86 (2012) 2961–2966.

[9] O.A. Al-Hartomy, R.K. Gupta, A.A. Al-Ghamdi, F. Yakuphanoglu, High performance organic-on-inorganic hybrid photodiodes based on organic semiconductor-graphene oxide blends, *Synth. Met.* 195 (2014) 217–221.

[10] N.M. Gabor, J.C. Song, Q. Ma, N.L. Nair, T. Taychatanapat, K. Watanabe, T. Taniguchi, L.S. Levitov, P. Jarillo-Herrero, Hot carrier–assisted intrinsic photoresponse in graphene, *Science (80-.)*. 334 (2011) 648–652.

[11] P. Wei, W. Bao, Y. Pu, C.N. Lau, J. Shi, Anomalous thermoelectric transport of dirac particles in graphene, *Phys. Rev. Lett.* 102 (2009) 166808.

[12] Z. Sun, H. Chang, Graphene and graphene-like two-dimensional materials in photodetection: Mechanisms and methodology, *ACS Nano.* 8 (2014) 4133–4156.

[13] K.J. Tielrooij, J.C.W. Song, S.A. Jensen, A. Centeno, A. Pesquera, A. Zurutuza Elorza, M. Bonn, L.S. Levitov, F.H.L. Koppens, Photoexcitation cascade and multiple hot-carrier generation in graphene, *Nat. Phys.* 9 (2013) 248–252.

[14] E. Ulrich Stützel, T. Dufaux, A. Sagar, S. Rauschenbach, K. Balasubramanian, M. Burghard, K. Kern, Spatially resolved photocurrents in graphene nanoribbon devices, *Appl. Phys. Lett.* 102 (2013) 43106.

[15] M. Buscema, M. Barkelid, V. Zwiller, H.S.J. van der Zant, G.A. Steele, A. Castellanos-Gomez, Large and tunable photothermoelectric effect in single-layer MoS2, *Nano Lett.* 13 (2013) 358–363.

[16] M.C. Lemme, F.H.L. Koppens, A.L. Falk, M.S. Rudner, H. Park, L.S. Levitov, C.M. Marcus, Gate-activated photoresponse in a graphene p–n junction, *Nano Lett.* 11 (2011) 4134–4137.

[17] L. Banszerus, M. Schmitz, S. Engels, M. Goldsche, K. Watanabe, T. Taniguchi, B. Beschoten, C. Stampfer, Ballistic transport exceeding 28 μm in CVD grown graphene, *Nano Lett.* 16 (2016) 1387–1391.

[18] A. De Sanctis, J.D. Mehew, M.F. Craciun, S. Russo, Graphene-based light sensing: Fabrication, characterisation, physical properties and performance, *Materials (Basel).* 11 (2018) 1762.

[19] J. Yan, M.-H. Kim, J.A. Elle, A.B. Sushkov, G.S. Jenkins, H.M. Milchberg, M.S. Fuhrer, H.D. Drew, Dual-gated bilayer graphene hot-electron bolometer, *Nat. Nanotechnol.* 7 (2012) 472–478.

[20] B. Chitara, S.B. Krupanidhi, C.N.R. Rao, Solution processed reduced graphene oxide ultraviolet detector, *Appl. Phys. Lett.* 99 (2011) 113114.

[21] B. Chitara, L.S. Panchakarla, S.B. Krupanidhi, C.N.R. Rao, Infrared photodetectors based on reduced graphene oxide and graphene nanoribbons, *Adv. Mater.* 23 (2011) 5419–5424.

[22] H. Yang, Y. Cao, J. He, Y. Zhang, B. Jin, J.-L. Sun, Y. Wang, Z. Zhao, Highly conductive freestanding reduced graphene oxide thin films for fast photoelectric devices, *Carbon N. Y.* 115 (2017) 561–570.

[23] V.K. Singh, S. M. Yadav, H. Mishra, R. Kumar, R.S. Tiwari, A. Pandey, A. Srivastava, WS2 quantum dot graphene nanocomposite film for UV photodetection, *ACS Appl. Nano Mater.* 2 (2019) 3934–3942.

[24] V. Mišeikis, S. Marconi, M.A. Giambra, A. Montanaro, L. Martini, F. Fabbri, S. Pezzini, G. Piccinini, S. Forti, B. Terrés, I. Goykhman, L. Hamidouche, P. Legagneux, V. Sorianello, A.C. Ferrari, F.H.L. Koppens, M. Romagnoli, C. Coletti, Ultrafast, Zero-Bias, Graphene photodetectors with polymeric gate dielectric on passive photonic waveguides, *ACS Nano.* 14 (2020) 11190–11204.

[25] L. Clinckemalie, D. Valli, M.B.J. Roeffaers, J. Hofkens, B. Pradhan, E. Debroye, Challenges and opportunities for CsPbBr3 perovskites in low- and high-energy radiation detection, *ACS Energy Lett.* 6 (2021) 1290–1314.

[26] X. Li, D. Yu, F. Cao, Y. Gu, Y. Wei, Y. Wu, J. Song, H. Zeng, Healing all-inorganic perovskite films via recyclable dissolution–recyrstallization for compact and smooth carrier channels of optoelectronic devices with high stability, *Adv. Funct. Mater.* 26 (2016) 5903–5912.

[27] Z. Ji, G. Cen, C. Su, Y. Liu, Z. Zhao, C. Zhao, W. Mai, All-inorganic perovskite photodetectors with ultrabroad linear dynamic range for weak-light imaging applications, *Adv. Opt. Mater.* 8 (2020) 2001436.

[28] Y.-H. Chang, W. Zhang, Y. Zhu, Y. Han, J. Pu, J.-K. Chang, W.-T. Hsu, J.-K. Huang, C.-L. Hsu, M.-H. Chiu, T. Takenobu, H. Li, C.-I. Wu, W.-H. Chang, A.T.S. Wee, L.-J. Li, Monolayer MoSe2 grown by chemical vapor deposition for fast photodetection, *ACS Nano.* 8 (2014) 8582–8590.

CHAPTER 17

Nanostructured Perovskites for Light-Emitting Diodes

Felipe M. de Souza and Ram K. Gupta

Pittsburg State University

CONTENTS

17.1 INTRODUCTION

The LEDs are components made of semiconductor elements that function based on the phenomenon of electroluminescence (EL), which occurs when there is an emission of light when the material is exposed to an electric field. Likewise, in a p–n junction, a light-emitting diode (LED) emits light when under voltage. In that sense, the LED's structure is usually based on inorganic semiconductors with crystalline or polycrystalline morphology containing p-type and n-type of materials. Hence, when the electric field is applied there is a recombination of the electrons (e^-) and holes (h^+) from the n-type and p-type materials, respectively, leading to the emission of a photon with the same energy as the bandgap [1]. The mechanism of this process is presented in Figure 17.1.

The effect of materials emitting radiation when exposed to an electric field was first observed for GaAs, AlGaAs along with AlAs which can radiate infrared (IR) and near IR (NIR) radiations under an electric field. This phenomenon took place due to their high electron mobility along with small bandgap. Despite their satisfactory EL properties, the fabrication of LEDs can be expensive due to the requirement of systems under very low pressure and high temperatures, which implies a higher cost. As an alternative to that, another class of emitting diodes that appeared were the organic LEDs (OLEDs), and quantum LEDs (QLEDs). They function similarly to traditional LEDs. However, OLEDs may present lower thermal stability as well as luminescence. On the other hand, QLEDS present

DOI: 10.1201/9781003185109-17

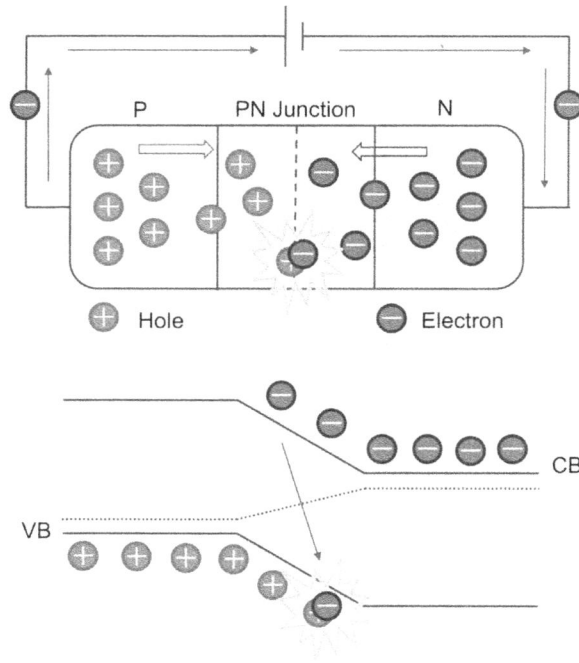

FIGURE 17.1   Diagram of a p-n junction that promotes the EL in LEDs. (Reproduced with permission [1]. Copyright (2021), American Chemical Society.)

more attractive properties when compared to OLEDs which include a luminescence with higher intensity, and it can be synthesized through solution-based approaches which are cheaper and relatively more efficient. Some of the most common nanomaterials used for QLEDS are CdSe, PbS, and InP, at which the quantum dots of the core-shell structure of CdSe/ZnSe demonstrated satisfactory properties in terms of external quantum efficiency (EQE). Despite the advantageous properties of quantum dots, one of their inherent issues lies in their toxicity and environmental issues due to the presence of heavy metals such as Cd or Pb, for instance. One of the attempts to address those challenges led to the development of InP and $CuInS_2$. However, these nanomaterials are intolerant towards defects, which hinders their large-scale application now. Yet, another class of material that appears as a viable alternative to these materials is the metal halide perovskite LEDs. These materials are semiconductors which is the general formula of ABX_3 at which A can be monovalent cations located at the corner of the cubic cell, B are divalent cations located at the crystal's body center and X can be halides that are located at the center of the cubic cell. An example of a $BaTiO_3$ perovskite cell structure is displayed in Figure 17.2 [2].

The inherent properties of metal halide perovskite are attractive for the development of LEDs which include high photoluminescence quantum yield (PLQY), long diffusion lengths for the excitons, and high absorption coefficient. On top of that, perovskites possess a broad range of tunability as changing the A or B components can lead to materials that can cover the whole visible and NIR spectra [3,4]. An example of these properties can

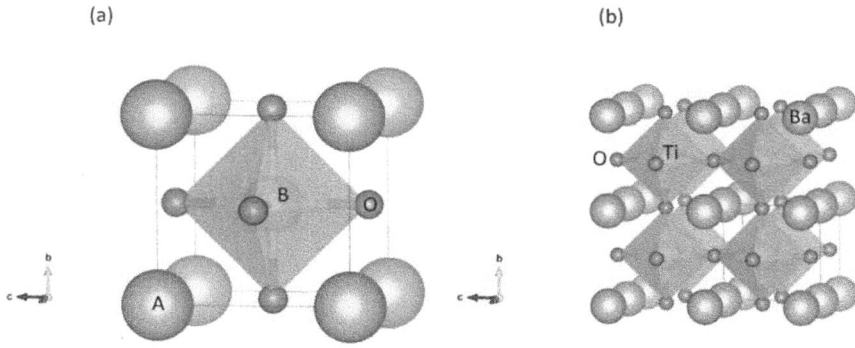

FIGURE 17.2 Stick and ball structure of BaTiO₃ perovskite cell displayed as (a) Unit cell of an ideal cubic perovskite and (b) Supercell structure. (Reproduced with permission [2]. Copyright (2021), The Authors, Some rights reserved. Licensee MDPI, Basel, Switzerland. This article is licensed under a Creative Commons Attribution 4.0 International License.)

be observed through $CsPbCl_3$, $CsPbBr_3$, and $CsPbI_3$ which have emissions at the violet, green, and NIR [5,6]. To study the properties and efficiency of a LED there are several parameters such as EQE, power efficiency (lm/W), turn-on voltage (V_t), thermal and/or color stability, and current efficiency (cd/A). In addition to that, the perovskite NIR LEDs can receive three structural classifications which are 3D, nanocrystalline, and layered. The 3D perovskite can be fabricated into a film through an in-situ process which usually displays a relatively high conductibility. Nanocrystalline perovskites can be synthesized either through hot-injection or ligand-assisted reprecipitation approaches which leads to nanostructures with satisfactory quantum confinement properties and high stability with the cost of low conductibility. Layered perovskite, likewise, 3D perovskite, can also be synthesized through in-situ thin film. In that regard, there are two phases that quasi-2D perovskite structures that can be obtained which are named as Ruddlesden-Popper (RP) and Dion-Jacobson (DJ) which can present optimized properties in terms of conductibility, quantum confinement, and stability [7].

17.2 METHODS TO SYNTHESIZE PEROVSKITES

Perovskite materials can be synthesized in several forms such as polycrystals, single crystals, thin films, or powders. Also, they can be generally divided into two categories which are the all-inorganic perovskite with the ABX_3 general formula, as previously mentioned, and the organic-inorganic hybrid perovskites which are composed of a cationic organic segment, a divalent metal, and a halide. Based on that, the all-inorganic perovskites nanocrystals class has been explored as means to serve as a more stable nanomaterial when compared to organic-inorganic hybrid perovskite. In that sense, the methylamine cations that have a diameter of 217 pm are replaced with Cs cations that have a smaller diameter of 167 pm, which can lead to $CsPbX_3$ structures with higher phase and photostability. Yet, the smaller size for the cation makes the cubic phase of $CsPbI_3$ unstable when compared to the orthorhombic phase. To address that, amine ligands can be used as they act as a template enabling the perovskite to maintain its structure under room temperature. Within that line, approaches to synthesize perovskites based on cesium lead halides ($CsPbX_3$) have been developed through the hot injection approach.

For that, Kovalenko et al. [8] used cesium-based salts that were converted into cesium oleate and placed to react with lead halide in octadecylene (ODE) at a temperature between 140 and 200°C. In addition, stoichiometrically equal quantities of oleic acid (OA) and oleic acetate (OAc) were added to stabilize the Pb precursors in the system. It has been observed that this approach yielded a better phase purity when the reaction occurred at higher temperatures, which led to a predominance of cubic phase [6]. Alongside that, the nanocrystal's size varied from 5 to 14 nm depending on the temperature. In addition, the $CsPbX_3$ also presented satisfactory properties by presenting a quantum yield with the highest of 90% along with a narrow line width emission that was in between 12 and 42 nm. In another approach performed by Zeng et al. [9] nanocrystals of a $CsPbX_3$ perovskite were synthesized through a facile method of a mixed solution of CsX and PbX in dimethyl sulfoxide (DMSO) or N, N-dimethylformamide (DMF) under room temperature. Then, OA and OAc were added to stabilize the system. This method proposed a feasible approach for the synthesis of all-inorganic perovskites as the nanocrystals were formed around 10 s after the precursor's solution was injected into toluene. On top of that, the reaction was performed at room temperature and displayed a PLQY with the highest of 95%, making it comparable with similar approaches performed at higher temperatures.

Based on these examples, it is noticeable that facile synthetical approaches can be used to obtain all-inorganic perovskites while allowing control of the nanomaterial's size which can lead to variation in the optical properties. This reaction control was explored by Akkerman et al. [10] which synthesized $CsPbBr_3$ nanoplatelets by using Cs-oleate as a precursor along with OA and oleylamine used as ligands. Then, HBr was added to a solution of Cs-oleate precursor which contained Cs_2CO_3 with 1-octadecene and OA. After that, a solution of $PbBr_2$ and DMF was added. The system turned into a turbid solution after 10 s. Lastly, acetone was added to finish the reaction which caused the solution's color to become greenish. After precipitating the nanoplatelets under centrifugation and redispersing them into toluene the solution's color changes to blue. The authors were able to define the nanoplatelets thickness based on the concentration of HBr as higher amounts led to thinner films. On top of that, an anion exchange process could also be performed by adding the precursors of the desired halides into a dispersion containing the nanoplatelets that were vigorously shaken to perform the anion exchange process. Yet, it is worth noting that the layered properties were improved when compared to the bulky counterparts of $CsPbBr_3$. In addition, films that were 3–5 monolayers thick were obtained through the spin coating technique containing a precursor solution of $PbBr_2$ and CsBr in DMSO. During the measurement of optical bandgap, it was observed that the number of monolayers caused an increase in the excitonic transitions when thinner monolayers were obtained along with a shift toward smaller wavelengths (blue shifts), which was attributed to quantum confinement effects.

As discussed, the synthesis of these inorganic perovskite LEDs can be performed in a facile manner. However, one of the concerns lies in the use of Pb-based materials which hold environmental concerns and high toxicity. An alternative that has been researched in that regard lies in the use of Sn-based materials that present lower toxicity along with a broad spectral response in emission in the NIR. One example of that was performed by Jellicoe et al. [11] that synthesized $CsSnX_3$ perovskite nanocrystals where X=I, Br, Cl, $Br_{0.5}I_{0.5}$, $Cl_{0.5}Br_{0.5}$. It is worth noting that Sn-based perovskite cannot be synthesized in the same manner as Pb-based as it led to the precipitation of CsX salts [12]. To avoid that, the Sn precursor was dissolved in tri-n-octylphosphine, which acted as a coordinating and reducing solvent. Then, this solution was added into another solution containing Cs_2CO_3 precursor along with OA and oleylamine at 170°C to obtain stable $CsSnX_3$ nanocrystals. The optical properties of $CsSnX_3$ could be tuned along with visible and NIR spectral region which was performed by reducing particle size to enable quantum confinement and incorporating different halides in the perovskite's structure. The latter process could be performed also through an anion exchange process. Within that line, two luminescent decay channels were observed and attributed to slow radiative e^-–h^+ recombination at defect sites and fast band-to-band emission, which were concluded after analyzing both spectral and temporal evolution of the Sn-based perovskite photoluminescence effect.

Another type of halide perovskite aside from the all-inorganic is the organic-inorganic hybrids. The synthesis of this type of perovskite has been performed by Pérez-Prieto et al. [13] which consisted of the addition of a lead bromide precursor, CH_3NH_3Br, into a solution of ODE, OAc, and OA. After a short time, the solution acquired a yellow color, and its particles were precipitated after the addition of acetone. The presence of acids and ammonium moieties aid in the stability of the perovskite quantum dots by preventing their precipitation. In addition, the 6 nm size nanoparticles presented a quantum yield increase up to 80% when the molar ratio of ammonium, as OA+MA and $PbBr_2$ were 4:1. Such effect was likely due to the more compact passivation layer of capping ligand's excess. However, phase impurities were observed, which led to a decrease in crystallinity. Also, organic-inorganic perovskite is generally unstable at a high temperature which requires them to be prepared in temperatures lower than 80°C. Yet, synthesis at higher temperatures has been performed for the synthesis of these nanomaterials that were performed through an in-situ process by using a perovskite precursor such as CsI or PbI_2, methylammonium iodide (MAI) and, formamidinium iodide (FAI) which can be dissolved in organic polar solvents such as DMSO or DMF under inert atmosphere. Then, the solution containing the precursors can be spin-coated and followed by an annealing process at around 100°C. Through that, grain sizes can range from 100 up to 1,500 nm. Another facile approach involves preparing a solution of $FAPbI_3$ with γ-butyrolactone (γBL) that is submitted to an oil-bath for 3 h at 100–105°C [14]. These few examples demonstrate that perovskite LED can be obtained through relatively facile solution-based methods which can provide a feasible way to produce them on a larger scale as the processing techniques become simpler and more efficient. However, some requirements must be met such as efficient light generation through proper tuning of the e^-–h^+ to allow a balance of charge carriers that are injected into the perovskite's active layer.

Another challenge commonly faced during the synthesis of organic-inorganic perovskites is the poor solubility of precursors in ODE. For that, Zhong et al. [15] performed a ligand-assisted reprecipitation (LARP) approach to synthesize $CH_3NH_3PbX_3$ which procedure is described in Figure 17.3. This method consisted of the use of polar solvents such as N,

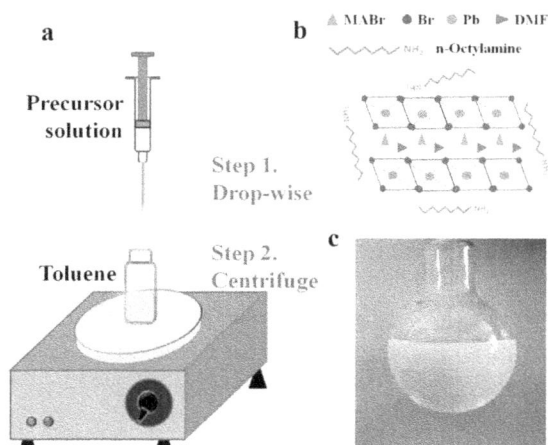

FIGURE 17.3 Scheme for (a) The LARP synthetical approach. (b) Precursor solution containing the starting materials. (c) Photocopy of the $CH_3NH_3PbBr_3$ perovskite nanocrystals dispersion. (Reproduced with permission [15]. Copyright (2015), American Chemical Society.)

N-dimethylformamide instead of ODE. Then, the clear solution containing the precursors was added into a toluene and acetone solvent system and stirred vigorously to form the nanocrystals. It was observed that the nanocrystal's size was dependent on the size of the organic chain ligand which led to 3.3 nm of diameter along with 70% of quantum yield. It was also observed that changing the capping ligand's and methylammonium's molar ratio without regarding the total amount of organic parts directed the formation of layered structures.

Huang et al. [16] demonstrated that perovskite nanocrystals could be precipitated with the addition of either tert-butanol or acetone as it caused a demulsification process. In that sense, the number of demulsification agents could influence the size of the nanocrystals from 2 to 8 nm. This approach deems versatile as it can be applied for both organic-inorganic as well as all-inorganic perovskite as for $CH_3NH_3PbX_3$ it yielded a PLQY of 92% whereas for $CsPbBr_3$ quantum dots yielded 60%. The scheme for the emulsion synthesis is presented in Figure 17.4.

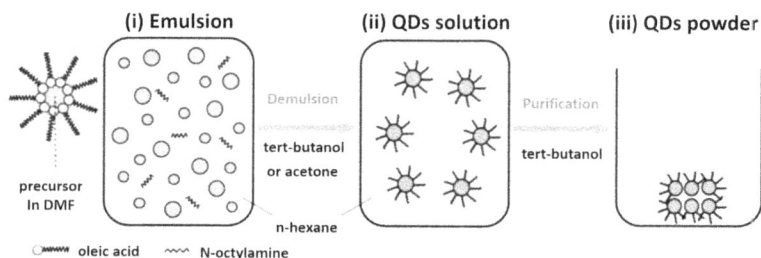

FIGURE 17.4 Scheme for the emulsion synthetic approach to obtain $CH_3NH_3PbBr_3$ organic-inorganic perovskite quantum dots. (Adapted with permission [16]. Copyright (2015), American Chemical Society.)

17.3 WORKING PRINCIPLE OF LEDs

Light emitting diodes (LEDs) are semiconductor-based devices that can emit light when an electric field is applied to them. For that, an n and p-type junction are necessary for their function. The phenomenon of light emission occurs because the application of a forward bias causes the e^- from the n-type component and the h^+ from the p-type materials to recombine which leads to the emission of a photon. In addition, the light color is based on the E_g of the semiconductor material. The practical uses of this technology culminated in the development of LED back in 1960 which are currently being updated to white LEDs (WLEDs), which are composed of traditional LEDs, mostly GaN, coated with phosphor emitters. The WLEDs have attractive properties such as long usage term, high efficiency, and small size [17,18]. Despite these advantages, GaN has some drawbacks such as production through epitaxy-based techniques performed over a sapphire wafer substrate which leads to a considerable increase in cost. Also, GaN requires a guiding plate since it can shine a light in all directions. Finally, its lack of flexibility can prevent some other fields of applications. Hence, OLEDs, QLEDs, and perovskite LEDs may provide an answer to these issues. Based on these aspects, the metal halide perovskites emerged as compelling active photoelectronic nanostructures for LEDs as they can be synthesized through solution-based approaches and can emit high-intensity lights at room temperature along with its relatively lower cost in terms of starting materials. The perovskite metal halides can be obtained in several nanostructures such as 3D, 2D, quasi-2D, and quantum dots (0D).

As previously mentioned, the 3D metal halide perovskite, as well as quantum dots, can be represented as ABX_3. On the other hand, 2D and quasi-2D metal halide perovskite can be shown are $C_mA_{n-1}B_nX_{3n+1}$ at which C can be a mono ($m=1$) or divalent ($m=2$) large-sized cation [19,20]. Within this line, a quasi-2D type of structure stays in between a 3D and 2D structure as it has a deductible thickness value. Based on that, a quasi-2D perovskite structure can exist in two different phases which are either RP or DJ. The main difference between these two phases is related to their intermolecular interaction. In that sense, RP perovskites interact through van der Waals attraction whereas DJ interacts through hydrogen bonding. The principle of quantum confinement according to the nanocrystal's size is also applied to both RP and DJ phases [7].

The objective of reducing the size of its nanocrystals is to enhance the quantum confinement effect as this property only appears in sufficiently small nanostructures, meaning that as the nanomaterial's size approaches that of an exciton's Bohr radius its optical properties enhance. In this sense, it promotes the entrapment of e^-–h^+ (excitons) throughout the dimensions. Hence, when the exciton is confined in one dimension it can move in one plane, which is named quantum wells. When confined in two dimensions the excitons can move along a line, also named quantum wire. Lastly, the exciton can be confined in all three dimensions, which prevents its movement in all directions [21]. As the length of the confined space decreases there is an increase in interaction between the e^-–h^+ which creates an additional electrostatic force that leads to an increase in E_g. Based on that, the nanocrystal perovskite's optical properties can be tuned [22,23]. In that sense, the higher the confinement degree, the larger the E_g, and emissions with a blue shifting (smaller wavelengths) are observed. Other phenomena that take place in LEDs are radiation and non-radiation recombination. To analyze that process, one should first note that within a semiconductor there are a number of e^- and h^+ that are in equilibrium. When the system within the semiconductor is in a non-equilibrium state there are two mechanisms at which the recombination of e^- and h^+ can take place. In one way the excess of energy in an electron can be released in the form of a photon, which is the desirable mechanism for LEDs named radiative recombination.

In the other way, the electron's excess of energy can also be transferred to the structure's lattice to increase its vibration or be given to the charge carriers to increase their kinetic energy named non-radiative recombination. The latter process is known as Auger recombination, which is non-desirable for LEDs since there is no emission of a photon. The radiation recombination can be increased in a few ways such as transforming a p-n junction from a homojunction to a heterojunction which leads to the creation of an active region that has a higher tendency to emit photons rather than absorb them. This effect is also enhanced when the active region's thickness gets close to the Broglie wavelength. Yet, despite the considerable progress and structural variety that surrounds the perovskite nanocrystals, there are some drawbacks that the scientific community is currently facing. One of those that hinder the implementation of perovskite LEDs in the market at the moment is based on their low lifespan when compared to the traditional LEDs. In that sense, it has been reported that perovskite LEDs can lose half of their EQE or initial luminance (T50) after 694 h. On the other hand, both QLED and OLED can last more than 1 million h [24,25]. It has also been observed that red LEDs last longer than green LEDs and blue LEDs present the lowest lifespan. The occurrence of this effect is likely due to the wider E_g of the semiconductor's LED which requires an operation at higher biased voltage. Also, because of the higher energy demand, the device can operate at a higher temperature.

In addition, a wider E_g facilitates the charge carrier accumulation at the interface of the semiconductor which can lead to degradation [26,27]. The color emission of perovskite LEDs is also influenced by the halide that composes its structure. Hence, a considerable range of emissions within the visible and NIR range can be obtained. Yet, the presence of different halides can lead to phase separation causing instability in the light emission [28]. Despite that, there has been a considerable improvement in the perovskite LEDs, mostly on the NIR range, which started with an efficiency of 0.76 and went up to 21% within less than 10 years of research. Thus, even though there are still some issues related to stability the perovskite LEDs show great potential for large-scale applications [29,30].

17.4 PEROVSKITES FOR LEDs

This session covers some of the current studies in the literature regarding LED perovskite by discussing its optoelectronic properties and the tunability strategies that can be employed during the experiment design. Based on that, Protesescu et al. [12] synthesized perovskite nanocrystals of $CsPbX_3$ where X = I, Br, and Cl which covered a broad color emission range within the visible spectrum along with the high intensity of light emission. One of the initial challenges with all-inorganic perovskite is their decreased solubility in commonly used solvents when compared to organic-inorganic perovskite metal halides ($MAPbX_3$). Despite that, the authors were able to obtain colloidal nanomaterials with the size of 4–15 nm $CsPbX_3$ with the geometry of cube and cubic perovskite. By varying the nanocrystal's size, the E_g could be tuned which allowed wavelength emission through 410–700 nm (visible spectral region). The photoluminescence of $CsPbX_3$ was characterized by emission line widths in a narrow range between 12 and 42 nm along with quantum yields from 50 and up to 90% and 1–29 ns of radiative lifetimes. The synthesis consisted of a controlled precipitation process of the ions involved (X^-, Pb^{2+}, and Cs^+) in the system. For that, the metallic precursors of PbX_2 and Cs-oleate reacted in octadecene, a solvent with a high boiling point, at around 140–200°C. It was observed that the reaction took place in between 1 and 3 s and through that, the nanocrystal size could be tuned. Also, the high-temperature synthesis worked as a driving force to obtain $CsPbX_3$ in the cubic phase.

In that regard, it is worth mentioning that obtaining a mono and stable phase system are important factors to maintain the optical-electronic properties. That effect correlates with nanocrystals of $CsPbI_3$ that were synthesized through the solution method at 305°C which yielded cubic phase nanocrystals with 100–200 nm. That causes the material to have an E_g of 1.74 eV (714 nm), which demonstrates the relevance of particle size to obtain a stable phase that leads to appreciable optical properties. Based on that, when the solution method synthesis of $CsPbI_3$ was performed around 140–200°C to obtain 4–15 nm nanocrystals there was a higher E_g which was attributed to the quantum confinement effect. In addition, the nanocrystals remained stable for months before they were recrystallized. The property analysis of these perovskite $CsPbX_3$ nanocrystals is presented in Figure 17.5.

As discussed, the variation of nanocrystal size through temperature was one of the parameters that promoted a larger variation of wavelength. Alongside that, varying the ratio of halides in terms of Br/I and Cl/I also increases the range of colors emitted by the $CsPbX_3$ nanocrystals. Another aspect that attracts the research of $CsPbX_3$ lies in their

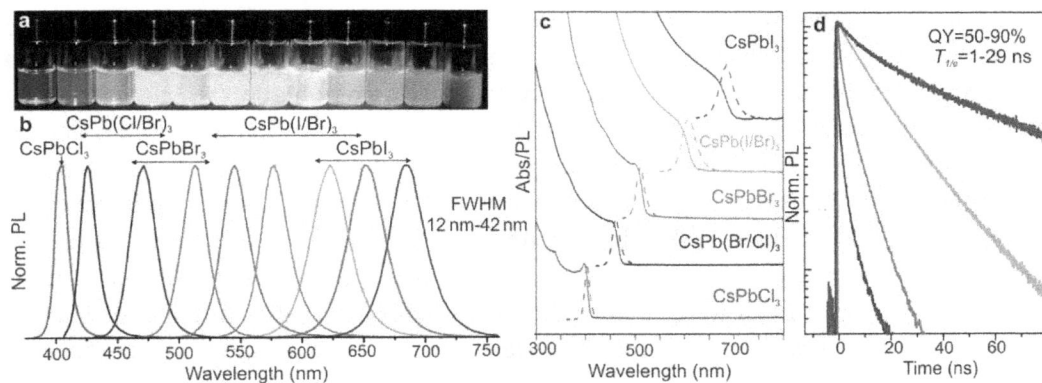

FIGURE 17.5 (a) Dispersion of $CsPbX_3$ in toluene and exposed to a UV lamp with a wavelength of 365 nm. (b) Photoluminescence spectra with a full width at half maximum (FWHM) from 12 to 42 nm and (c) Absorption in function of wavelength for the $CsPbX_3$. (d) Time-based photoluminescence decay for the samples presented in (c) With exception of $CsPbCl_3$. (Reproduced with permission [12]. Copyright (2015), American Chemical Society.)

high ionic character which makes them stoichiometrically organized even though Pb^{2+} and Cs^+ cations present considerably different radii. This level of organization is not necessarily observed in some quaternary metal chalcogenides such as $CuZnSnS_2$ for instance, as they can have a considerable disordered and heterogeneous distribution of ions along with their structure. Such effect is likely attributed to the ionic sites which are mostly tetrahedral. Also, the smaller degree of stoichiometric organization can lead to more defects in the structure, such as interstices and vacancies, leading to a variation in acceptors and donors state within the E_g. Such effects can, at some point, cause weakening and broadening of light emissions among other effects [31]. Yet, within the same line of comparison between $CsPbX_3$ and metal chalcogenides, it was observed that the former obtained quantum yield values that were comparable to the latter (50%–90%). In addition to that, the $CsPbX_3$ provided a wide range of pure colors which suggested promising applications for commercial displays. Aside from these properties, the perovskite nanocrystal should also present appreciable miscibility with both inorganic and organic matrixes to increase its applicability. For that, the nanocrystals can be incorporated into a polymeric matrix such as poly-(methylmetacrylate) (PMMA). It was noted that only the emission intensity remained high but also led to an increase in the photopolymerization rate performed at the monomer, methylmetacrylate. Such an effect could be caused by the reabsorption of the perovskite's emission by the photoinitiator itself which could perhaps increase the population of free radical species that could promote the photopolymerization. The color emission scale and the luminescence effect of $CsPbX_3$ and its incorporation into the PMMA matrix are presented in Figures 17.6a and b, respectively.

Enabling facile ways to synthesize materials for high-end applications such as LEDs is one of the key factors that can allow their applications in large-scale processing. This concept was explored by Li et al. [9] who was able to synthesize $CsPbX_3$ perovskite nanocrystals at room temperature, without the requirement of inert gases or injection process. On top of that, the reaction took place in a few seconds while yielding considerably high quantum yields in the range of 70%, 80%, and 95% along with FWHM of 18, 35, and 20 nm for the blue, red, and green emissions, respectively. The nanocrystals also presented a high generation of excitons along with satisfactory radiative recombination which was related to the quantum well band alignment of the $CsPbX_3$, self-passivation of halogen, and its exciton binding energy of 40 meV. Under the scope of these satisfactory properties is worth noting the synthetical process which consisted of the dissolution of ion sources based on CsX and PbX_2 into DMF or DMSO. In the same vessel, oleylamine and OA were added to function as surface ligands. An important aspect of this approach is that the concentrations of the ion precursors were smaller than their saturation point in DMF or DMSO. Hence, to promote a rapid crystallization process, toluene was added to the solution. Both ion precursors have extremely poor solubility in toluene. In addition, toluene itself is soluble in DMF and DMSO. This solubility discrepancy led to a drastic drop in solubility of the ion precursors causing a supersaturated media that induced fast recrystallization which is represented in Equation 17.1.

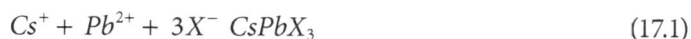

$$Cs^+ + Pb^{2+} + 3X^- \quad CsPbX_3 \tag{17.1}$$

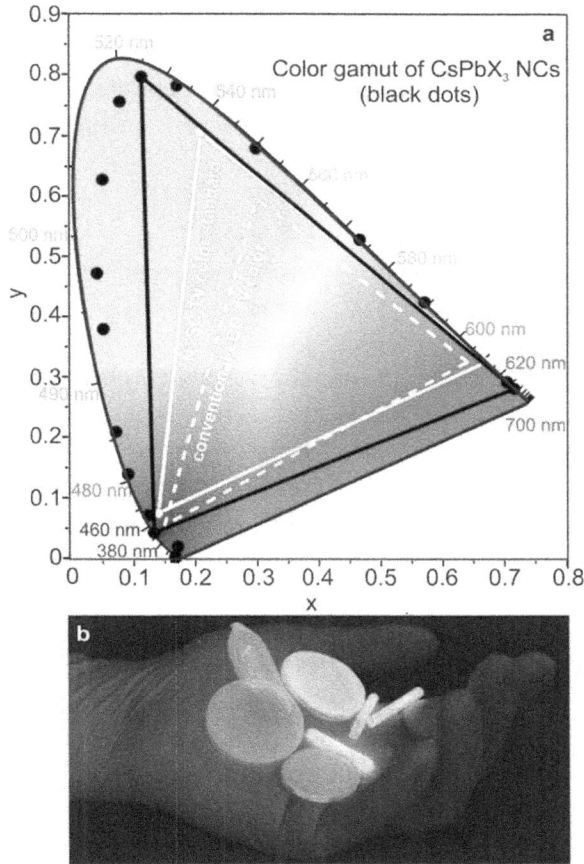

FIGURE 17.6 (a) Color emission of the CsPbX$_3$ (black dots) and compared to both NTSC (white solid triangle) and LCD TV (white dashed triangle). (b) Luminescence effect of CsPbX$_3$ in PMMA exposed to a 365 nm wavelength. (Adapted with permission [12]. Copyright (2015), American Chemical Society.)

The reaction's system was kept under stirring and both oleylamine and OA aid on the nanocrystal's dispersity by promoting surface functionalization. This facile approach for the synthesis of the inorganic perovskite quantum dots suggested an alternative to cadmium-based quantum dots. In addition to that, there was an improvement in the overall luminescence as the perovskite quantum dots presented a narrower linewidth as well as higher quantum efficiency when compared to cadmium quantum dots. Similarly, to the previous example, the variation of halide concentration led to different optical properties, yet without causing phase separation. In that sense, it was observed the higher content of higher concentrations of Br$^-$ in the CsPbX$_3$ led to a characteristic green color, whereas a higher concentration of I$^-$ led to color shifts at longer wavelengths and for Cl$^-$ at shorter ones. A thermal study was also performed to understand the variance in optical properties. Through that, it was noted that as temperature increased there was a blueshift of photoluminescence peaks. Such behavior was similar to PbS quantum dots and organic-inorganic perovskite such as $CH_3NH_3PbX_3$, however not yet fully investigated [9]. Simultaneous to that effect there was also a consistent decrease in the photoluminescence intensity. Based on this study the binding energy of the exciton was found to be 40 meV based on Equation 17.2 at which E_b is the exciton binding energy, k_b is the Boltzmann constant and I_0 is the intensity at 0 K.

$$I(T) = \frac{I_0}{1 + Ae^{-\frac{E_b}{k_b}/T}} \qquad (17.2)$$

Such values of exciton binding energy were considerably higher than thermal disturbance energy at room temperature of around 26 meV as well as that of traditional LEDs such as GaN with 25 meV. Based on these results it is inferred that $CsPbX_3$ can generate excitons at room temperature along with high e^-–h^+ rate recombination [32]. One strategy that is often performed on the synthesis of cadmium-based quantum dots lies in its surface functionalization to prevent excited electrons to get trapped because of surface defects. This process can be performed by coating the quantum dots' surface with a proper ligand such as the case of CdSe core with ZnS shell structure. For this case, this effect can be performed by increasing the amount of halogens ratio in the structure. Within that line, Br^-, for instance, can play a similar role as it can interact with cationic species to avoid electron trapping. Hence, it can improve the efficiency of the excited carrier, ultimately leading to an enhancement of photoluminescence and quantum yield. Such an effect can be named self-passivation [33]. In addition to that, the Br over the perovskite's surface should be Cs leading to the formation of $PbBr_x$ analogs that possess a large E_g of around 4 eV. Through that, a core-shell structure of CsPbBr@Br quantum dots is formed [34,35].

The organic-inorganic hybrid perovskite is another interesting class of nanocrystals that can serve as an alternative to the decrease in the use of scarce metals such as Cs by replacing it with an organic cationic group. In addition, this type of perovskite can be usually synthesized under milder conditions when compared to most inorganic perovskite. Despite these factors, there are several challenges on the line such as improving their thermal and colloidal stability, along with reaching monodispersity and proper control of morphology. As means to propose an approach to fix some of these issues, Patra et al. [36] synthesized formamidinium lead bromide ($FAPbBr_3$) which by varying the concentration of the A^+ site from ABX_3-type of structure led to the formation of quasi-spherical, dodecahedrons, and monodispersed cubes. Such a process was achieved partially by using a lower amount of FA^+ against Pb^{2+}. Also, the use of phenacyl bromide to function as both halide and ligand along with mild temperatures of 50°C were an important factor for the synthesis' success. The schematics for the approach employed in the study are demonstrated in Figure 17.7. It was observed that synthesizing $FaPbBr_3$ using lower concentrations of FA^+ in comparison to Pb^{2+} played a major role in terms of controlling the nanocrystal facet. In addition, the decrease in temperature led to the nanocrystal's size becoming larger reaching around 20 nm. This approach promoted some degree of control over the organic-inorganic perovskite along with satisfactory optical properties such as PLQY close to 100% in all the

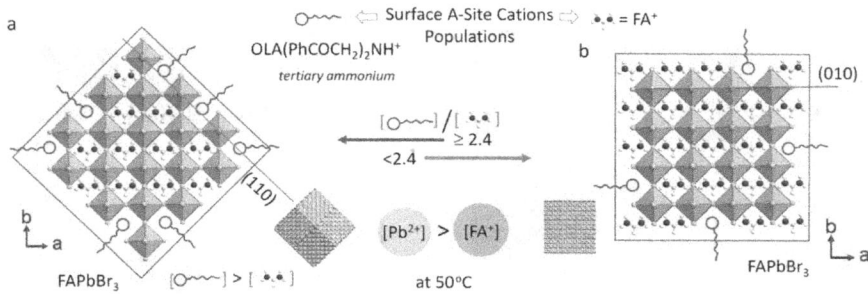

FIGURE 17.7 Scheme for the variation of FA$^+$ concentration and the influence of tertiary ammonium ions derived from the reaction between oleylamine and phenacyl bromide leading to different shapes of the FAPbBr$_3$ nanocrystals. Atomic model of (a) Suggested (110) group of facets with a predominance of tertiary ammonium ions bonded at the FAPbBr$_3$ dodecahedron nanocrystals and (b) Suggested (010) group facets with a predominance of FA$^+$. Each octahedral structure represents a unit of PbBr$_6$$^{4-}$. (Reproduced with permission [36]. Copyright (2021), American Chemical Society.)

obtained shapes. Yet, both temperature and mostly concentration of organic cation seemed to have played a major role in the nanocrystals morphology.

Another important aspect regarding LED perovskite is that a considerable amount of Pb is employed for their synthesis which implies the risk of contamination with heavy metals that can lead to bioaccumulation along with environmental concerns. Hence, it deems important to develop an alternative route that can avoid the use of Pb. One of the options proposed by the latest literature relies on the replacement of Pb with Sn. Based on that, Sn-based perovskite can present narrower E_g values when compared to Pb analogs, along with long lengths of carrier diffusion, and low exciton binding energy [37,38]. However, the Sn-based perovskite turned out to not be stable in air, making the requirement of use of inert gas to perform the synthesis. Such effect was attributed to the oxidation states of Sn^{2+}/Sn^{4+} that collapsed the nanocrystal's structure [39–41]. To address that it has been proposed the use of Sn^{4+} as its highest oxidation state would be more stable against O$_2$. Based on that, relatively stable Cs$_2$SnI$_6$ has been synthesized as a more stable counterpart when compared to CsSnI$_3$ for instance [42,43]. Within this line, Wang et al. [44] synthesized perovskite nanocages of CsSnBr$_3$. The synthetical approach was based on a facile hot-injection method by using ODE as a solvent along with oleylamine and OA as ligands. Yet, tri-n-octylphosphine, a stronger coordinating solvent, was also required to properly dissolve and coordinate with SnO. The source of Sn was obtained through stannous 2-ehtylhexanoate, which is less toxic along being a lower cost option compared to trioctylphosphine SnBr$_2$ (TOP-SnBr$_2$). For the source of bromide, MgBr$_2$ was used as a simpler option in terms of procedure when compared to the use of halogenated amines. It is worth noting that even though Mg^{2+} would be present in reaction media it would not be incorporated into the perovskite's structure, based on the tolerance factor theory which describes that Mg^{2+} would not be able to penetrate the nanocrystals, as it acts as a cation that enables the transport of Br$^-$ into the crystal. The authors confirmed the absence of

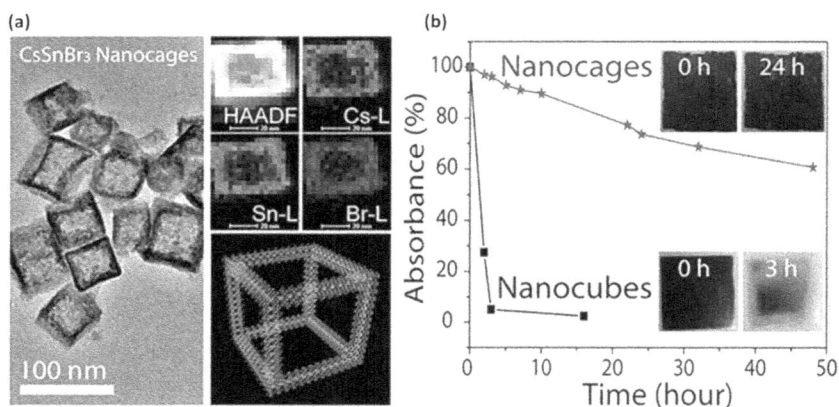

FIGURE 17.8   (a) TEM for the $CsSnBr_3$ nanocages along with EDS mapping with Scanning TEM-High Angle Annular Dark Field (STEM-HAADF) which corresponds to the Cs, Sn, Br, and elemental maps, respectively. (b) Stability study based on the change in absorbance over time while comparing $CsSnBr_3$ nanocages and nanocubes. (Adapted with permission [44]. Copyright (2017), American Chemical Society.)

Mg through X-ray spectroscopy (EDS) and X-ray photoemission spectroscopy. A solution was prepared with these compounds and Cs oleate was added, leading to the formation of a dark red solution within a few seconds which points to the formation of $CsSnBr_3$. It was found that an increase in reaction temperature led to an increase in crystal size, hence allowing some degree of tunability.

As mentioned, Sn-based perovskite can deteriorate quickly due to its susceptibility to react with O_2 and moisture. Yet, the presence of Sn^{2+} may be required to form the crystal. To work around this situation an SnF_2 additive can be incorporated into the system as it can suppress the conversion of Sn^{2+} into Sn^{4+}, thus improving stability. In that sense, it was suggested that the presence of F^- acts as a stronger electron withdrawer when compared to Br^- leading to a stronger bonding with Sn^{2+}, hence preventing its oxidation [45,46]. Based on this principle, perfluorooctanoic acid was employed to stabilize the Sn^{2+}. Through that, not only the perovskite nanocrystals provided higher stability when exposed to air, but also maintained their inherent optical properties. Throughout this discussion, the $CsSnBr_3$ perovskite nanocage structures are demonstrated through TEM along with its stability in terms of absorption in function of time as can be seen in Figure 17.8. Thus, despite the remaining challenge in improving Sn-based perovskite's stability there has been considerable progress into using this type of material as optic electronic components for devices in the future as they can propose a safer and likely cheaper alternative.

17.5 CONCLUSION

Throughout the concepts and discussions provided in this chapter, it was notable that there has been considerable progress within the field of developing nanocrystal perovskites for LEDs and optic electronic devices alike. Based on that, one of the core aspects remains on the size-related properties to promote quantum confinement as this effect plays a major role in the electronic properties of the nanomaterial by generating excitons that can be recombined to emit photons. Alongside that effect, inducing the quantum confinement within the desired dimension is also an important aspect for the optimization of properties. The processing is also an important aspect taken into consideration to place material into the market. Based on that, the perovskite LEDs can be synthesized through considerably facile and fast procedures based on solution techniques that can be performed at room temperature without the need to immerse in inert gas. Also, reaction times can be as short as 10 s. These aspects are attractive points for the research and manufacturing of these materials as they can be widely and quickly produced. On the other hand, traditional LEDs require dry-based and high-temperature techniques, which are relatively more costly due to specific instrumentation along with proper control of reaction conditions. Another attractive feature of perovskite LEDs lies in their facile tunability as optical properties can be changed with a small change on the procedure's parameters such as temperature, time, precursors concentrations and so, which enables a myriad of materials that can be synthesized under similar conditions. Yet, there must be an effort input to improve the stability of perovskite LEDs as they are not yet able to compete with traditional LEDs. Another important aspect is that both LED and perovskite LED require relatively scarce and toxic chemicals the former concerning Cd-based nanomaterials and the former Pb-based materials. Within this line, there has been considerable progress in both areas as OLEDs employ the use of organic groups which can be cheaper and lower in cost alongside organic-inorganic hybrid perovskite that offers a different type of material that requires a smaller amount of metallic non-renewable resources. On top of that, there has been an effort into the development of Pb-free perovskite, which is still in its infant phase, yet it is showing progress. Thus, based on this discussion it is possible that perovskite LED-based material can potentially be a viable option alongside traditional LEDs to compose high-end optic electronic devices.

REFERENCES

1. Chen J, Xiang H, Wang J, Wang R, Li Y, Shan Q, Xu X, Dong Y, Wei C, Zeng H (2021) Perovskite white light emitting diodes: progress, challenges, and opportunities. *ACS Nano* 15:17150–17174.
2. Aïssa B, Ali A, El-Mellouhi F (2021) Oxide and organic–inorganic halide perovskites with plasmonics for optoelectronic and energy applications: a contributive review. *Catalysts* 11: 1057.
3. Vashishtha P, Veldhuis SA, Dintakurti SSH, Kelly NL, Griffith BE, Brown AAM, Ansari MS, Bruno A, Mathews N, Fang Y, White T, Mhaisalkar SG, Hanna JV (2020) Investigating the structure–function relationship in triple cation perovskite nanocrystals for light-emitting diode applications. *J Mater Chem C* 8:11805–11821.
4. Protesescu L, Yakunin S, Kumar S, Bär J, Bertolotti F, Masciocchi N, Guagliardi A, Grotevent M, Shorubalko I, Bodnarchuk MI, Shih C-J, Kovalenko M V (2017) Dismantling the "red wall" of colloidal perovskites: highly luminescent formamidinium and formamidinium–cesium lead iodide nanocrystals. *ACS Nano* 11:3119–3134.
5. Davis NJLK, de la Peña FJ, Tabachnyk M, Richter JM, Lamboll RD, Booker EP, Wisnivesky Rocca Rivarola F, Griffiths JT, Ducati C, Menke SM, Deschler F, Greenham NC (2017) Photon reabsorption in mixed CsPbCl3:CsPbI3 perovskite nanocrystal films for light-emitting diodes. *J Phys Chem C* 121:3790–3796.

6. Stoumpos CC, Malliakas CD, Kanatzidis MG (2013) Semiconducting tin and lead iodide perovskites with organic cations: phase transitions, high mobilities, and near-infrared photoluminescent properties. *Inorg Chem* 52:9019–9038.

7. Vashishtha P, Bishnoi S, Li C-HA, Jagadeeswararao M, Hooper TJN, Lohia N, Shivarudraiah SB, Ansari MS, Sharma SN, Halpert JE (2020) Recent advancements in near-infrared perovskite light-emitting diodes. *ACS Appl Electron Mater* 2:3470–3490.

8. Vybornyi O, Yakunin S, Kovalenko MV (2016) Polar-solvent-free colloidal synthesis of highly luminescent alkylammonium lead halide perovskite nanocrystals. *Nanoscale* 8: 6278–6283.

9. Li X, Wu Y, Zhang S, Cai B, Gu Y, Song J, Zeng H (2016) CsPbX3 quantum dots for lighting and displays: room-temperature synthesis, photoluminescence superiorities, underlying origins and white light-emitting diodes. *Adv Funct Mater* 26:2435–2445.

10. Akkerman QA, Motti SG, Srimath Kandada AR, Mosconi E, D'Innocenzo V, Bertoni G, Marras S, Kamino BA, Miranda L, De Angelis F, Petrozza A, Prato M, Manna L (2016) Solution synthesis approach to colloidal cesium lead halide perovskite nanoplatelets with monolayer-level thickness control. *J Am Chem Soc* 138:1010–1016.

11. Jellicoe TC, Richter JM, Glass HFJ, Tabachnyk M, Brady R, Dutton SE, Rao A, Friend RH, Credgington D, Greenham NC, Böhm ML (2016) Synthesis and optical properties of lead-free cesium tin halide perovskite nanocrystals. *J Am Chem Soc* 138:2941–2944.

12. Protesescu L, Yakunin S, Bodnarchuk MI, Krieg F, Caputo R, Hendon CH, Yang RX, Walsh A, Kovalenko MV (2015) Nanocrystals of cesium lead halide perovskites (CsPbX3, X = Cl, Br, and I): novel optoelectronic materials showing bright emission with wide color gamut. *Nano Lett* 15:3692–3696.

13. Gonzalez-Carrero S, Galian RE, Pérez-Prieto J (2015) Maximizing the emissive properties of CH3NH3PbBr3 perovskite nanoparticles. *J Mater Chem A* 3:9187–9193.

14. Han Q, Bae S-H, Sun P, Hsieh Y-T, Yang Y (Michael), Rim YS, Zhao H, Chen Q, Shi W, Li G, Yang Y (2016) Single crystal formamidinium lead iodide (FAPbI3): insight into the structural, optical, and electrical properties. *Adv Mater* 28:2253–2258.

15. Zhang F, Zhong H, Chen C, Wu X, Hu X, Huang H, Han J, Zou B, Dong Y (2015) Brightly luminescent and color-tunable colloidal CH3NH3PbX3 (X = Br, I, Cl) quantum dots: potential alternatives for display technology. *ACS Nano* 9:4533–4542.

16. Huang H, Zhao F, Liu L, Zhang F, Wu X, Shi L, Zou B, Pei Q, Zhong H (2015) Emulsion synthesis of size-tunable CH3NH3PbBr3 quantum dots: an alternative route toward efficient light-emitting diodes. *ACS Appl Mater Interfaces* 7:28128–28133.

17. Yao E-P, Yang Z, Meng L, Sun P, Dong S, Yang Y, Yang Y (2017) High-brightness blue and white LEDs based on inorganic perovskite nanocrystals and their composites. *Adv Mater* 29:1606859.

18. Jiang C, Zou J, Liu Y, Song C, He Z, Zhong Z, Wang J, Yip H-L, Peng J, Cao Y (2018) Fully solution-processed tandem white quantum-dot light-emitting diode with an external quantum efficiency exceeding 25%. *ACS Nano* 12:6040–6049.

19. Liu X-K, Xu W, Bai S, Jin Y, Wang J, Friend RH, Gao F (2021) Metal halide perovskites for light-emitting diodes. *Nat Mater* 20:10–21.

20. Karlsson M, Yi Z, Reichert S, Luo X, Lin W, Zhang Z, Bao C, Zhang R, Bai S, Zheng G, Teng P, Duan L, Lu Y, Zheng K, Pullerits T, Deibel C, Xu W, Friend R, Gao F (2021) Mixed halide perovskites for spectrally stable and high-efficiency blue light-emitting diodes. *Nat Commun* 12:361.

21. Takagahara T, Takeda K (1992) Theory of the quantum confinement effect on excitons in quantum dots of indirect-gap materials. *Phys Rev B* 46:15578–15581.

22. Butkus J, Vashishtha P, Chen K, Gallaher JK, Prasad SKK, Metin DZ, Laufersky G, Gaston N, Halpert JE, Hodgkiss JM (2017) The evolution of quantum confinement in CsPbBr3 perovskite nanocrystals. *Chem Mater* 29:3644–3652.

23. Vashishtha P, Metin DZ, Cryer ME, Chen K, Hodgkiss JM, Gaston N, Halpert JE (2018) Shape-, size-, and composition-controlled thallium lead halide perovskite nanowires and nanocrystals with tunable band gaps. *Chem Mater* 30:2973–2982.

24. Won Y-H, Cho O, Kim T, Chung D-Y, Kim T, Chung H, Jang H, Lee J, Kim D, Jang E (2019) Highly efficient and stable InP/ZnSe/ZnS quantum dot light-emitting diodes. *Nature* 575:634–638.

25. Bi C, Hu J, Yao Z, Lu Y, Binks D, Sui M, Tian J (2020) Self-assembled perovskite nanowire clusters for high luminance red light-emitting diodes. *Adv Funct Mater* 30:2005990.

26. Zhao L, Roh K, Kacmoli S, Al Kurdi K, Jhulki S, Barlow S, Marder SR, Gmachl C, Rand BP (2020) Thermal management enables bright and stable perovskite light-emitting diodes. *Adv Mater* 32:2000752.

27. Guo Y, Jia Y, Li N, Chen M, Hu S, Liu C, Zhao N (2020) Degradation mechanism of perovskite light-emitting diodes: an in situ investigation via electroabsorption spectroscopy and device modelling. *Adv Funct Mater* 30:1910464.

28. Mao W, Hall CR, Bernardi S, Cheng Y-B, Widmer-Cooper A, Smith TA, Bach U (2021) Light-induced reversal of ion segregation in mixed-halide perovskites. *Nat Mater* 20:55–61.

29. Zhao X, Tan Z-K (2020) Large-area near-infrared perovskite light-emitting diodes. *Nat Photonics* 14:215–218.

30. Xu W, Hu Q, Bai S, Bao C, Miao Y, Yuan Z, Borzda T, Barker AJ, Tyukalova E, Hu Z, Kawecki M, Wang H, Yan Z, Liu X, Shi X, Uvdal K, Fahlman M, Zhang W, Duchamp M, Liu J-M, Petrozza A, Wang J, Liu L-M, Huang W, Gao F (2019) Rational molecular passivation for high-performance perovskite light-emitting diodes. *Nat Photonics* 13:418–424.

31. Zhang W, Zhong X (2011) Facile synthesis of ZnS–CuInS2-alloyed nanocrystals for a color-tunable fluorchrome and photocatalyst. *Inorg Chem* 50:4065–4072.

32. Wang Y, Li X, Song J, Xiao L, Zeng H, Sun H (2015) All-inorganic colloidal perovskite quantum dots: a new class of lasing materials with favorable characteristics. *Adv Mater* 27:7101–7108.

33. Liu Y, Chen P, Wang Z-H, Bian F, Lin L, Chang S-J, Mu G-G (2009) Efficient two-photon absorption of CdSe-CdS/ZnS core-multishell quantum dots under the excitation of near-infrared femtosecond pulsed laser. *Laser Phys* 19:1886–1890.

34. Ng T-W, Thachoth Chandran H, Chan C-Y, Lo M-F, Lee C-S (2015) Ionic charge transfer complex induced visible light harvesting and photocharge generation in perovskite. *ACS Appl Mater Interfaces* 7:20280–20284.

35. Kulbak M, Cahen D, Hodes G (2015) How important is the organic part of lead halide perovskite photovoltaic cells? Efficient CsPbBr3 cells. *J Phys Chem Lett* 6:2452–2456.

36. Patra A, Bera S, Nasipuri D, Dutta SK, Pradhan N (2021) Tuning facets and controlling mono-dispersity in organic–inorganic hybrid perovskite FAPbBr3 nanocrystals. *ACS Energy Lett* 6:2682–2689.

37. Ma L, Hao F, Stoumpos CC, Phelan BT, Wasielewski MR, Kanatzidis MG (2016) Carrier diffusion lengths of over 500 nm in lead-free perovskite CH3NH3SnI3 films. *J Am Chem Soc* 138:14750–14755.

38. Huang L, Lambrecht WRL (2013) Electronic band structure, phonons, and exciton binding energies of halide perovskites $CsSnCl{}_{3}$, $CsSnBr{}_{3}$, and $CsSnI{}_{3}$. *Phys Rev B* 88:165203.

39. Song T-B, Yokoyama T, Stoumpos CC, Logsdon J, Cao DH, Wasielewski MR, Aramaki S, Kanatzidis MG (2017) Importance of reducing vapor atmosphere in the fabrication of tin-based perovskite solar cells. *J Am Chem Soc* 139:836–842.

40. Li W, Li J, Li J, Fan J, Mai Y, Wang L (2016) Addictive-assisted construction of all-inorganic CsSnIBr2 mesoscopic perovskite solar cells with superior thermal stability up to 473 K. *J Mater Chem A* 4:17104–17110.

41. Wang F, Ma J, Xie F, Li L, Chen J, Fan J, Zhao N (2016) Organic cation-dependent degradation mechanism of organotin halide perovskites. *Adv Funct Mater* 26:3417–3423.

42. Kaltzoglou A, Antoniadou M, Kontos AG, Stoumpos CC, Perganti D, Siranidi E, Raptis V, Trohidou K, Psycharis V, Kanatzidis MG, Falaras P (2016) Optical-vibrational properties of the Cs2SnX6 (X=Cl, Br, I) defect perovskites and hole-transport efficiency in dye-sensitized solar cells. *J Phys Chem C* 120:11777–11785.

43. Saparov B, Sun J-P, Meng W, Xiao Z, Duan H-S, Gunawan O, Shin D, Hill IG, Yan Y, Mitzi DB (2016) Thin-film deposition and characterization of a Sn-deficient perovskite derivative Cs2SnI6. *Chem Mater* 28:2315–2322.

44. Wang A, Guo Y, Muhammad F, Deng Z (2017) Controlled synthesis of lead-free cesium tin halide perovskite cubic nanocages with high stability. *Chem Mater* 29:6493–6501.

45. Koh TM, Krishnamoorthy T, Yantara N, Shi C, Leong WL, Boix PP, Grimsdale AC, Mhaisalkar SG, Mathews N (2015) Formamidinium tin-based perovskite with low Eg for photovoltaic applications. *J Mater Chem A* 3:14996–15000.

46. Xing G, Kumar MH, Chong WK, Liu X, Cai Y, Ding H, Asta M, Grätzel M, Mhaisalkar S, Mathews N, Sum TC (2016) Solution-processed tin-based perovskite for near-infrared lasing. *Adv Mater* 28:8191–8196.

Index

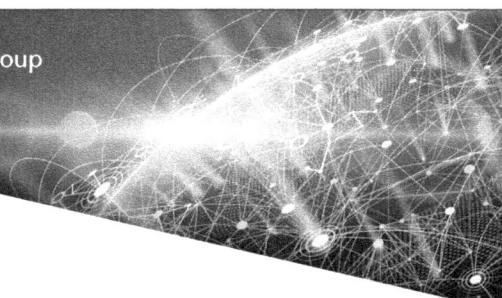

For Product Safety Concerns and Information please contact our EU
representative GPSR@taylorandfrancis.com
Taylor & Francis Verlag GmbH, Kaufingerstraße 24, 80331 München, Germany

www.ingramcontent.com/pod-product-compliance
Lightning Source LLC
Chambersburg PA
CBHW080924220326
41598CB00034B/5668

* 9 7 8 1 0 3 2 0 2 7 7 2 2 *